Hi-tech Crop Production and Pest Management

The Editors

Dr. Wajid Hasan, Ph.D. PDF devoted his academic career to Agriculture, particularly in Agriculture Entomology as a Research and Extension. Dr. Hasan is working as a Subject Matter Specialist (Entomology) at KVK Jehanabad, Bihar Agricultural University, Bihar. He was selected for Dr. D.S. Kothari UGC-Post Doctoral Fellow (P.D.F.) in 2010 by the University Grant Commission of India. Dr. Hasan did his doctorate in Agricultural Entomology (2009) from G.B. Pant University of Agri & Tech., Pantnagar. He has made a vibrant link with farmers by advising them plant protection measures when needed for which he has received 11 awards from several societies in the outstanding contribution of the relevant discipline, published 48 research papers, 20 books, 25 book chapters, 38 popular articles, 6 success stories, 2 Practical Manuals, 6 success stories and 47 delivered lecture in different seminar/symposia/conference/workshop and 7 radio talks. He has organized 11 International Conference as Chief-Organizing Secretary, Co-Organizing and Executive Organizing Secretary. Dr. Hasan organized 244 (9896 trainees) Training for Practicing Farmers, Rural Youths, Extension Functionaries, Two Skill training (200hrs), 26 On-Farm Trials (OFT) conducted. Dr. Hasan organized 394 (17896 trainees) Training for Practicing Farmers, Rural Youths, Extension Functionaries, Three Skill training (200hrs), 34 On-Farm Trial (OFT) conducted. Dr. Hasan has a life membership of 5 scientific societies. He is working as Editor-in-Chief in the International Journal of Agricultural and Applied Sciences (IJAAS) and in the Editorial board in 5 scientific International Journals.

Dr. Md. Motiar Rohman is a plant breeder working at Plant Breeding Division of Bangladesh Agricultural Research Institute. He is expertized in crop improvement through conventional and biotechnological approaches. He completed his PhD from United Graduate School of Agricultural Sciences, Ehime University, Japan. He developed 10 crop varieties suitable for unfavourable ecosystem like drought and salinity. Besides, he has been serving as research supervisor of MS and PhD students of different agricultural universities of Bangladesh. So far, he has supervised 15 MS and 5 PhD students. He has published more than 100 scientific papers and a good number of book chapters in national and International journals and books. He has also received international awards from international conference and workshop. Dr. Rohman has also been serving as editor of journals, and organizer of conferences at home and abroad. For his outstanding contribution, he has been serving as an executive member in different position of Plant Breeding and Genetics Society of Bangladesh.

Dr. Mehjabeen is a Science Associate in the Asia Pacific Association of Agricultural Research Institution (APAARI) based in Bangkok. She is responsible for the outreach of few global research projects like the ASTI and, many other research projects in its complete research cycle. She is also an editor for a monthly journal, Just Agriculture. She is a board member for a monthly magazine, Urban Transport New which, is circulated and read by the readers of more than 97 countries. She is one of the Board of Directors for the Hi limit group based in Ghana. She has served as an assistant professor for many years for different academic institutions. She served as a managing editor for monthly newsletters for colleges she worked for along with her teaching career. She stayed as a Subject matter expert and helped with dissertation and thesis for students across the globe. She holds PhD in economics with a specialized area, agricultural economics, Masters in Economics, Masters in Business Administration and Masters in Mass communication. She has severed as a consultant for the Indian Council of Agricultural Research for the mega project, Soil Health Card. She was involved in drafting the reports and working papers and other documentation processes. She authored and published many research papers with both national and international collaborations. She stayed as a University topper and she has been awarded as the Best Research Scholar for her contribution in the field of Agricultural Economics.

Dr. Payel Panja is presently working as a Teaching Associate in the School of Agriculture and Allied Sciences, The Neotia University, West Bengal, India. She has completed her M.Sc. and Ph. D in Post Harvest Technology from Bidhan Chandra Krishi Viswavidyalaya, Kalyani, West Bengal. She has qualified ICAR-NET in Vegetable Science. She has received different 6 award from several societies in the outstanding contribution of the relevant discipline, published 18 research papers, 2 books, 12 book chapters, 3 popular articles and 15 delivered lecture in different national/international seminar/symposium/conference/workshop. She is the member in several scientific societies. She has also acted as editorial Board member in one journal and reviewer in many journals. She has organized 3 International Conference as Co-Organizing and Joint Organizing Secretary.

Hi-tech Crop Production and Pest Management

– Editors –

Dr. Wajid Hasan
Subject Matter Specialist (Entomology),
Krishi Vigyan Kendra (BAU Sabour),
Jehanabad, Bihar, India

Dr. Md. Motiar Rohman
Senior Scientific Officer
Plant Breeding Division, Bangladesh Agricultural Research Institute,
Bangladesh

Dr. Mehjabeen
Science Associate
Asia Pacific Association of Agricultural Research Institution (APAARI)
based in Bangkok

Dr. Payel Panja
Teaching Associate
School of Agriculture and Allied Sciences,
The Neotia University, West Bengal, India

BIOTECH

2022

BIOTECH BOOKS®

ISBN 978-81-7622-500-7

Published by: **BIOTECH BOOKS®**
4762-63/23, Ansari Road, Darya Ganj,
New Delhi - 110 002
Phone: 09899295259, 011-43003222
E-mail: biotechbooks@yahoo.co.in
Website: www.biotechbooks.in

Typeset at:**Classic Computer Services**
Delhi - 110 035

Printed at:**Replika Press Pvt. Ltd.**

PRINTED IN INDIA

Preface

Technological progress in the recent years have increased the intensity and scale of agricultural land use. We have to deal with a very wide and fast-growing scope of technology in agriculture sector which are necessary for optimizing production, improving productivity, improving quality and profitability along with meeting challenges of climate change and environment. Intensification in agriculture follows a pattern of declining marginal output and increasing complexity. Earlier farming was based on the concept of subsistent agriculture and when Green revolution started in 1950s, the main focus of agricultural research was on crop production to increase the food grain productivity. Recently, the trend of agriculture research has been towards higher production with improved productivity and maintaining the quality of environment must be addressed at every aspects of farming. Therefore, strategies are needed to develop farming systems to achieve optimization of profitability of crop production while maintaining soil quality and reducing environmental degradation. Solution of these new challenges requires consideration of how numerous components interact to effect plant growth. To achieve this goal crop modelling can play a major role, future agriculture research will require considerably with more effort and resource than present research activity. Nanotechnology can be effectively used for the basement of IPM with advances in monitoring, soil assessment and quality control by nano-based sensors.

Resource management has become a formidable task with loss of cultivable land to urbanization. Coercion to increase production has resulted in adoption of intense farming. This however has resulted in significant increase in the pest problem. Despite ten-fold increase in insecticide use during the last 80 years, losses inflicted by pests have nearly doubled. Unlike chemical method, which kills non target species, cause detrimental health effects to human beings and pollute environment. The use of pesticides still dominates the management of insect pests and is a health hazard for humans and the environment. Control of insect pests and plant disease by chemical can be spectacular but the accumulation of harmful chemical residue sometimes causes serious ecological problems. Application of synthetic pesticides is a usual practice to ward off infestation of insect pests and diseases from field crops. sustain productivity, conserve natural resources, reduce production costs, improve environmental health, maintain biodiversity, and reduce agrochemical use for crop production/protection. The development of resistance and resurgence to these chemical insecticides has made the management of insect pests extremely difficult. The other environmental concerns are also associated with insecticides like residue in food and feed, health hazard to humans reduction in non-target organisms population and environmental pollution. So the plant protectionist all over the world trying to find eco-friendly and sustainable alternatives for management of insect pests and disease.

The book is structured into 24 chapters and primarily for the degree, post-graduate students, and for researchers of the field. The chapters will clearly define the Hi-tech Crop Production and Pest Management and their impacts on agricultural productivity. We wish to express a deep sense of gratitude to those who helped us directly or indirectly during the preparation of the manuscript of this text. We hope that the book is useful and interesting to readers, teachers, and students and would create in them the urge to know more about recent researchers going related to Smart Agriculture. We are highly thankful to all authors who contribute their research/ideas to enhance the utility of the book.

Wajid Hasan

Contents

Hi-tech Crop Production

Chapter 1

Crop Modelling for Pioneering Sustainable Agriculture

Mukesh Kumar Mehla[*1], Khyati Singh[2]*

[1]Department of Soil and Water Engineering, College of Technology and Engineering, MPUAT, Udaipur – 313001, Rajasthan
[2]Department of Vegetable Science, College of Agriculture, CCS Haryana Agricultural University, Hisar – 125004, Haryana

Crop models are mathematical equations that represent a real-world system, reactions that occur within the plant and interactions between the plant and its environment. Unlike in the fields of physics and engineering, universal models do not exist within the agricultural sector. Models are built for specific purposes and the level of complexity is accordingly adopted. There are various models used in agriculture like DSSAT, CropSyst, EPIC, AQUACROP and APEX, etc. They are structured for varying subsystems and these models can be built to simulate a specific crop or a specific aspect of the crop production system. These models are based on comprehensive climatic, soil and physiological data which makes them capable being adopted universally as they support extrapolation to alternative cropping cycles, crop and locations, etc. Also capable of quantifying temporal and spatial variability. They give acceptable results in relatively short time span and at comparatively low costs enabling the researchers to investigate effect of a large number of management strategies which would be very cumbersome and costly by using traditional methodologies. Owing to the complexity of the system and lack of availability of proper datasets, it becomes hard to model the agricultural systems. Regardless of its limitations, crop modelling approach remains a promising means of assessing different crop and field management strategies, yield estimation, crop planning, access the impact climate change, environmental impact assessment and formulation of policies etc.

Introduction

Technological progress in the recent years have increased the intensity and scale of agricultural land use. We have to deal with a very wide and fast-growing scope of technology in agriculture sector which are necessary for optimizing production, improving productivity, improving quality and profitability along with meeting challenges of climate change and environment. Intensification in agriculture follows a pattern of declining marginal output and increasing complexity. Earlier farming was based on the concept of subsistent agriculture and when the Green revolution started in 1950s, the main focus of agricultural research was on crop production to increase the food grain productivity. Recently, the trend of agriculture research has been towards higher production with improved productivity and maintaining the quality of environment must be addressed at every aspects of farming. Therefore, strategies are needed to develop farming systems to achieve optimization of profitability of crop production while maintaining soil quality and reducing environmental degradation. Solution of these new challenges requires consideration of how numerous components interact to effect plant growth. To achieve this goal crop modelling can play a major role, future agriculture research will require considerably with more effort and resource than present research activity.

Crop production is based on making right decisions at right time in a right way. The decision for a crop production system was derived from traditional experience based on agricultural research, in which the correlation and regression of crop yields are analysed for certain defined variables. Analysis of correlation and regression has assisted qualitative and quantitative understanding of various variables and their interactions with each other and the environment that were involved in cropping systems and have contributed to the progress of agronomy (Asseng *et al.*, 2014). However, this information attained from such model analysis are very much site specific. The correlation studied can only be reliable when applied to other site where parameters like climate, soil and other crop management are similar to those used in developing the original functions. Regression based crop models have limited applications in decision-making as they are of very crude nature. Agriculture models are very rudimentary representations of the real-world systems because of inherent complexities involved with the systems. The sound understanding of model structure, its scope and limitation is necessary for a judicious application of such model. Crop modelling is a new branch of learning with lesser available back ground literature. Therefore, the aim of this chapter is to provide some basic information about crop modelling.

Modelling in Agricultural Systems

A model can be defined as a simplified representation of a system and the study of the system and its behaviour is done using any simulation model. A system consists of a set of components, abstract or real. The simple representation of a crop is referred to as Crop model which are used to study crop growth and its responses to the changes in environment. Agricultural systems are characterized by having many organizational levels. From the individual components within a single plant, through constituent plants, to farms or a whole agricultural region or nation, lies a whole range of agricultural systems. Since the core of agriculture is concerned with plants, the level that is of main interest to the agricultural modeller is the plant. Reactions and interactions based on plant biological processes are combined to form a big picture of the plant which is then extrapolated to the crop and their output.

Agricultural models are mathematical equations that represent the reactions that occur within the plant and the interactions between the plant and its environment. Owing to the complexity of the system and the incomplete status of present knowledge, it becomes impossible to completely represent the system in mathematical terms and hence, agricultural models' images of the reality. Unlike in the fields of physics and engineering, universal models do not exist within the agricultural sector (McCauley, 2020). Models are built for specific purposes and the level of complexity is accordingly adopted. Inevitably, various models are structured for varying subsystems and these models can be built to simulate a specific crop or a specific aspect of the crop production system.

Crop modelling is important research tool that can help in accomplishing the goal of sustainable agriculture around the world (Ramankutti *et al.*, 2018). Remote sensing data obtained from satellite sensors can be used for crop modelling at larger scales (Jin *et al.*, 2018; Bajocco *et al.*, 2019). There is a lot of uncertainty involved in the crop modelling that can arise from different factors that influences it (Confalonieri *et al.*, 2016; Salder *et al.*, 2016). Scale of the study and the data used have significant effect on the accuracy and applicability of model predictions and crop modelling study are conducted at different scales varying from field scale to regional and large area studies (Faivre *et al.*, 2004; Ewert *et al.*, 2011; Hoffmann *et al.*, 2016; Malek *et al.*, 2017; Mourtzinis *et al.*, 2017; Müller *et al.*, 2017).

The main objective of constructing crop models is to evaluate the harvestable (economic) yield. The models with different levels of complexity are developed based on the amount of data and knowledge that is available within a particular field. The most pertinent aspects of crop models are described below:

Different Types of Crop Models

Empirical Model

Empirical models are used for direct descriptions of the collected data and are typically expressed as regression equations (with one or a few factors) and evaluate the final yield. For examples the model may include the effect of fertilizer application on crop yield, or the association between leaf size and leaf area in a given plant species. The limitation of this model is that it is site specific and cannot used universally.

Mechanistic Model

A mechanistic model describes the conduct of the system in respect of lower level attributes. They are based on some mechanism or explanation at the lower levels. These models have the capacity to imitate relevant biological, physical or chemical processes and gives explanation for the working and a particular response.

Static and Dynamic Models

A static model does not have time as a variable even if the end-products of cropping systems are accumulated over time, *e.g.*, the empirical models. On the other hand, dynamic models directly incorporate time as a variable and most of these models are expressed as differential equations.

Deterministic and Stochastic Models

A deterministic model functions by proposing definite predictions for quantities (*e.g.*, crop yield, rainfall or animal liveweight) without any associated variance, probability distribution or random element. However, there are always irregularities and variations due to presence of heterogeneity in the system and recorded data. These deviations are inherent to both biological and agricultural systems. In some of the cases, deterministic models may be adequate despite these inherent variations while in other cases they might be unsatisfactory such as in rainfall prediction. The deterministic models become more inadequate with higher uncertainty in the system and in contrast to this stochastic model appears.

Simulation and Optimizing Models

Simulation models are comprised by group of models that are designed to mimic the behaviour of a system. They function as mechanistic model and in the mostly they are deterministic in nature. These models are structured to imitate the system at less time intervals and the aspect of variability related to daily change in parameters like weather conditions and soil characteristics are featured.

The short simulation time-step requires large amount of input data (climate, soil and crop parameters) for the model to work and generally provides the possibility of specifying management possibilities and they can aid to examine a wide range of management practices at low costs. Most crop models that are used to estimate crop yield fall within this category (Mulla *et al.*, 2020).

Optimizing models have the specific aim to create the best opportunity in terms of management inputs for practical functioning of the system. For deriving solutions, they use decision rules that are consistent with some optimising algorithm. This brings some rigidity into their structure leading to interruptions in representing stochastic and dynamic aspects of agricultural systems. Linear and non-linear programming were used initially at farm level for enterprise selection and resource allocation. Later, applications to assess long-term adjustments in agriculture, regional competition, transportation studies, integrated production and distribution systems as well as policy issues in the adoption of technology, industry re-structuring and natural resources have been developed. Optimising models do not allow the incorporation of many biological details and may be poor representations of reality. Using the simulation approach to identify a restricted set of management options that are then evaluated with the optimising models has been reported as a useful option.

Table 1.1: Important Crop Models

Software/Model	*Applicability Details*
SLAM II	Forage harvesting operation
SPICE	Whole plant water flow
REALSOY	Soyabean
MODVEX	Model development and validation system
IRRIGATE	Irrigation scheduling model
COTTAM	Cotton
APSIM	Modelling framework for a range of crops
GWM	General weed model in row crops
MPTGro	Acacia spp. and Leucaena Spp.
GOSSYM-COMAX	Cotton
CropSyst	Wheat and other crops
SIMCOM	Crop (CERES crop modules) and economics
LUPINMOD	Lupin
TUBERPRO	Potato and disease
SIMPOTATO	Potato
WOFOST	Wheat and maize, Water and nutrient
WAVE	Water and agrochemicals
SUCROS	Crop models
ORYZA1	Rice, water
SIMRIW	Rice, water
SIMCOY	Corn
CERES-Rice	Rice, water
GRAZPLAN	Pasture, water, lamb
EPIC	Erosion Output Impact Calculator
CERES	Series of crop simulation models
DSSAT	Framework of crop simulation models includingmodules of CERES, CROPGRO and CROPSIM
QCANE	Sugarcane, potential conditions
AUSCANE	Sugarcane, potential and water stress conditions, erosion
CANEGRO	Sugarcane, potential and water stress conditions
APSIM-Sugarcane	Sugarcane, potential growth, water and nitrogen stress
NTKenaf	Kenaf, potential growth, water stress

Some Important Crop Models

AQUACROP Model

AquaCrop, developed by the Land and Water Division, FAO, is a crop growth model used to evaluate environmental and management impacts on crop production. This system is simple, accurate and widely applicable which simulates yield response to irrigation, and specifically deals with limited water conditions in crop production. It simulates on relatively small number of parameters and performs simple determination methods which requires mostly-intuitive input-variables, while the computation is based on complex biophysical processes to assure a truthful simulation of the crop response in plant-soil systems. It requires crop data, meteorological data, soil characteristics and management practices adopted as Inputs and give output in the terms of crop canopy, biomass and yield response to the stresses originating from water and salt balances. It is available as standalone windows-based software, code and as plug-in to GIS software (Raes *et al.*, 2011; Lorite *et al.*, 2015; Foster *et al.*, 2017).

Crop Growth and Development

AquaCrop simulate crop growth and development by its growth engine. Which simulate crop growth based on temperature and water availability which is also a limiting factor. This process and its limitations can be understood by means of four step calculation and their formulas. First step is to simulate the fraction of the green canopy of the total surface area which is given in terms crops canopy cover (CC) factor (Eq. 1). After that a logistic function simulate CC from the initial value (CC_0) to the maximum (CC_x) while considering a growth coefficient (c_{gc}) in the early season and a decline coefficient (c_{dc}) in the late season. In second step, it calculates crop transpiration (T_r) from a reference evapotranspiration (ET_0) and a crop transpiration coefficient ($K_{c.TR}$) which is proportional to the CC (Eq. 2). After that biomass (B) is calculated based on the crop transpiration and a normalized crop water productivity (wp*) factor (Eq. 3). Lastly, the biomass is converted into crop yield (Y) by multiplying it with a harvest index (hi) factor considered for the crop (Eq. 4). Calculation for Crop canopy cover, Transpiration, Biomass and dry on any day (i) is given by

$$CC_i = \frac{\text{Soil covered by green canopy}}{\text{Unit ground surface area}} \tag{1}$$

$$T_{r_i} = K_{S_i} K_{C.TR_i} ET_{o_i} \tag{2}$$

$$B = wp^{\bullet} \times \sum_{i=1}^{n} K_{s.b_i} \frac{T_{r_i}}{ET_{o_i}} \tag{3}$$

$$Y = h_i \times B = f_{HI} \times h_{io} \times B \tag{4}$$

where,

CC: Crop canopy cover (m^2/m^2)

T_r: Crop transpiration (mm/day)

ET_{oi}: Crop reference evapotranspiration (mm/day)

K_{CTRi}: Crop transpiration coefficient which is proportional to CC

K_{si}: Soil water and salinity stress coefficient

K_{sbi}: Cold stress coefficient

B: Total biomass production (g/m^2)

wp*: Normalised crop water productivity (g/m^2)

Y: Dry yield (g/m^2)

h_i: Harvest index (g/g) a product of the reference harvest index (h_{io}, g/g) adjusted for water and temperature stress with f_{HI}

n: Number of simulation days per growing period.

Soil Water Balance

AquaCrop model simulates a daily soil water content (S) calculates soil water balance from water fluxes entering (rain, irrigation, capillary rise) and leaving the crop rootzone (surface runoff, deep percolation, soil evaporation, crop transpiration) with the help of Eq. 5. Soil water balance in the crop rootzone on any day (i) is given by,

$$S_i - P_i + I_i + C_i - SO_i - D_i - E_i - T_i \qquad (5)$$

where,

S: Soil water content (mm)

P: Precipitation(mm)

I: Irrigation (mm)

C: Capillary rise (mm) which depends on the soil type and groundwater table depth

SO: Surface runoff (mm) estimated by Curve Number (CN) method (Eq. 6)

D: Deep percolation (mm) estimated from drainability (m^3/m^3/day) based on the soil type

E: Soil evaporation (mm) from Eq. 7

T: Crop transpiration (mm) from Eq. 2

$$RO_i = \frac{(P_i - 0.2S_i)^2}{P_i + S - 0.2S_i} \qquad (6)$$

$$E = K_{r_i} \times K_{e_i} \times ET_{o_t} \qquad (7)$$

where,

S_1: Maximum potential storage (mm) which depends on soil type

K_r: Evaporation reduction coefficient

K_e: Evaporation coefficient proportional to the soil fraction not covered by the crop (1-CC)

DSSAT Ecosystem Model

It was designed as an integrated crop modelling platform (Jones *et al.*, 1998). The DSSAT approach allows the user to prepare individual input files that define the experiment needed to run the model and analyse the output after conducting the simulation. Also allows comparison with observed data and strategic analyses for model scenarios. This approach was utilised as DSSAT Version 2.1 (IBSNAT, 1989) and DSSAT Version 3.0 (IBSNAT, 1993). The latest version of the model is DSSAT Version 4.7.5 (Hoogenboom *et al.*, 2019) and future version is implemented in the version jDSSAT (Resenes *et al.*, 2019) which has simpler features for easier simulation of crop growth, development and yield for different crops of same field.

Crop Water Modelling

There are different models that simulate water uptake, water dynamics in the soil and wetting patterns in root zones. They are generally based on analytical or numerical solutions of equations which descried these phenomena. They include WetUp, HYDRUS 2D/3D, So-WaM, Neuro-Drip, CoupModel, Drip-Irriwater and DIDAS (Cook *et al.*, 2003; Simunek *et al.*, 2008; Wesseling *et al.*, 2009; Hinnell *et al.*, 2010; Jansson, 2012; Arbat *et al.*, 2013; Friedman, 2016). These models account for the different soil and the root system properties, climate, and plant/emitter configuration to simulate soil water in root zone. DIDAS model is used evalvating different irrigation systems and schedules. This software can be especially useful finding optimal emitter spacing for close growing crop like onion which needs closely spaced drip lines without costly and time consuming field trails and experiments (Mehla and Singh, 2019).

DIDAS

It is based on analytical solutions of the Richards equation. This is available in the form of computer program which solves Richards equation for both steady and unsteady water flow in point and line source emitters based type of source, the geometry of the domain, and the initial and boundary conditions. This software consists of three modules first is drip-system design module which uses data about root-zone, plant resistance, the configuration of the water source and its location, and soil hydraulic parameters to simulate water uptake processes and determine effect of geometric attributes on water use efficiency. Second is diurnal pattern module which simulate diurnal patterns of plant-atmosphere resistance to water uptake and evaporation, it is primary step for evaluating irrigation scheduling. Third module is irrigation-scheduling optimization tool which used to simulate effect of different irrigation schedules but doesn't calculate direct effect of irrigation schedules in terms of yield but calculate RWUE (Relative Water use efficiency) for optimizing irrigation scheduling. For our study we are mainly concerned with drip-system design module.

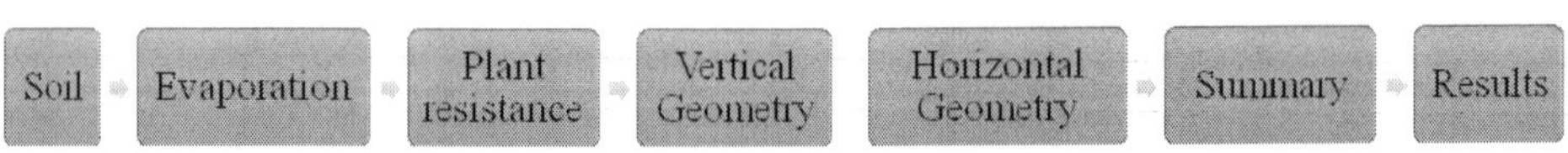

Figure 1.1: Flow Chart of Various Steps Involved in Simulating a Ccenario in Drip System Design Module in DIDAS.

Data Requirement for Modelling

Crop modelling requires a number of parameters to simulate crop, study effect of management and environment on the crop. Meteorological observations are most important data required for any modelling study, climate data is basis of the growth calculations which are simulated at a certain time step based crop parameters and climatic data. These data can be obtained from the national meteorological organizations, they provide weather data from different stations and observatories across the globe. Along with point based weather data there are options of grid based data which can be useful for regional analysis. Weather data can also be obtained from satellites observations or combination of satellite and ground data. There are different climate models which give historic and future climate data projections which needs basis correction and downscaling to be used for analysis (Duveiller *et al.*, 2015). These are based on general circulation models (GCM) and regional circulation models (RCM) *etc.*, developed by different organizations at different levels (Baron *et al.*, 2005). There is a need for high precision and accuracy of the data. The data obtained from surface observatories has proved to be excellent. It gained the confidence of the people across the globe for decades.

Modelling Climate Change Impacts

There limited land and water resources available to us, while population of the globe is increasing rapidly day by day there is need to efficient planning and management of these limited resources and develop alternatives to support this ever growing population. This creates a major problem for agriculture sector which has to provide for the masses and the production (productivity) must be increased to meet rapidly growing demands while natural resources must be protected. At current level of production, we will not be able to meet future demand by doubling crop production by the year 2050 (Ray *et al.*, 2013). In recent years there has been a growing concern that climate change and its impact on agriculture and our environment as whole. The climate change has led to dramatic change in local and regional weather patterns, melting glaciers and weather extremes are causing drastic changes around the world having severe negative impacts across the globe which have been reason of concern for many. Farmers needs to adapt to climate change-through changes in farming practices, cropping patterns, and use of new technologies to reduce the impact of climate change on crop production. Crops models can be applied for studying the potential impact of climate change and climate variability on agriculture and analyse different potential options to tackle climate change impacts of different provides a direct link between models, climate and the concerns of the society (Challinor *et al.*, 2014; Donatelli *et al.*, 2015; Challinor *et al.*, 2018).

Different Stages of Model Development

Strength of Model

The strength of models is indicated by its ability to deliver a robust framework for defining a system and develop and understanding how different variables influence various components, study long-term impact of certain interventions on the system, analysis of risks involved with adopting a particular strategy and achieve better results as a cheap alternative to traditional methods of experimentation.

Model Calibration

Calibration of model is the process of adjustment of the system parameters to adjust simulation results and get desired result output. In many instances, even if a model is based on observed data, simulated values do not exactly comply with the observed data and minor adjustments have to be made for some parameters.

Model Validation

The model validation stage confirms that the calibrated model is a close representative of the real situation. The procedure consists of a comparison of simulated output and observed data that have not been previously used in the calibration stage. Ideally, all mechanistic models should be validated both at the level of overall system output and at the level of internal components and processes. The latter is an important aspect because due to the occurrence of feedback loops in biological systems, good prediction of system's overall output could be attributed to compensating internal errors. However, the insufficiency of detailed datasets makes it difficult to validate all the components and only the determinant components are adopted for validation. For example, in a soil-water-crop model, as biomass accumulated is highly determined by factors like leaf area and extractable water, hence it is imperative to validate these components. The approach of model validation is still elementary. The main reason is that, in contrast to the case of disciplinary experiments, the model is testing a set of multiple hypotheses simultaneously.

At present, the complications in the validation of system simulation models is due to uncertainty of field data that makes the validation complex. Practical execution of model reveals that soil, plant and climatic conditions are rarely accurate and variation in the data may arise from nearby sites. Many times, such parameters that were not routinely estimated can be significant and are then measured arbitrarily. The parameters evaluated also differ due to the diversity arising from the inherent soil diversity over relatively short distances and the effects of husbandry practices on soil properties.

Crop data over the growing period reflect soil diversity as well as variation in environmental factors. The sampling errors also adds to the inaccuracies in observed data. The validation procedures require comparisons of qualitative and quantitative data. It is recommended that before initiating the quantitative tests, qualitative assessment of time-trends of simulated and collected data for both internal variables and systems outputs. To adjust the inadequate predictions of model outputs can be corrected by "re-fitting" of the regression curves or fine-tuning of one or more internal variables. This practice should be performed with caution as arbitrary changes can make possible the changes in model structure that may restrict the application of the model as a forecasting too. In some cases, it is better to obtain more reliable data by using further experimentation than to apply extensive modifications to model parameters to obtain an acceptable fit for the suspected data.

Applications of Crop Modelling

Simulation modelling is gaining importance in field of farm and resource management, research and teaching, production forecasts and policy analysis. These models can be used for wide range of applications in research, crop system management and policy analysis. Major applications of crop modelling are as described:

- **Improve research understanding:** Crops models can help in ensuring the integration of research understanding from experimentations, identify of the major factors driving the system and knowledge gaps. Thus, crop modelling can contribute to targeted and efficient research planning.
- **Integration of knowledge**: Crops modelling adopts a modular approach which allows the scientist to independently pursue research in different disciplines and later integrate the acquired knowledge. Model can be altered to cater for various application for different processes, crops and climate *etc.* For example, APSIM model allows the integration of different crops.
- **Experiment documentation and data organization:** Simulation model development, testing and application requires technical and observational data in large quantity supplied in given units and in a particular order. Model needs good data organisation and database systems for proper data handling. This enhances the efficiency of data manipulation and documentation of experiments.
- **Genetic improvement**: With the improvement of simulation models, they can mimic the system more thoroughly and more detailed information about the effect of various genetic traits on economic yields can be integrated into genetic improvement programs. For *e.g.*, researchers used NTKenaf model to design crop ideotypes for specific environments.
- **Yield analysis:** The use of multiple simulation models can be done to assess climate-determined yields in various crops. For *e.g.*, the CANEGRO model is used in the South African sugar industry and using this modelling approach, the amount of yield reduction is possible due to non-climatic reasons (*e.g.*, soil fertility, delay in sowing, pests and diseases). Due to changing weather trends, improved technology and genetic improvement, simulation models have been shown to be useful in differentiating yield gains.
- **Cultural and input management:** Management decisions regarding cultural practices and inputs have a major impact on yield. Simulation models, that allow the specification of management options, offer a relatively inexpensive means of evaluating a large number of strategies that would rapidly become too expensive if the traditional experimentation approach were to be adopted. Based on local conditions, models for water requirement evaluation can be created for command areas. Many publications are available describing the use of simulation models with respect to cultural management (planting and harvest date, irrigation, spacing, selection of variety type) and input application (water and fertiliser).

- **Risks assessment and investment support:** Using a combination of simulated yields and gross margins, economic risks and weather-related variability can be assessed. These data can then be used as an investment decision support tool.
- **Site-specific farming:** Crop modelling can help in determining the crop production capacity of such lands where crop was not sown before or in a long time. Profit maximisation may be achieved by managing farms as sets of sub-units and providing the required inputs at the optimum level to match variation in soil properties across the farm. By coupling simulation models with geographic information systems (GIS) we can obtain predicted yield over the farm and use such maps for decision making.
- **Best management practices**: The proper handling of agronomic practices is critical for achieving high returns from any crop production program. Simulation can be done to strategize best crop management activities for sustainable agriculture *i.e,* without affecting the soil quality, soil fertility, ecological balance of nature and reducing environmental degradation. Models having chemical leaching or erosion components can be used to determine the best practices over the long-term. The EPIC model has been used to appraise erosion risks due to cropping practices and tillage.
- **Yield prediction:** Pre-harvest yield estimating for industries over large areas is important to the producer (harvesting and transport), the processing agent (milling period) as well as the marketing agency. It also helps the researcher in terms of planning. The technique uses actual weather records at the time of crop production together with forecast data to estimate yield across the industry.
- **Introduction of a new crop:** Agricultural research is linked to the prevailing cropping system in a particular region. Data concerning the growth and development of a new crop in that region would be lacking. It may many years (5-15 years) to develop a new crop variety using varying breeding methods suitable for different crops based on their mode of reproduction, morphological features and other characteristics. The modelling approach can be used for conducting multi-location field trails to understand genotype and environment interaction (G x E).and provide estimates of yield probability in target environments. Developing a simulation model based on scientific data collected elsewhere and a few datasets collected in the new environment helps in the assessment of temporal variability in yield using long-term climatic data. Running the simulations with meteorological data in a balanced network of locations also helps in locating the industry.
- **Global climate change and crop production:** The increasing diversity in climatic conditions and extreme weather calamities has been a major concern of scientists across the world. Increased levels of CO_2 and other greenhouse gases are contributing to global warming with associated

changes in rainfall pattern. Crop cultivation is highly dependent on climate therefore variation in global climate affect the crop productivity as several morphological and physiochemical processes like plant growth, photosynthesis, respiration, evapotranspiration, *etc.* are affected by changing temperature, relative humidity, precipitation, solar radiation, *etc.* Assessing the effects of these changes on crop yield is important at the producer as well as at the government level for planning purposes.

Limitations of Crop Modelling

Agricultural scientists have developed the models for their application be the user-group which includes the agronomists, breeders, agricultural scientists, policy-makers, extension workers, and farmers. Since different users have different expertise in the modelling field, models can be misused. As crop models are not universal, the user has to select the most appropriate model in accordance with their objectives. While making a prudent choice of model, it is necessary that the limitations of model are taken into consideration, such as the study of modelling to put it into proper perspective and achieve successful applications (Patricia *et al.*, 2013).

Since agricultural systems are complex to fully understand, the modeller usually adopts a simplified equation to describe the broader sub-system based on the model objective and the developer's views. Models that do not originate from strong interdisciplinary collaboration are often advantageous in the developer's area of expertise, but weak in other areas.

The performance of model can be related to the quality of scientific information used in model development, calibration and validation. When a model is implemented in a new state (*e.g.*, replacing a new variety), the calibration and validation steps are important for the correct simulation. Model performance can only be examined based on the quality of input data. It is easier to record data for the above-ground crop growth and development but for evaluation of below-ground is difficult to examine the soil characteristics and root growth. Using approximations may cause inaccurate results.

Most simulation models require reliable and complete meteorological data. However, meteorological sites may not completely represent all the weather parameters at a chosen location, for instance, data may be accessible for only one or a few parameters (rainfall or temperature) while data for solar radiation may not be available. In such cases, the modeller sometimes has to rely on the generated data or fill gaps for incomplete data that affect the performance of the model.

Before applying any model, the users should have the understanding of the model structure, its requirements, assumptions and limitations for application and decision-making purposes. Model developers can raise expectations of users beyond its capabilities. For *e.g.*, A model similar to QCANE that has been developed for sugarcane growth in non-limiting conditions may lead to inaccurate analysis and result if it is applied to simulate under nitrogen stress conditions or water.

Conclusion

Crop modelling can contribute significantly as a tool for research and decision-making, tools for education, training and technology-transfer. Research community has benefited mostly from crop/soil models so far using them wide range of applications. These models are based on comprehensive climatic, soil and physiological data which makes them capable being adopted universally as they support extrapolation to alternative cropping cycles, crop and locations *etc.* Also capable of quantifying temporal and spatial variability. They give acceptable results in relatively short time span and at comparatively low costs enabling the researchers to investigate effect of a large number of management strategies which would be very cumbersome and costly by using traditional methodologies. These models are used as primarily tools for organizing and supplementing knowledge gained in experimentation. We need to strengthen our research infrastructure and integrate modelling with field experiments to use these models for solving real world problems more relevant to the current times and effectively disseminate the results from work using models to potential beneficiaries. Their importance as research tools, model development and application can contribute significantly to identify gaps in our knowledge, facilitating more efficient and targeted research planning. Simulating crop models efficiently simulate crop production. Hence these models can be used to assess irrigation and real time water requirement. In addition, studying and correlating crop yield response in different irrigation areas over several years. Calibration of a system helps in obtaining research data of several years with less time and resource.

REFERENCES

Arbat, G., Puig-Bargués, J., Duran-Ros, M., Barragan, J., Ramirez de Cartagena, F. 2013. Drip-Irriwater: computer software to simulate soil wetting patterns under surface drip irrigation. Comput. Elect. Agric. 98, 183–192.

Asseng, S., Zhu, Y., Basso, B., Wilson, T., and Cammarano, D. 2014. *Simulation Modeling: Applications in Cropping Systems* (N. K. B. T.-E. of A. and F. S. Van Alfen (ed.); pp. 102–112). Academic Press.

Bajocco S, Ferrara C, Alivernini A, Bascietto M, Ricotta C. 2019. Remotely-sensed phenology of Italian forests: going beyond the species. Int J Appl Earth Obs 74: 314–321

Baron C, Sultan B, Balme M, Sarr B, Traore S, Lebel T, Janicot S, Dingkuhn M. 2005. From GCM grid cell to agricultural plot: scale issues affecting modelling of climate impact. Philos T R Soc B 360: 2095

Challinor AJ, Watson J, Lobell DB, Howden SM, Smith DR, Chhetri N. 2014. A meta-analysis of crop yield under climate change and adaptation. Nat Clim Change 4(4): 287–291

Challinor AJ, Mueller C, Asseng S, Deva C, Nicklin KJ, Wallach D, Vanuytrecht E, Whitfield S, Ramirez-Villegas J, Koehler A.K. 2018. Improving the use of crop models for risk assessment and climate change adaptation. Agric Syst 159: 296–306

Confalonieri R, Orlando F, Paleari L, Stella T, Gilardelli C, Movedi E. 2016. Uncertainty in crop model predictions: what is the role of users? Environ Modell Softw 81: 165–173

Cook, F.J., Thorburn, P.J., Fitch, P., Bristow, K.L., 2003. WetUp: a software tool to display approximate wetting patterns from drippers. Irrig. Sci. 22, 129–134.

Deryng D, Elliott J, Folberth C, Müller Pugh CTAM, Boote KJ, Conway D, Ruane AC, Gerten D, Jones JW, Khabarov N, Olin S, Schaphoff S, Schmid E, Yang H, Rosenzweig C 2016. Regional disparities in the beneficial effects of rising CO_2 concentrations on crop water productivity. Nat Clim Chang 6(8): 786–790

Donatelli M, Confalonieri R 2011. Biophysical models for cropping system simulation. In: Flichman G (ed) Bio-economic models applied to agricultural systems, Springer, p 59–87.

Donatelli M, Srivastava AK, Duveiller G, Niemeyer S, Fumagalli D 2015. Climate change impact and potential adaptation strategies under alternate realizations of climate scenarios for three major crops in Europe. Environ Res Lett 10: 075005

Duveiller G, Donatelli M, Fumagalli D, Zucchini A, Nelson R, Baruth B.2015 A dataset of future daily weather data for crop modelling over Europe derived from climate change scenarios. Theor Appl Climatol (3–4): 573–585

Ewert F, van Ittersum MK, Heckelei T, Therond O, Bezlepkina I, Andersen E. 2011. Scale changes and model linking methods for integrated assessment of agri-environmental systems. Agr Ecosyst Environ 142: 6–17

Faivre R, Leenhardt D, Voltz M, Bnoit M, Papy F, Dedieu G, Wallach D. 2004. Spatialising crop models. Agronomie 24: 205–217

Foster, T., Brozoviæ, N., Butler, A.P., Neale, C.M.U., Raes, D., Steduto, P., Fereres, E. and Hsiao, T.C. 2017. AquaCrop-OS: An open source version of FAO's crop water productivity model. *Agricultural water management* **181**: 18-22.

Friedman, S.P., Communar, G., and Gamliel, A. 2016. DIDAS - User-friendly software package for assisting drip irrigation design and scheduling. Comput. Elect. Agric., 120: 36-52.

Hinnell, A.C., Lazarovitch, N., Furman, A., Poulton, M., Warrick, A.W., 2010. Neuro Drip: estimation of subsurface wetting patterns for drip irrigation using neural networks. Irrig. Sci. 28, 535–544.

Hoffmann H, Zhao G, Asseng S, Bindi M, Biernath C, Constantin J. 2016. Impact of spatial soil and climate input data aggregation on regional yield simulations. PLoS ONE 11(4): e0151782

Hoogenboom, G., Porter, C. H., Shelia, V., Boote, K. J., Singh, U., White, J. W., Hunt, L. A., Ogoshi, R., Lizaso, J. I., Koo, J., Asseng, S., Singels, A., Moreno, L. P. and Jones, J. W. 2019. Decision Support System for Agrotechnology Transfer (DSSAT) Version 4.7. DSSAT Foundation, Gainesville, FL.

IBSNAT 1989. The Decision Support System for Agrotechnology Transfer Version 2.1 (DSSAT v2.1) User's Guide. Dept. of Agronomy and Soil Science, COTA and HR, University of Hawaii, Honolulu, HI

IBSNAT. 1993. The IBSNAT Decade. Dept. of Agronomy and Soil Science, COTA and HR, University of Hawaii, Honolulu, HI, p. 178.

Jansson, P.E., 2012. CoupModel: model use, calibration, and validation. Trans. ASABE 55, 1337–1346.

Jones, J. W., Tsuji, G. Y., Hoogenboom, G., Hunt, L. A., Thornton, P. K., Wilkens, P. W., Imamura, D. T., Bowen, W. T. and Singh, U. 1998. Decision support system for agrotechnology transfer: DSSAT v3. Springer, Dordrecht, the Netherlands.

Jin X, Kumar L, Li Z, Feng H, Xu X, Yang G, Wang J 2018. A review of data assimilation of remote sensing and crop models. Eur J Agron 92: 141–152

Lorite, I.J., Garcia-Vila, M. and Fereres, E., 2015. Reference manual AquaCrop-GIS version 2.1. FAO. Rome, Italy.

Malek K, Stockle C, Chinnayakanahalli K, Nelson R, Liu M, Rajagopalan K, Barik M, Adam JC 2017. VIC–CropSyst-v2: a regional-scale modeling platform to simulate the nexus of climate, hydrology, cropping systems, and human decisions. Geosci Model Dev 10: 3059–3084

McCauley, D. J. 2020. Predictive Agriculture: Crop Modeling for the Future. *CSA News*, *65*(5), 3–9.

Mehla, M. K., and Singh, K. 2019. Optimizing drip Irrigation system design for Onion crop using DIDAS software. *Bulletin of Environment, Pharmacology and Life Sciences*, *8*(2), 64-69.

Mourtzinis S., Edreira JIR, Conley SP, Grassini P. 2017. From grid to field: assessing quality of gridded weather data for agricultural applications. Eur J Agron 82: 163–172

Mulla S., Singh S.K., Singh K.K. and Praveen B. 2020. Climate Change and Agriculture: A Review of Crop Models. In: Venkatramanan V., Shah S., Prasad R. (eds) Global Climate Change and Environmental Policy. Springer, Singapore.

Müller C, Elliott J, Chryssanthacopoulos J, Arneth A, Balkovic J, Ciais P. 2017. Global gridded crop model evaluation: benchmarking, skills, deficiencies and implications. Geosci Model Dev 10: 1403–1422

Patricia O.D., Yeboah S., Sylvester A., Shadrack A. and Owusu D. E. 2013. Crop modeling: A tool for agricultural research – A review. E3 Journal of Agricultural Research and Development. 2. 1-6.

Raes, D., Steduto, P., Hsiao, T.C. and Fereres, E. 2011. Reference manual AquaCrop version 4.0. FAO. Rome, Italy.

Ramankutti N, Mehrabi Z, Waha K, Jarvis L, Kremen C, Herrero M, Rieseberg L.H. 2018. Trends in global agricultural land use: implications for environmental health and food security. Ann Rev Plant Biol 69: 789–815

Ray, D.K., Mueller, N.D., West, P.C., Foley, J.A. 2013. Yield trends are insufficient to double global crop production by 2050. PLoS ONE 8: e66428

Resenes, J., Pavan, W., Holbig, C., Fernandes, J. M. and Hoogenboom, G. 2019. jDSSAT: a JavaScript modules for DSSAT-CSM integration. SoftwareX 10.

Sadler EJ, Barnes EM, Batchelor WD, Paz J, Irmak A. 2016. Addressing spatial variability in crop model applications. In Agricultural system models in field research and technology transfer. CRC Press, Boca Raton, pp 258–269

Simunek, J., van Genuchten, M.T., Sejnac, M., 2008. Development and applications of the HYDRUS and STANMOD software packages and related codes. Vadose Zone J. 7, 587–600.

Wesseling, J.G., Ritsema, C.J., Oostindie, K., Stoof, C.R., Dekker, L.W., 2009. A new, flexible and widely applicable software package for the simulation of onedimensional moisture flow: SoWaM. Environ. Model. Software 24, 1127–1132

Chapter 2

Applications of Plant Biotechnology in Crop Improvement

Tanu Shiri[1]*, Arvind Kumar[1], S.S. Gaurav[1], S.K. Singh[2] and S. Jain[3]

[1]Department of Genetics and Plant Breeding, C.C.S. University Campus, Meerut – 250004, Uttar Pradesh,
[2]Department of Agriculture Botany, CCR (PG) College, Muzaffarnagar – 251001, Uttar Pradesh,
[3]Department of Bioscience, Shri Ram College, Muzaffarnagar – 251001, Uttar Pradesh,

Recombinant DNA innovation has essentially enlarged the traditional yield improvement, and has an incredible guarantee to help plant raisers to satisfy the expanded food need anticipated for the 21st century. Emotional advancement has been made in the course of recent decades in controlling qualities from various and colorful sources, and embeddings them into microorganisms and yield plants to give protection from creepy crawly irritations and diseases, flexibility to herbicides, dry spell, soil saltiness and aluminum harmfulness; improved post-collect quality; upgraded supplement take-up and healthful quality; expanded photosynthetic rate, sugar, and starch creation; expanded viability of biocontrol specialists; improved comprehension of quality activity and metabolic pathways; and creation of medications and antibodies in crop plants. In spite of the various and broad valuable utilizations of biotechnology items, there stays a basic need to introduce these advantages to the overall population in a genuine and reasonable manner that invigorates an impartial and mindful open discussion.

The turn of events, testing and arrival of rural items produced through biotechnology-based cycles ought to be ceaselessly advanced dependent on the latest encounters. This will require a dynamic and smoothed out administrative structure, unmistakably strong of the advantages of biotechnology, however exceptionally to the prosperity of people and condition.

Keywords: *Recombinant DNA, Microorganisms and Antibodies.*

Introduction

Plant tissue culture includes a great significance in plant biotechnology specially within the crop improvement programmes. The term tissue culture could also be defined because the process of in-vitro culture of explants (pieces of living differentiated tissues) in nutrient medium under sterile conditions. However, the tissue culture incorporates the term tissue culture just as cell culture, organ culture and suspension culture too. The results of plant biotechnology are being moved quickly from research facilities to the fields (http://plant tissue culture). Notwithstanding this, the tissue culture contributes enormously for understanding the examples and dependable elements of development, digestion, morphogenesis and separation of plants. Tissue Culture: The in-vitro culture of the tissue for example Callus culture.

- **Cell Culture:** Denotes the in-vitro culture of one or more number of cells (https://www.docsity.com/en/pharmacognosy-4th-semester-).
- **Organ Culture:** This term is used for in-vitro culturing of organs like embryo, root or shoot apices (https://www.scribd.com/document/451534653/TISSUE-CULTURE).
- **Suspension Culture:** Defined as the culture of cell and cell aggregates suspended in a liquid medium (https://allnoteseasy.weebly.com/strategies-for-enhancement-in-food-production.html).
- **Ex plant:** The excised piece of distinguished tissue or the organ that is used for culture is termed as explant. *e.g.*, embryos, young leaf, bud, *etc.*

Agriculture today is on the verge of a technological period, in a manner it has never been seen before (https://thoughcopyforest2.live/6011780307/). As we have a tendency to enter the new millennium, one key development that comes to our mind is that the emergence of biotechnology, that offers a number of the best opportunities and solutions to a number of the more problems faced by us. Biotechnology is a group of biological techniques that share two things in common; they manipulate living cells and their molecules and have a large rang of uses that can improve our lives (http://balli0406.zzux.com/627.html). The major techniques of biotechnology are genetic engineering, cell culture, tissue culture, bioprocessing, protein engineering *etc.* (http://163.172.173.229/45.html). These technique of producing modified plants of known necessary agriculture qualities, in large numbers of plants from small pieces of plant in relatively short periods of time (http://def.amima.id/317FC0B/plant-tissue-culture.html). It is a technique of fast propagation under controlled disease-free conditions. Entire crop population with premium qualities are often created from a single elite specimen plant (https://vdocuments.mx/tissue-culture-banana-cultivation-bird-tissue-culture-banana-cultivation-6-2.html).

Depending on the species, the plants become prepared for plantation within the field. This technique of plant propagation greatly reduces the labour and area requirement, for producing new varieties and can also markedly enhance propagation rates. Methods of plant propagation and formation are of particular interest to our country and work on a wide range of vegetables, fruit crops and trees are in progress. Several scientists have been experimenting to extend the application of tissue culture to make plant species commercially vital. Such as, experiments on crops like coconut, date palm. Cashew, Mango, Orange *etc.* are being made in the different research laboratories. With the appearance of plant tissue technology, it is now possible to propagate fine varieties of flowers, forest, and fruit trees by tiny plantlets. Commercialization of these crops has already taken place. In Ornamental crops, Orchids, Carnation, Gladiolus, Gerbera, Anthurium *etc.* has been commercially grown. In Fruit crops, Banana, Sugarcane *etc.* has been commercially grown. In Forest trees such as Teak, Eucalyptus *etc.* has been commercially grown (https://vdocuments.mx/tissue-culture-banana-cultivation-bird-tissue-culture-banana-cultivation-6-2.html). Tissue culture method used Medicinal plants are also being experimented and soon would be ready for commercial plantation. Biotechnology is an area with a tremendous potential in finding basic issues of food, fiber, fuel and medicine particularly in developing Asian countries (https://mj.vsevchoko.ru/184).

History of Plant Tissue Culture

The German Botanist Guttlieb Haberlandt first proposed the significance of plant tissue and cell culture in seclusion, in 1902. He is viewed as the dad of plant tissue culture. He utilized tissue of *Lamium Puroureum* and *Eichhornia crassipes,* the epidemis of Ornithooalium and epidermal hairs of Pulmonaria Mollissima. He developed them on a specific salt arrangement with sucrose and watched clear development in the cells. The cells stayed alive for as long as multi month. They developed in size, changed shape; thickening of cell dividers happened and starch showed up in the chloroplasts, which at first needed it. Nonetheless, none of the cells separated. The disappointment was that he was taking care of profoundly separated cells and the present a day development hormone, vital for actuating division in develop cells, were not accessible to him. Hanning (1982) had started another line of examination, which later formed into a significant applied region of in-vitro methods. Hanning extracted almost develop undeveloped organisms of certain plants like Raphanus Sativus and effectively developed them to development on mineral salts and sugar arrangement. Van Overbeck (1941) and associates showed just because the stimulatory impact of coconut milk, which was like incipient organism sac liquid, on incipient organism improvement and callus development in Datura. This demonstrated a defining moment in the field of incipient organism culture, for it empowered the way of life of youthful undeveloped organisms which neglected to develop on a blend of mineral salts, nutrients, amino acids and sugar. Ensuing definite work by Raghavan and Torrey (1963), Norstog (1965) and others prompted the advancement of Synthetic media for the way of life of more youthful incipient organisms.

Laibach (1925, 1929) showed the down to earth application of Plant Tissue Culture Course Material, Modern College 13 undeveloped organism culture in the field of plant reproducing. He disconnected incipient organisms from nonviable seeds of a specific plant and raised them to development on a supplement medium. In 1922, working autonomously Robbins (USA) and kotte (Germany) detailed some accomplishment with developing disengaged root tips. White made the principal fruitful report of consistently developing tomato root tips in 1934. During 1939 - 1950 broad work on root culture was embraced by Street to comprehend the function of nutrients embed development and shoot-root relationship. Gautheret (1983) effectively refined cells of Salix, Nicotiana-Hybrid and carrot on engineered media. They, just because, showed that development controllers and nutrients whenever added to media upgraded the development framing jumble of cells called callus. Skoog (1944), and Miller (1961) exhibited the enlistment of divisions in secluded, develop and separated cells by utilizing manufactured just as common mixes. Muir (1958) built up a method of developing single cells into fluid medium if there should be an occurrence of *Tenetes Erecta* and *Nicotiana Tabacum*. It raised entire plants beginning from single cells of tobacco. Skoog and Miller (1961) demonstrated that changing the general convergences of the two substances in the medium could manage the organ separation. The primary reports of some incipient organism arrangement from Carrot tissue showed up in 1958-59 by Reinert (Germany) and Steward (USA). Ball (1946) effectively raised entire plants of Lupinus and Tropaeolum by refined shoot tips. Morel and Martin (1951), just because, recouped infection free Dahlia plants from contaminated people by refined their shoots. Murashige (USA) utilized this strategy to mu Ui ply plants in enormous number for a few animal types going from greeneries to foliage, bloom and organic product plants. Guha and Maheshwari (1966) showed the chance of raising huge quantities of plantlets from dust grains of Dhatura. In 1972 Carlson and others created the primary substantial half and half between two plants by fiising their protoplasts.

Objective of Tissue Culture

The essential target of tissue culture task could be proliferation of enormous amount of good quality planting material from tip top mother plants inside brief timeframe, space with least expense per plant.

Conceptual Foundation of Plant Tissue Culture Technology

Plant tissue culture alludes to the development "In-Vitro" of all plant parts, regardless of whether a solitary cell, a tissue or organ, under ailment free conditions on supplement medium. In the existence pattern of any creature, two gametes of other gender circuit to frame a solitary cell-zygote. From this single-celled zygote starts the whole multicellular and multi organed body of a higher creature. In a blooming plant, for instance, structures as practically assorted as underground roots, green leaves and blossoms all emerge from the single-celled zygote through a large number of divisions of cells. Plant Tissue Culture Course Material, Modern

College Theoretically, hence, all the phones in a plant body, in the case of dwelling in the blossoms, leading tissues or root tips, ought to have gotten a similar genetical material as initially present in the zygote. There must be some different elements superimposed on the hereditary qualities of cells, which achieve this immense variety communicated by the hereditarily indistinguishable cells. The cycle of varieties is called separation. This separation is really gone before by certain cell and sub cell changes. An inquiry that emerges at this stage is whether the cell changes fundamental separation of different kinds of cells are lasting and irreversible or whether there is simply a versatile change to suit the utilitarian need of creature when all is said in done. During the ordinary life pattern of a plant, it is accepted, that the occasions prompting separations are of lasting nature. In any case, the investigations of Vouching on extremity in cuttings proposed something else. He had seen that all cells along the length are equipped for shaping roots just as shoots however their thickness is chosen by their general situating of the cutting. The most ideal approach to respond to this question and see more about the interrelationship between various cells of an organ and various organs of a creature would anyway be to eliminate them from the impact of their neighboring cells and tissue and develop them in detachment on supplement media. This has prompted the establishment of another part of science as 'Cell and Tissue Culture'. It is pertinent to both plant and creature cells. Plant tissue culture has procured numerous handy applications in agribusiness, cultivation and ranger service. It is progressively getting famous as a piece of ongoing field of Biotechnology.

Applications of Plant Tissue Culture

In the mid-fifties it was seen that plant cells are amiable to concoction controls in the medium whereby they can be actuated to frame composed structures and complete plants. This revelation is viewed as significant for the use of cell and tissue culture techniques to defeat a few issues associated with agribusiness, agriculture and plant reproducing.

1. The method gives an approach to quick increment of appealing and exceptional plants. The ~1,00,000 plants/year/bud in Eucalyptus and twenty thousand plants/year/bud in turmeric were found.
2. As the trials uncover, infection tainted plants additionally contain some sound stocks as such they can be acquired by isolating shoot tips for their in-vitro proliferation. This has given victories in Strawberries and Sugarcane.
3. The advancement of haploids through the strategy of anther culture has an expected hugeness in essential and applied hereditary qualities and plant rearing. During the previous 20 years the method has been effectively stretched out to around 20 plant species including some monetary plants.
4. The incipient organism culture has been valuable in defeating seed lethargy. It is additionally used for creating practical plants from crosses, which typically bomb because of the passing of juvenile incipient organisms. Trials were effective in the event of Jute and Rice.

5. The undeveloped organism tissue culture is likewise applied for the spread of uncommon plants. In certain examinations, coconuts grew delicate, strong and greasy tissue instead of the fluid endosperm (Mohan Ram, 1976). These are uncommon and over the top expensive, served uniquely at exceptional feasts in Philippines. Under ordinary conditions the coconut seeds neglect to grow. Utilizing the procedure of in-vitro culture of extracted incipient organisms De Guzman (1969) prevailing with regards to making plantlets from makapuno nuts (https://www.google.com/search).
6. Another significant utilization of undeveloped organism culture is found in acquiring some uncommon half and halves. It is conceivable to raise total crossover plants through undeveloped organism culture. This strategy has been productively utilized for some interspecific crosses of harvests like Tomato, Papaya and Cotton (https://www.google.com/search).
7. It is conceivable to segregate and culture single cells of plants. This aides in freak choice comparable to trim improvement, as done in Tobacco, Datura and so forth. The strategy is additionally helpful in the creation of some concoction substances in the business. Sometimes, cell societies contained multiple times more synthetic substance than the roots.
8. Recently tissue culture is utilized in protoplast culture of various assortments of plants and these protoplasts are utilized for physical hybridization.
9. A couple of elite harvest assortments have been broadly embraced, bringing about the vanishing of an enormous number of more established assortments. The woods, which house the wild races of the vast majority of the yields, are being cut for enormous scope. Consequently, tissue culture can be utilized to save germplasm for example tissue protection of these plants should be possible indistinguishably (http://shodhganga.inflibnet.ac.in/bitstream/10603/110292/12/12chapter per cent 202. pdf).

Procedure of Plant Tissue Culture

1. **Medium Preparation and sterilization (Autoclave)**: In the commercial laboratory, the medium is prepared for initiation, subculture and rooting purposes. The medium includes of the nutrients such as Micro and Macro Nutrients, Vitamins, Irons and growth regulators required for the growth of the plant. The medium is disinfected in the electrically worked Autoclave at Fifteen pounds temperature. This cleansing is essential to stay away from the bacterial sullying. This medium is then filled the way of life bottles and these containers are again cleaned in the autoclave. At that point the medium is cooled and is prepared for immunization (https://fliphtml5.com/szqfe/cimw/basic).
2. **Mother plant Selection and sterilization:** The commercial laboratory decides the plant species, which are to be multiplied in the laboratory. In this manner, the mother plants are chosen from the infection free

Figure 2.1: Autoclave Used in Sterilization Process of Glass Wares and Nutrient Media.

territories. Sound, illness and infection free plants are chosen as mother plants. The genuine plant part, which is called as ex-plant is chosen for immunization. This plant part is washed with water, fluid cleanser and germicide arrangement. This ex-plant is washed again with the concoction answer for evade any contagious defilement originating from the field condition. To eliminate the remainders of the synthetics it is altogether washed by refined water.

3. **Inoculation of explants on sterilized medium of known composition:** Inside the laminar stream station, the ex-plant is dealt with again with a disinfectant Chemical called Sodium Hypo chloride. After this treatment the ex-plant is washed triple with twofold refined water to wipe out the substance extras. The explants are immunized on the sanitized medium. The vaccination happens in the Inoculation room, on the Laminar Air Flow station, which keeps a persistent air current, which keeps this seat without the danger of pollution. The master specialists in the research center does the immunization.
4. **Transfer of cultures to the growth room:** Immunized societies are then moved to the development room where a run of the mill temperature of 15o to 20o C with the assistance of forced air systems and glass 1000 clean air is kept up. The fake lighting course of action is likewise made for the development of the plant.
5. **Transfer of cultures for Subculture:** After 3-6 weeks from immunization, the vaccinated ex-plant shows development in various shoots. These shoots are relocated on the subculture or duplication mode for development of the plants. For subculture, separate development medium is readied. A

few subculture cycles are done in the lab for large scale manufacturing from the ex-plant. (Figure 2.2)

Figure 2.2: Transfer of Cultures for Subculture.

6. **Partition of in-vitro shoots and establishing:** The shoots are isolated in the laminar wind stream stations and these completely developed shoots are moved to the pulling vehicle for root age. Contingent on the species, attaching expects One to Three weeks (Figure 2.2).
7. **Move to greenhouse for solidifying**: Rooted shoots are eliminated from the lab and are moved to the greenhouse for solidifying. In the solidifying method the plants are first kept in the mugginess chambers for acclimatization and afterward are moved to the greenhouse.

Commercialization of Plant Tissue Culture Technique

In Western Europe, business miniature proliferation began as late as mid 1980's and there were all out 200 and forty business plant tissue culture research centers delivering a great many plants for each year before the finish of 1988. Indeed, even Holland began business miniature engendering in 1980's and before the finish of 1988 Holland was creating 62 million plants with the assistance of 67 tissue culture research centers. Indeed, even Israel had five business tissue culture labs creating 5,000,000 plants for every year before the finish of 1988. Additionally, in different nations numerous business tissue culture research facilities were set up and created a huge number of plants for their nation. These nations included, Poland, Yugoslavia, and Soviet Union and so on In America business miniature engendering began in 1965 and there were around hundred business tissue culture labs at that point.

Indian Scenario

In 1980's while every one of these nations were creating a large number of plants, India had just four business tissue culture research facilities. At last the research facilities expanded, yet they couldn't create the amount that agribusiness and cultivation market required. Numerous business plant tissue culture research centers-initiated tasks in 1990's. At present the plant tissue culture is well investigations, tested and Biotechnologies for Agriculture and Aqua Culture. India has accomplished an achievement in this method by leading innovative work with exceptional examination labs like; Indian Council for Agriculture Research, Delhi; Indian Institute of Horticulture Research, Bangalore and National Chemical Laboratory, Pune. Presently the need is to make this procedure solid in the business territory of creation. Notwithstanding of the help of the public level Research Institutes, Agricultural Universities and Government Agriculture Department, business tissue culture is as yet confronting diverse issues.

Help for Plant Tissue Culture Business from Different Sources

It was discovered that distinctive sort of monetary and specialized help was accessible for business plant tissue culture research centers. The business research facilities didn't recognize these sources. Following are the distinctive help plans accessible for business tissue culture research facilities.

1. Assistance from Government Agencies Various Central and State Government organizations

These organizations have been making a decent attempt to support agribusiness business and farming preparing area in the State. The lacking component is, a discourse between the ranchers, farming industry and the Government improvement offices. Government offices have concocted various plans. The vast majority of the business research centers didn't know about these plans. Business research facilities should examine these plans and utilize them in making their business practical. Plans actualized by Central and State Government for advancement of Agri Business and Agro Processing Sector:

1.1. Name of the Scheme

To build up Tissue Culture Laboratory in private area. Plan executing organization: Central Government Nature of Assistance: Assistance is accessible for one unit for each year for private area for setting up tissue culture research center. Appropriation accessible is Rupees Ten Lakhs.

1.2. Name of the Scheme

Help for beginning Greenhouses. Plan executing office: Central Government Nature of Assistance: Under this plan for setting up Greenhouses for innovative farming after help is accessible.

1.2.1. G.H.I

Greenhouse edge and U.V. Film would be financed up to Fifty per cent with a constraint of rupees Thirty-1,000 200 and fifty just for one greenhouse.

1.2.2. G.H.2

For incompletely controlled nurseries utilizing Fan and Pad, endowment at the pace of 40 per cent with the restriction of rupees One lakh is accessible. Both the plans of nursery are given the help just for 500 sq. mt. of region for every recipient.

1.3. Name of the Scheme

Support for Drip Irrigation for High Value Crops. Plan executing organization: Agriculture Department, Government of Maharashtra Nature of Assistance: Under Centrally supported plan help is accommodated trickle water system framework for organic products, blossoms and vegetables. The pace of endowment is 90 per cent for B.C., S.T., little and negligible ranchers and ladies. For different ranchers the pace of appropriation is 70 per cent. Appropriation is accessible for a zone up to limits endorsed under Agricultural Land Ceiling Act. The pace of sponsorship is rupees 25 thousand for every hector as the most extreme breaking point.

1.4. Name of the Scheme

Property charge on Greenhouse Scheme actualizing organization: Government of Maharashtra Nature of Assistance: Greenhouses and Poly houses work for cutting edge cultivators of Vegetables, Flowers and Nursery plants won't be charged property charge by town panchayats.

1.5. Name of the Scheme

To give power at a concessional rate for innovative. horticulture. Plan executing organization: Maharashtra State Electricity Board. Nature of Assistance: Electricity rate for cutting edge agribusiness like tissue culture, greenhouses in Private and Public area will have concessional rate as follows: Rupees two and 25 paise per unit for innovative horticulture, which would incorporate tissue culture and mushroom development. To pick up the advantage from Maharashtra State Electricity Board for a concession in power charges, the research center should be situated external the Industrial zone.

2. Help from Micro Engendering Innovation

Help from Micro spread innovation parks to advance tissue culture exercises, the Department of Biotechnology, Ministry of Science and Technology, Government of India has created Micropropagation Technology Parks. At present there are two Micropropagation Technology Parks in being. One is situated in 25 Maharashtra at the National Chemical Laboratory, Pune and the other is at Tata Energy Research Institute, New Delhi. The innovation parks are created to give a viable stage to move of demonstrated advancements to the business people in the field of business plant tissue culture. The innovation stops additionally go about as an interface between the Research Institutes and the tissue culture business for quickening commercialization of the tissue culture innovation. Micropropagation innovation parks offer after administrations to the tissue culture business.

2.1. Innovation Move

Transfer of demonstrated advancements to the clients, preparing, usage and appropriation of the innovation at the customer's site.

2.2. Agreement Research

Development of cycle for the more current yields, refinement of existing convention.

2.3. Specialized Help for Creation of Plants

An indigenously planned profoundly refined research center and greenhouse exists to deliver plants for enormous scope. A solid gathering of researchers and profoundly experienced staff are accessible with the unit and give specialized help for creation of plants for estate projects or field preliminaries.

2.4. Preparing of the Staff

Training programs are coordinated on broad preparing for plant tissue culture and concentrated preparing on explicit plant of revenue.

2.5. Setting Up and Authorizing of Tissue Culture Lab and Greenhouse

Designing of the research center and greenhouse, preparing the lab and the greenhouse.

2.6. Warning Consultancy for Generally Speaking Running of the Research Facility

Consultancy accommodated running a tissue culture unit, which incorporates all the perspectives, directly from washing of crystal till creation and solidifying of plants in the greenhouse.

2.7. Turn-key Venture Dependent on the Customer's Need

All the above administrations can be taken independently or in blends according to the requirements. Turn-key venture is likewise offered which incorporates all the above administrations together and some other related angles or issues.

3. Help from Financial Institutes

Monetary help is accessible to tissue culture business houses from the State Bank of India, Bank of Baroda, Central Bank of India, Dena Bank and NABARD. The terms and states of help vary from organization to establishment.

4. Assistance for Marketing

Public Horticulture Board, Ministry of Agriculture, Government of India was set up in 1984 for coordinated advancement of cultivation in the nation. Public cultivation board energizes and advances the improvement of agriculture industry in the nation. The board helps in expanding creation and showcasing of cultivation produce. Public Horticulture Board's developing project in introduction: Development of promoting of agriculture produces through investment of Soft Loan. Objective: To build stream new and prepared agricultural produce to focused homegrown and outside business sectors. Products: Fruits, vegetables of business significance Assistance: The credit backing will be made upto a little less than half of the advance part upto a constraint of rupees hundred lakhs for each venture @ four per cent overcharge for every annum, repayable in five portions after a ban of three to five years.

5. Benefit from Biotech

Biotech Consortium India Limited Biotech Consortium India Limited was set up with the target of giving the linkages to encourage quickened commercialization of biotechnology. This association was consolidated as a public restricted organization in 1990 under the Indian Companies Act 1956. It is advanced by the Department of Biotechnology, Government of India. Biotech Consortium India Limited has been busy with development improvement, advancement move, adventure consultancy, finance association, information dispersal, work planning and circumstance related to biotechnology. It has helped more than One hundred and twenty clients including scientists, technologists, research foundations, schools, unique finance managers, the corporate region, government, banks and money related associations.

Advtanges of Plant Tissue Culture

Mass Multiplication of Elite Clones

Micropropagation allows the making of tremendous amounts of plants from small amounts of the mother plant. The creation requires decently short periods of time to create plants. Dependent upon the species under creation, a singular ex-plant can be expanded into two or three thousand plants in less than one year.

Predictable with Type creation: Large number of reliable with the sort plants could be incited inside a short period of time and space and that too reliably. For example, it very well may be possible to spread Two to Four lakhs of tissue refined plants from a singular growth or rose against 10 to 15 plants by standard techniques. Similarly, it may take around Two to Four months to convey a sound planting material by tissue culture infers, however at any rate Six to Eight months is needed for most species by the latest procedure for plant causing.

- ☆ **Higher Yields:** Tissue Culture Plants may have extended spreading and blossoming, more vital force and better return, essentially on account of likelihood of removal of illnesses.
- ☆ **Flexible method:** The versatility of nurseries can be improved. As the capital theory on mother plant is diminished to close to zero, it very well may be less difficult to conform to advancing conditions. Besides, an unrivaled program of creation is possible, because of the more noticeable plant consistency and the openness in the mass at whatever point.
- ☆ **Innovation of new combinations:** Tissue culture can be utilized for raising new arrangements.

Conclusion

Expansion by tissue culture offers extraordinary business prospects in expand plants, vegetables and natural item plants, where assessment of the thing is high. In India the tissue culture method has supposedly been effective in excess of a hundred types of plants. It has been assessed that in India, in excess of Three hundred and fifty million tissue refined plants are being created yearly through tissue culture strategy. Plant Tissue Culture has come to remain as an instrument in plant science. Plant Tissue Culture can possibly resolve the issues of test science, which in any case through customary techniques is hard to handle. Sooner rather than later this method will assume a noticeable part in hereditary designing, rearing and afforestation programs. Tissue culture strategy is a shelter for the farming and the agriculture business in view of its various points of interest. Tissue culture can create a few quantities of sound, infection free and consistent with type plants. The benefit of this strategy is with the end goal that these plants can be planted whenever during the year, which tackles ranchers' occasional manor issues. Illness free plants, duplicated through tissue culture, produce more significant returns than tainted ones. This biotechnology application can be utilized for both, customary and new assortments. Painstakingly observed creation would guarantee clean plantlets for circulation. Tissue culture action is taking a state of an industry the same number of ranchers are planting tissue culture developed plantlets; agro-brokers are purchasing and selling tissue culture developed plantlets while some are sending out either the plantlets or the produce of plants developed by tissue culture, particularly assortments of blossoms like roses, anthuriums and gerberas are getting a charge out of high benefits. Tissue culture is the best progression in plant rearing. Agribusiness and agriculture industry should exploit this strategy in arriving at the best statures at public and worldwide levels. In India, tissue culture is quickly turning into a business strategy for engendering new and uncommon species and hard to-proliferate plants. As of now, the interest for miniature proliferated plants is more prominent than the stockpile with certain plants. A few producers spend significant time in just the miniature proliferation of plantlets, leaving the developing on for example solidifying movement to other people. Numerous cultivators are coordinating a tissue culture lab into their general activity. While many plant tissue culture labs are coming up, a portion of the research centers are being shut down because of different reasons and a large number of the current labs are discovered to be griping about the issues they are confronting. Business utilization of plant tissue culture strategy has tremendous business potential if tissue culture research centers are liberated of their requirements, especially monetary and advertising. They are the organizations that would empower India hold its independence on farming creation front. Account being the core of any business, funds should be controlled and managed foundationally in tissue culture business also. A careful investigation of the monetary feasibility and business possibilities of the tissue culture adventure should be completed by the new contestant. It is the need of the time with the setting of a time of globalization and the development of World Trade Organization.

REFERENCES

Carlson, P.S., Smith, H.H. and Dearing, R.D., 1972. Parasexual interspecific plant hybridization. *Proceedings of the National Academy of Sciences*, *69*(8), pp.2292-2294.

Gautheret, R.J., 1983. Plant tissue culture: A history. *The botanical magazine= Shokubutsu-gaku-zasshi*, *96*(4), pp.393-410.

Gupta, G.R., Guha, S.I.P.R.A. and Maheshwari, S.C., 1966. Differentiation of buds from leaves of *Nicotiana tabacum* L. in sterile culture. *Phytomorphology*, *16*, pp.175-182.

Hanning, G.E. and Conger, B.V., 1982. Embryoid and plantlet formation from leaf segments of *Dactylis glomerata* L. *Theoretical and Applied Genetics*, *63*(2), pp.155-159.

http: / /plant tissue culture: history, methods and application www.biologydiscussion.com

https: / /www.docsity.com / en / pharmacognosy-4th-semester-1 / 5529632 /

https: / /www.scribd.com / document / 451534653 / TISSUE-CULTURE

https: / /allnoteseasy.weebly.com / strategies-for-enhancement-in-food-production.html

https: / /thoughcopyforest2.live / 6011780307 /

http: / /balli0406.zzux.com / 627.html

http: / /163.172.173.229 / 45.html

http: / /def.amima.id / 317FC0B / plant-tissue-culture.html

https: / /vdocuments.mx / tissue-culture-banana-cultivation-bird-tissue-culture-banana-cultivation-6-2.html

https: / /mj.vsevchoko.ru / 184

https: / /www.google.com / search

http: / /shodhganga.inflibnet.ac.in / bitstream / 10603 / 110292 / 12 / 12_chapter per cent 202.pdf

https: / /www.scribd.com / document / 417811331 / Plant-biotechology-for-crop-improvment-project-2-docx

https: / /fliphtml5.com / szqfe / cimw / basic

Laibach, F., 1925. das Taubwerden von Bastardsamen und die kunsiliche Aufzucht fruh absterbender Bastardembryonen. Z. Bot. 17: 417–459.

Laibach, F., 1929. Ectogenesis in plants. J. Hered. 20: 201–208.

Miller, C.O., 1961. Kinetin and related compounds in plant growth. *Annual Review of Plant Physiology*, *12*(1), pp.395-408.

Morel, G. and Wetmore, R.H., 1951. Tissue culture of monocotyledons. *American Journal of Botany*, pp.138-140.

Muir, W.H., Hildebrandt, A.C. and Riker, A.J., 1958. The preparation, isolation, and growth in culture of single cells from higher plants. *American Journal of Botany*, pp.589-597.

Norstog, K., 1965. DEVELOPMENT OF CULTURED BARLEY EMBRYOS I. GROWTH OF 0.1–0.4 mm EMBRYOS. *American Journal of Botany*, *52*(6Part1), pp.538-546.

Raghavan, V. and Torrey, J.G., 1963. Growth and morphogenesis of globular and older embryos of Capsella in culture. *American Journal of Botany*, *50*(6Part1), pp.540-551.

Skoog, F., 1944. Growth and organ formation in tobacco tissue cultures. *American Journal of Botany*, pp.19-24.

van Overbeek, J., Conklin, M.E. and Blakeslee, A.F., 1941. Chemical stimulation of ovule development and its possible relation to parthenogenesis. *American Journal of Botany*, pp.647-656.

Chapter 3

Torrefaction of Crop Residue as Biomass and its Application

Kanishk Verma, Parmod Sharma, Nitin Karwasra and Yadvika

Department of RBEE, COAE&T, CCSHAU, Hisar, Haryana

Torrefaction is a thermal conversion method of biomass in the low temperature range of 200-300°C. Biomass is pre-treated to produce a high quality solid biofuel that can be used for combustion and gasification. In this review the characteristics of torrefaction are described and a short history of torrefaction is given. Torrefaction is based on the removal of oxygen from biomass which aims to produce a fuel with increased energy density by decomposing the reactive hemicellulose fraction. Different reaction conditions (temperature, inert gas, reaction time) and biomass resources lead to various solid, liquid and gaseous products. The torrefaction process produces a brown to black uniform solid product, as well as condensable (water, organics, and lipids) and non-condensable gases (CO_2, CO and CH_4). Typically during torrefaction, 70 per cent of the mass is retained as a solid product, containing 90 per cent of the initial energy content, while 30 per cent of the lost mass is converted into condensable and non-condensable products. A short overview of the different mass and energy balances is presented. Finally, the technology options potential are described.

Keywords: *Torrefaction, Biofuel, Combustion and Gasification.*

Introduction

Energy security and environmental sustainability are the major emerging issues in the world that can only be addressed through the diversification in the energy resources and clean fuels. The promotion of indigenous renewable energy sources and the low carbon fuels could be a win-win solution while addressing issues of the global warming and climate change. Volatility of oil prices has encouraged the global community to reduce the dependence on oil and replace it with a clean and renewable energy resources (Deutmeyer *et al.*, 2012). Biomass energy – a renewable energy – could be a good candidate for replacement of fossil fuels. It can be used in three forms, such as solid (briquette, pellet, char), liquid (ethanol, biodiesel), and gaseous (producer gas, biogas) (Koh and Hoi, 2003). (Nhuchhen, Basu, and Acharya 2014)

Energy produced from biomass is considered carbon-neutral because the carbon dioxide released during conversion is already part of the carbon cycle. Energy from biomass can be produced from different thermochemical (combustion, gasification, and pyrolysis), biological (anaerobic digestion and fermentation), or chemical (esterification) processes, where direct combustion can provide a near-term energy solution. The growing interest in biomass as a solid fuel includes combustion to produce steam for electrical power and commercial plant uses, as well as gasification to produce a combustible gas (large partial pressure of nitrogen and CO_2, called producer gas) and syngas (carbon monoxide and hydrogen with low amounts of nitrogen and CO_2). Still, the use of either producer gas or syngas in modern reciprocating or gas turbines, or to produce higher value chemicals and fuels, is limited due to biomass feedstock preparation, accumulation logistics, and economics. (Tumuluru *et al.*, 2011)

Biomass can be transformed into gas or liquid fuels via a variety of methods, such as gasification, pyrolysis, anaerobic digestion, fermentation and transesterification. It can also be utilized as a solid fuel and burned directly for the generation of heat and power. However, biomass is characterized by its high moisture content, low calorific value, hygroscopic nature and large volume or low bulk density, which result in a low conversion efficiency as well as difficulties in its collection, grinding, storage and transportation. For those reasons, biomass is usually blended with coal for co-firing rather than used alone in power plants (Bergman and Kiel 2005). In the past, a number of biomass pretreatment methods have been developed to address the aforementioned disadvantages. Among the explored biomass upgrading methods, torrefaction and densification (or pelletization) are two noticeable routes for solid fuel production.(Chen, Peng, and Bi 2015)

Limitations of Biomass as Fuel

Some of the inherent problems with raw biomass materials compared to fossil fuel resources (low bulk density, high moisture content, hydrophilic nature, and low calorific value) render raw biomass difficult to use on a large scale. These limitations greatly impact logistics and final energy efficiency. Due to its low energy density compared to fossil fuels, very high volumes of biomass are needed, which compounds problems associated with storage, transportation, and feed handling at cogeneration, thermochemical, and biochemical conversion plants.

High moisture in raw biomass is one of the primary challenges, as it reduces the efficiency of the process and increases fuel production costs. High moisture content in biomass leads to natural decomposition, resulting in loss of quality and storage issues such as off-gas emissions. Another consequence of high moisture content is the uncertainty it causes in biomass's physical, chemical, and microbiological properties. Irregular biomass shapes constitute another issue, especially during feeding in a cofiring or gasification system. In addition, biomass has more oxygen than carbon and hydrogen, making it less suitable for thermochemical conversion processes. Considered collectively, these properties make raw biomass unacceptable for energy applications. To overcome these challenges and make biomass suitable for energy applications, the material must be preprocessed. One of the commonly used preprocessing operations is grinding, which helps to achieve a consistent particle size; however, the moisture content of the biomass limits the performance of many grinders. Furthermore, grinding can be very costly when smaller particle sizes are desired and, in some cases, impractical for biomass with high moisture content. High moisture content can also result in inconsistent particle sizes (especially when the particles are less than 2 mm), which may not react consistently, thereby reducing the efficiency and increasing the costs of the conversion process. Also, raw biomass is thermally unstable due to high moisture, which results in low CVs and inconsistent particle-size distribution issues when used in thermochemical processes such as gasification. This can lead to inconsistent products and the formation of condensable tars, which results in problems like gas-line blockage. A viable option is to pretreat the biomass before the end-use application. Pretreatment helps alter biomass's physical properties and chemical composition and makes it more suitable for conversion. The pretreatment can be a chemical, thermal, or mechanical process, like ammonia fiber explosion, torrefaction, and steam explosion, respectively. These pretreatment processes help alter the amorphous and crystalline regions of the biomass and bring significant changes in structural and chemical compositions.

Torrefaction, which is a thermal pretreatment process, is a viable technology that significantly alters the physical and chemical composition of the biomass. Torrefaction is defined as slowly heating biomass in an inert environment and temperature range of 200–300°C. This process improves the physical, chemical, and biochemical composition of the biomass, making it perform better for cofiring and gasification purposes. Many researchers have studied the effect of torrefaction process time and temperature on the physical and chemical composition. However, a detailed literature review is lacking on the torrefaction process in terms of biomass reactions (such as depolymerization, devolatilization, and carbonization) and product properties. The primary focus of this research is to conduct a detailed literature review on biomass torrefaction, which includes (a) biomass reactions, including chemical and structural changes, (b) torrefaction product yields in terms of condensable, noncondensable, and solid product, and (c) the solid torrefied product's physical, chemical, and storage properties for energy applications. (Tumuluru *et al.*, 2011)

Torrefaction Overview

Torrefaction is a method to improve biomass properties for energy generation. In literature, it is defined as a thermal treatment process through which biomass is heated between temperature of 200–300°C in an inert condition and at a relatively low residence time. Historically, torrefaction principle became known in relation to wood pretreatment in the 1930s in France (Bioenergy Update 2000) when the production of torrefied wood (TW) was researched for use in gasifier, not until the 1980s when there is an interest in substituting charcoal for TW in metallurgic processing plant that first torrefaction demonstration plant was built in France by a French company, Pechiney, to produce TW of 12,000 tons/acre. During torrefaction, the biomass properties are changed to better fuel characteristics for combustion and gasification applications. The torrefied products show relatively similar characteristics as coal. Torrefaction combined with densification provides an energy dense fuel of 20 to 25 GJ/ton (Bergman and JHA 2005).

Torrefied materials possess following characteristics:

1. Hydrophobic behaviour: TB has hydrophobic characteristics owning to the destruction of its O–H bond structure, hence making it incapable to retain or absorb moisture.
2. Inhibiting biological decomposition: stopping biologi- cal decomposition like rotten
3. Improved grindability: Torrefied biomass has improved grindability. This leads to more efficient co-firing in existing coal-fired power stations or entrained-flow gasification for the production of chemicals and transportation fuels. TB is more brittle owing to its higher C/H and C/O ratios, hence provides enhanced pulverize characteristics and requires far less energy for grinding compared to that of raw biomass (Bergman and Kiel 2005, Arias 2008, Bridgeman 2010).
4. Higher heating value: Torrefaction increases the calorific value of biomass and as a result increases their energy density (Bergman *et al.*, 2005, Sadaka and Negi 2009). Densification increases the bulk and volumetric density of biomass. Hence, a combination of torrefaction and pelletization processes produce torrefied pellets, which pilot-scale experiments have shown to have better handling than biomass pellets due to its hydrophobicity (Wang 2011). Torrefaction process causes dehydration that initiates and propagates cracks in the lignocellulosic structure (*e.g.* wood), as a result induces porosity and density changes.
5. Particle sizes and distribution of torrefied biomass: Pulverized torrefied biomass exhibit more uniform and smaller particle sizes compared to that of pulverized raw biomass (Phanphanich and Mani, 2010; Govin, 1988)

What is Torrefaction?

Although the definitions exhibit similarities in terms of torrefaction processes, the operating temperature range differs from studies to studies depending on the biomass types that were researched. Bergman and Chen (Bergman 2005, Chen and Kuo 2011] defined the Torrefaction temperature range from 200 to 300°C; Prins (Prins *et al.*, 2006) and Pimchuai (Pimchuai *et al.*, 2010) defined temperature between 230 and 300°C; meanwhile, Arias (Arias *et al.*, 2008) defined temperature range between 220 and 300°C. Studies have shown that biomass exhibit different behaviour to thermal treatment owing to their types, origin and properties (Bridgeman *et al.*, 2008); hence, the initiation of biomass decomposition depends on the type of biomass. In order to develop a more general definition of torrefaction, an experimental study on a range of biomass types will be required to determine the temperature at which a biomass sample is torrefied. This may be exemplified by hydrophobicity, *i.e.* the operating temperature and residence time when the torrefied biomass seizes to absorb water. Although the typical definition that mostly occur in published journals is "the thermal pretreatment method carried out between the operating temperature of 200°C and 300°C under inert condition and relatively short reactor residence time and slow heating rate less than 50°C/min" (Bridgeman *et al.*, 2008). It is carried out under conditions of atmospheric pressure and in the presence of a minimum amount of oxygen in order to avoid spontaneous combustion. Recently, a number of researchers including present authors have carried out Torrefaction research at different oxygen concentrations. P. Rousset (Rousset *et al.*, 2011) in their study showed that the different oxygen concentrations did not significantly affect the composition of the solid by-product for low temperatures. An oxygen concentration of 6 per cent apparently shows better characterisations on grindability and hydrophobicity tests of torrefied biomass (Sule 2012). Therefore, torrefaction of biomass can be defined as a thermochemical pre-treatment process in an oxygen reduced condition at a temperature range from 200 to 300°C for a shorter residence time that maximizes the solids content and enhances its hydrophobic characteristics. The torrefaction process involves the decomposition of biomass during which various types of volatiles are liberated, and the final product is a solid fuel generally called torrefied biomass or torrefied fuel. (Acharya, Sule, and Dutta 2012)

Torrefaction Process

The pre-conversion of biomass using torrefaction involves three main steps: chopping, drying and Torrefaction (roasting) (Yan *et al.*, 2009) as shown in Figure 3.1. During Torrefaction process, biomass is fed into a chopper to reduce them into fine or more uniform particles. The chopped biomass then goes through the drying section to remove the moisture and then fed into the torrefaction reactor.

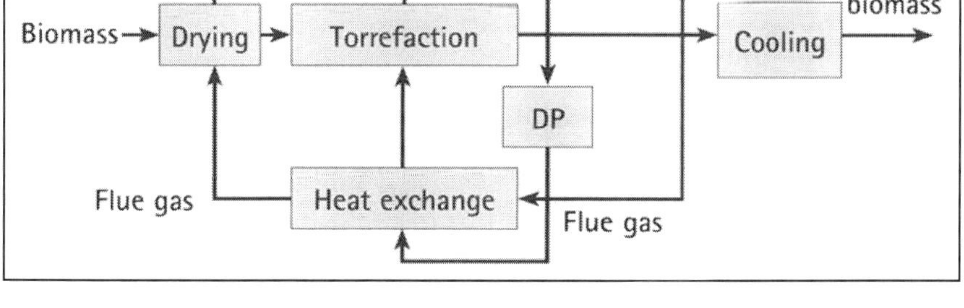

Figure 3.1: Basic Principle Concept for Directly Heated, Two Stage Torrefaction with Gas Recycling (Tumuluru *et al.*, 2011).

The moisture liberated during drying composed of both condensable and non-condensable gases and volatiles as stated in Figure 3.2.

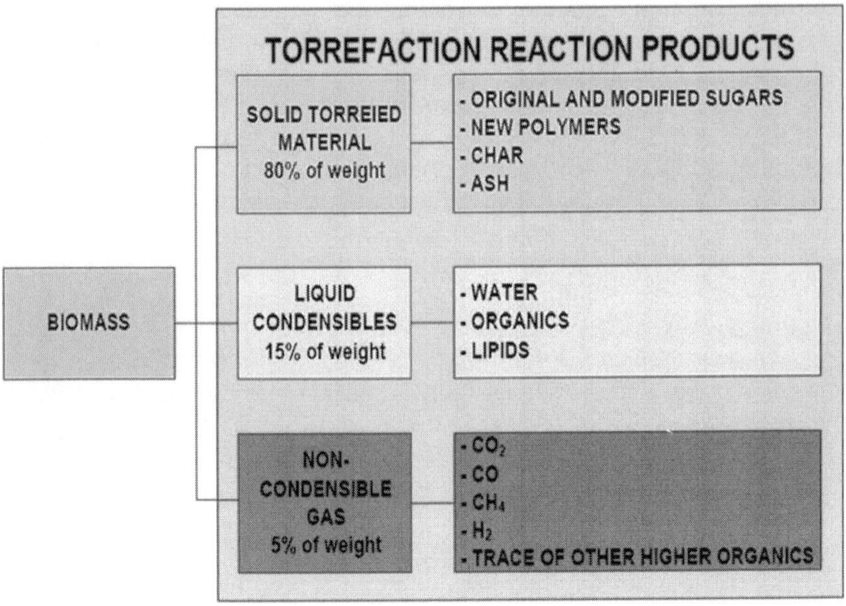

Figure 3.2: Products Formed during Torrefaction Process.

The higher the temperature of torrefaction, the higher the combustion heat of the waste volatiles gas liberated during the process. After a complete devolatilisation of the biomass, the final solid product that remains is often referred to as torrefied biomass. The improved combustible properties of biomass after torrefaction result in an attractive solid fuel for combustion and gasification processes. Furthermore, the improved grindability of torrefied biomass makes it advantageous for pelletization, which facilitates storage, transportation and co-combustion of biomass with coal (Bergman and Kiel 2005). During torrefaction process, biomass undergoes series of decomposition reactions that cause the liberation of gaseous products including volatile organic compounds. In particular, the C, H, O compositions of the biomass become altered. The decomposition of biomass polymer structure during torrefaction causes the destruction of its hydroxyl (OH) group and making it incapable to form hydrogen bond with water and hence loses its tendency to absorb water (Chen *et al.*, 2011). As a result, torrefied biomass is nonpolar molecular structure, which is practically hydrophobic. During torrefaction process, biomass undergoes two stage processes: drying and torrefaction. During drying, biomass loses majority of its moisture at temperature around 110°C, and further increase of treatment temperature initiates the decomposition of its polymeric structure, predominantly, hemicellulose. At torrefaction temperature between 250 and 300°C, high significance of decomposition occurs in hemicellulose and relatively slight decomposition in lignin and cellulose. Consequently, majority of biomass weight loss is attributable to the decomposition of hemicellulose into volatile compound. Since only slight devolatilisation occurs in lignin and cellulose, torrefied biomass retains majority of its energy content (Bergman *et al.*, 2005). According to Bergman *et al.*, the typical process of Torrefaction retains around 70 per cent of its mass which contains around 90 per cent of its initial; hence, around 30 per cent of the mass containing only 10 per cent of energy content of the biomass is converted into torrefaction gases (*i.e.* the volatile organic compounds released as flue gas). This is illustrated in Figure 3.3. Prior to torrefaction process, chopping of biomass feed- stock into uniform sizes may be required depending on the feedstock type and properties.

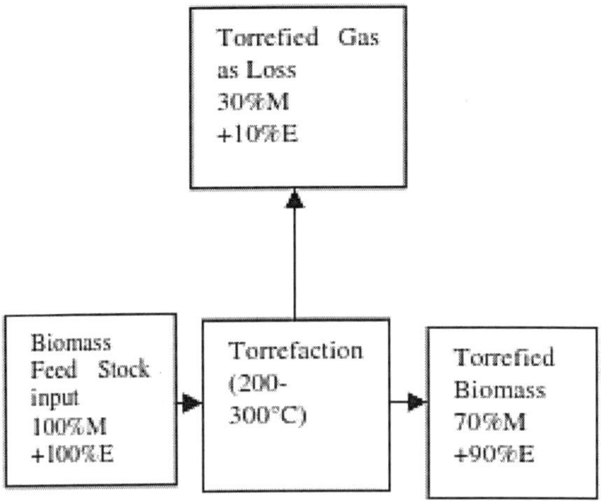

Figure 3.3: Mass and Energy and Energy Balance of a Typical Torrefaction Process (M = mass and E = energy).

Particle size has significant effect in torrefaction reactions according to Ciolkosz and Wallace (Ciolkosz and Wallace 2011), especially when large biomass feedstock is being processed. Although no definite sizes are recommended for biomass in torrefaction process by most studies, the sizes can be based on processing equipment and biomass properties. Most studies agree that temperature parameter has more significant effect in enhancing the combustible properties of biomass than residence time (Bridgeman *et al.*, 2008 and Bergman *et al.*, 2005).

Torrefaction Kinetics

The weight loss kinetic of torrefied wood and concluded that the kinetics of torrefaction occurs in two steps reactions: hemicellulose decomposition and cellulose decomposition. And since hemicellulose decomposition occurs faster than the cellulose decomposition, it contributes significantly towards the overall mass yield of torrefied wood. Due to these different fractions, biomass can decompose in different way under various conditions. Biomass undergoes four stages during torrefaction process: moisture evaporation, hemicellulose decomposition, lignin decomposition and cellulose decomposition (Van der Stelt *et al.*, 2011).

Torrefaction Mechanism

During torrefaction process, the thermal decomposition of biomass causes numerous reactions to occur through their polymer/cell structure. At low torrefaction temperatures, decomposition occurs in the hemicellulose structure by means of a limited devolatilisation and carbonization; meanwhile, in the lignin and cellulose structure, a minor decomposition occurred. It has been observed that the transition from one decomposition regime occurs at narrow temperature range for hemicellulose while the transitions for lignin and cellulose occur over at wide temperature range. Hence, it can be concluded that hemicellulose is the most reactive polymer constituent of biomass, and it is attributed to the significant mass loss in biomass during torrefaction (Rousset *et al.*, 2011).

Temperature has a significant effect in the degree of decomposition of biomass. In the process of heating lignocellulosic materials, the decomposition of the polymer structure of the material undergoes stages of decomposition regimes due to increasing temperature. At temperature between 200 and 280°C, only hemicellulose has undergone depolymerization with limited devolatilisation reactions while in lignin and cellulose, these reactions are still occurring. After majority of the moisture has been removed at temperature between 100 and 120°C, the significant weight loss of biomass is attributable to depolymerisation and limited devolatilisation of the hemicellulose during torrefaction process (between temperature of 200 and 300°C). Hence, since only slight depolymerization and devolatilisation reaction occur in lignin and cellulose during torrefaction, majority of the energy content remains in the torrefied products. Furthermore, in a study on the torrefaction impact on lignocellulosic structure of biomass, P. Rousset (Rousset *et al.*, 2011) concluded that the slight weight loss that occurred in biomass at temperature of 230°C was attributed to slight decomposition of hemicellulose, and at temperature around 260°C, severe decomposition of hemicellulose and slight of lignin contributed to massive biomass weight loss. Rousset went further to categorize the temperature range for thermal decomposition of hemicellulose as 150 to 350°C, cellulose as 275 to 350°C and lignin as 250 to 500°C. During decomposition of lignocellulosic polymer structure, other important parameter is the residence time, which accounts for the transition periods that exist from a decomposition regime to another.

Application

Torrefied biomass applications pulverized fuel combustion in coal-fired power stations and entrained-flow gasification are particularly interesting product outlets for biomass. In both applications, biomass has to be fed to the reactor as a powder, which is difficult, costly, and achievable only at very low capacity in classical coal mills. Due to this limitation, wood pellets are currently the state-of-the-art for cofiring, as they consist of sufficiently small particles. Consequently, wood pellets also have some limitations in terms of energy content and moisture uptake during storage and transportation. Torrefied biomass, because it is energy-dense and hydrophobic in nature, can be a good replacement for wood pellets in cofiring and gasification plants. The high fuel quality of torrefied biomass makes it very attractive for combustion and gasification applications. Due to high calorific values, the thermal energies of the combustion and gasification system can be improved significantly. However, data is lacking on milling, handling, storing, transporting, and combusting large amounts of torrefied biomass at a commercial scale. Common torrefied biomass applications include: (1) high-quality smokeless solid fuels for industrial, commercial, and domestic applications, (2) solid fuel for cofiring directly with pulverized coal at electric power plants, (3) an upgraded feedstock for fuel pellets, briquettes, and other densified biomass fuels, and (4) high-quality biomass solid fuel for advanced bioenergy application. (Tumuluru *et al.*, 2011)

Conclusion

Interest in biomass torrefaction has grown significantly in recent years, as has knowledge of its processes and properties. The process of torrefaction is dominated by the thermal breakdown of hemicellulose to a combination of gases, condensable liquids, and solid components that, together with the cellulose and lignin present in the raw material, comprise a product suitable for utilization as fuel. Torrefied products can substitute charcoal in a number of applications such as fuel for domestic cooking stoves, residential heating, manufacture of improved solid fuel products such as fuel pellets, compacted fireplace logs and barbecue briquettes for commercial and domestic uses.

Important advantage of torrefied wood compared to untreated wood is its uniformity. Torrefaction has the potential to become an important biomass pre-treatment technology which can improve the biomass to a high quality solid fuel with good characteristics in terms of energy density, homogeneity, grindability, and hydrophobic behaviour. The main advantage of torrefaction is the improvement of energy density and grindability.

Torrefied biomass most likely requires densification if it is to be handled successfully in a bioenergy supply chain. Characteristics such as pelletizing, biological degradation, and dust forming of the solid biomass need more attention. Due to the low moisture content of torrefied wood the transport cost is lower and the quality as a fuel better. It is easily packaged and transported, and thus constitutes an efficient fuel. The properties of torrefied biomass should lead to an improved operation in gasifiers for which the stability of the process is important. However, the production of TBP's appears to be a significant technological challenge at this point. Torrefied biomass can be also used as fuel for industrial applications. With a 30-35 per cent fixed carbon content, torrefied wood has a promising potential as a reducer. The process of torrefaction moves the chemical and physical properties of raw biomass close to those of bituminous coal, which allows co-utilization with high substitution ratios of biomass in the existing coal-fired boilers without major modifications.

Nowadays available torrefaction technologies are basically designed and tested for biomass, so further research is required, especially on reaction kinetics, to obtain the data needed for reactor design at large scale, because several gaps still exist in the understanding of torrefaction. Research about product characteristics is also recommended in order to better understand co-firing with coal, and there is a need for continued work to optimize this option.(Nunes, Matias, and Catalão 2014)

REFERENCES

Acharya, Bimal, Idris Sule, and Animesh Dutta. 2012. "A Review on Advances of Torrefaction Technologies for Biomass Processing." *Biomass Conversion and Biorefinery* 2 (4): 349–69. doi: 10.1007/s13399-012-0058-y.

Arias B, Pedida C, Fermoso J, Plaza MG, Rubiera F, Pis JJ (2008) Influence of torrefaction on the grindability and reactivity of woody biomass. *Fuel Process Technol J* 89(2): 169–175

Arias BR, Pevida CG, Fermoso JD, Plaza MG, Rubiera FG, and Pis-Martinez JJ. (2008) Influence of torrefaction on the grindability and reactivity of woody biomass. *Fuel Process Technol* 89(2), 169–175

Bergman PCA, Boersma AR, Zwart RWH, and Kiel JHA (2005) Torrefaction for biomass co-firing in existing coal-fired power stations. Report ECN-C-05-013, ECN

Bergman, PCA, and Kiel JHA (2005) Torrefaction for biomass upgrading. In: 14th European Biomass Conference and Exhibition,Paris, France, 17–21 October

Bergman, PCA, and Kiel JHA (2005) Torrefaction for biomass upgrading. In: 14th European Biomass Conference and Exhibition, Paris, France, 17–21 October

Bioenergy Update (2000) vol. 2 No. 4 https: //www.bioenergyupdate.com/magazine/security/NL0400/bioenergy_update_april_2000.htm

Bridgeman TG, Jones JM, Shield I, Williams PT (2008) Torrefaction of reed canary grass, wheat straw and willow to enhance solid fuel qualities and combustion properties. *J Fuels* 87: 844–856

Bridgeman TG, Jones JM, Williams A, Waldron DJ (2010) An investigation of the grindability of two torrefied energy crops. *Fuel J* 89(12): 3911–3918

Chen W H, Cheng W Y, Lu K M, Huang Y PA (2011) In evaluation on improvement of pulverized biomass property for solid fuel through torrefaction. *Appl Energy* 88: 3636–3644

Chen W, Kuo P (2011) Torrefaction and co-torrefaction characterization of hemicellulose, cellulose and lignin as well as Torrefaction of some basic constituents in biomass. *Energy J* 36(2): 803–811

Chen, Wei Hsin, Jianghong Peng, and Xiaotao T. Bi. 2015. "A State-of-the-Art Review of Biomass Torrefaction, Densification and Applications." *Renewable and Sustainable Energy Reviews* 44. Elsevier: 847–66. doi: 10.1016/j.rser.2014.12.039.

Ciolkosz D, Wallace R (2011) A review of torrefaction for bioenergy feedstock production. Biofuels, *Bioprod Biorefin* 5(3): 317–329

Deutmeyer, M., Bradley, D., Hektor, B., Tumuluru, J. S., Hess, R., Nikolaisen, L. and Wild, M. (2012). "Possible Effect of Torrefaction on Biomass Trade," IEA Bioenergy, Task: 40 Sustainable International Bioenergy Trade. Retrieved at http: //www.bioenergytrade.org/download s/t40-torrefaction-2012.pdf, 13th May 2013.

Govin A, Repellin V, Rolland M, and Duplan J (1988) Effect of torrefaction on grinding energy requirement for thin wood particle production [online materials]. http: //hal.archives-ouvertes.fr/docs/00/46/23/39/PDF/AG-SFGP09.pdf

Koh, M. P. and Hoi, W. K. (2003). "Sustainable Biomass Production for Energy in Malaysia," *Biomass and Bioenergy*, 25 517- 529

Nhuchhen, Daya, Prabir Basu, and Bishnu Acharya. 2014. "A Comprehensive Review on Biomass Torrefaction." *International Journal of Renewable Energy and Biofuels* 2014: 1–56. doi: 10.5171/2014.506376.

Nunes, L J R, J C O Matias, and J P S Catalão. 2014. "A Review on Torre Fi Ed Biomass Pellets as a Sustainable Alternative to Coal in Power Generation." *Renewable and Sustainable Energy Reviews* 40: 153–60. doi: 10.1016/j.rser.2014.07.181.

Phanphanich M, Mani S (2010) Impact of torrefaction on the grindability and fuel characteristics of forest biomass. *Biores Technol J* 102(2): 1246–1253

Pimchuai A, Dutta A, Basu P (2010) Torrefaction of agricultural residue to enhance combustible properties. *Energy Fuel J* 24 (9): 4638–4645

Prins MJ, Ptasinski KJ, Janssen FJJG (2006) More efficient biomass gasification via torrefaction. *Energy Fuels J* 31(15): 3458–3470

Rousset P, Davrieux F, Macedo L, Perré P (2011) Characterisation of the torrefaction of beech wood using NIRS: combined effects of temperature and duration. Biomass Bioenergy J 35(3): 1219–1226

Sadaka S, Negi S (2009) Improvements of biomass physical and thermochemical characteristics via torrefaction process. Environ Prog Sustain Energy, AlChE J 28(3): 427–434

Sule I (2012) Torrefaction behaviour of agricultural biomass and their commercial feasibilities, M.A.Sc. Thesis, University of Guelph

Stelt, M. J C van der, H. Gerhauser, J. H A Kiel, and K. J. Ptasinski. 2011. "Biomass Upgrading by Torrefaction for the Production of Biofuels: A Review." *Biomass and Bioenergy* 35 (9). Elsevier Ltd: 3748–62. doi: 10.1016/j.biombioe.2011.06.023.

Tumuluru, Jaya Shankar, Shahab Sokhansanj, J Richard Hess, Christopher T Wright, and Richard D Boardman. 2011. "A Review on Biomass Torrefcation Process and Product Properties for Energy Applications." *Industrial Biotechnology* 7 (5): 384–401. doi: 10.1089/ind.2011.0014.

Van der Stelt MJC, Gerhauser H, Kiel JHA, Ptasinski KJ (2011) Biomass upgrading by torrefaction for the production of biofuels: a review. *Biomass Bioenergy* 35: 3748–e3762

Wang G, Luo Y, Deng J, Kuang J, Zhang Y (2011) Pretreatment of biomass by torrefaction. *Chin Sci* Bull 56: 1442–1448

Yan W, Acharjee TC, Coronella CJ, Vásquez VR (2009) Thermal pretreatment of lignocellulosic biomass. *Environ Prog Sustain Energy* 28: 435–440

Chapter 4

Response of Phytohormones under Abiotic Stress

G. Babithraj Goud[1], V.G. Ishwarya Lakshmi[2], Halavath Sai Kumar[3], B. Akhila[4] and G. Sharanya[5]

[1]Department of Crop Physiology, PJTSAU, Hyderabad – 500030, Telangana
[2]Department of Genetics and Plant Breeding, PJTSAU, College of Agriculture, Rajendranagar, Hyderabad – 500030, Telangana
[3]Department of Genetics and Plant Breeding, SHUATS, Uttar Pradesh
[4]Department of Fruit Science, SKLTSHU, Rajendranagar, Hyderabad – 500010, Telangana
[1]Department of Analytical Chemistry, PJTSAU, Hyderabad, Telangana

A stress signal is first detected by the receptors located on membrane of plant cells. The signal information is then transduced down-stream, resulting in energizing of various stress-responsive genes, plants exposed to one stress may become more tolerant to another. This is known as cross-tolerance. Abiotic stress conditions lead to biosynthesis of signaling molecules, including phytohormones, which have important functions as main messenger sign in signal transduction, regulating cell metabolism. Thus, it regulates many genes, that leads to metabolic effect, hormones are endogenously bio-active content formed in the plants, which plays crucial roll in growth and development. These phytohormones include auxins, gibberellins, cytokinins, abscissic acid and ethylene. Recently, JA, SA, and BRs were reported to be-playing a vital part in plant metabolism and stress responses, have also been classified as phytohormones. In this study, the mechanisms responsible for the perception, signal transduction of phytohormones, complexity of signal cross-talk and controlling downstream abiotic stress responses were discussed.

Keywords: *Phytohormones, Abiotic stress, Signaling, Gene expression.*

Introduction

The adaptive responses enforced by salt, cold, heat and drought stress are mainly governed by phytohormones at extremely low concentrations by signaling pathways. There were significant reports in the circumstance of regulating other procedures in plants, includes abiotic stress tolerance, defense signaling pathways of SA, JA, and ET. However, the physiological, biochemical, and molecular mechanisms stimulated by phytohormones via which plants integrate adaptive responses under abiotic stress are largely unknown (Hubbard *et al.,* 2010).

Phytohormones in Plant Response different Stresses Perception Plants receive environmental signals through specific receptors, which trigger acascade of events leading to variation of cellular or metabolic activity, including regulation of the expression of specific genes. Ion channels, intracellular signalling proteins, and second messengers plays a vital role in these signal transduction cascades. The external stress signal is identified at membrane level by the membrane receptors, including receptor-like kinase (RLK) or histidine kinase (HK) and this putative membrane-localized receptors like protein might plays a vital role in osmotic stress. The two component HK was initially found in plants due to its role in perception of various environmental signals. However, when the extra-cellular sign senses a signal in the cytoplasmic histidine residue self-phosphorylated histidine is approved to an aspartate receiver in a regulator which is constitute of protein sensor. These two components sensors may attached with mitogen-activated downstream Kinases (MAPKs) targeted cellular responses Stress Signaling Transduction lead to perception of stress like cold, heat, water stress, and salinity stress. Ca2+ releasing early events cells to many environmental stimuli are known to involve in membrane depolarization and elevations of cytosolic Ca2+. Membrane fluidity and reorganization of the cytoskeleton are essential for cold-induced cytosolic Ca2+oscillations.

The cyclic ADP-ribose-gated cADPR-gated Ca2+channels are involved in ABA-induced expression of cold-regulated genes. Inositol 1,4,5-trisphosphate (IP3) gated Ca2+channels dehydration of salt induces Ca2+elevations. Calcium may in turn regulate a host of enzyme activities via calcium-dependent protein kinases and calmodulin. Activators and inhibitors are found that calcium involved in long term signaling in abiotic stress. The stress signal perceived by the membrane enter inside the nucleus to induce many stress responsive genes, the products which leads to plant adaptation to stress tolerance directly or indirectly.

On the whole, the stress response would be action of many genes, which might cross-talk with each other. The stress involved genes such as phytohormones ABA, ET, SA, JA, *etc.,* intracellular signaling are governed by protein kinases and phosphatases. Transduction up-on receives the sign from membrane receptors cells, signalling modes are Protein phosphorylation and de-phosphorylation which regulate various cellular processes like, enzyme activation, protein localization and degradation.

Most plant protein that play major role in protein phosphorlayse. AAPK is known to block ABA-induced stomata closure by eliminating ABA activation of plasma membrane anion channels. In a stress signaling cascade, inactivation of phosphor-proteins is usually accomplished by de-phosphorylation. Among the four major subgroups of proteins, phosphatases the serine/threonine phosphatase PP2A is a component of the stress signaling pathway which participates and governed by abiotic stress responses.

In response to water deficit, OsPP2A-1-5 in rice was up-regulated in response to high salinity in leaves whereas TaPP2Ac - 1 revealed that improves drought tolerance in wheat (País *et al.*, 2010). In plants, the MAPK pathways are intracellular signal modules, mediator tosignal transduction, are involved growth development PCD and in responses to a range of environmental stimuli including cold, heat, drought, UV, and reactive oxygen species (ROS). MAPK cascades shown to be concerned in SA, ABA, auxin and CKs signal transduction. Abscissic acid signaling ABA is an important phytohormone that plays vital role various stress signals (Peleg and Blumwald 2011).

As a result, ABA has become the stress responsive hormone. The main stress function of ABA involve the regulation of water balance, in plants, abiotic stress such as drought, salt, and cold stress. These different stresses sharing some common features and all induce dehydration stress to the plant cells, which activate ABA generation to various extents, that induces a common set of genes involved in ABA biosynthesis, which include zeaxanthinepoxidase (ZEP, known as ABA1 in Arabidopsis), 9-cisepoxycarotenoiddioxygenase (NCED), ABA, ABA3, also known as LOS5. ABA sign transduction transduction have been extensively studied.The flowering time control protein FCA G-protein-coupled receptor 2 (GCR2) GCR-type G-protein 1 (GTG1) and GTG2 and cytosol/nucleuslocalized).

Abscissic Acid and Drought Stress drought is often caused by long-term water deficit in the soil that cannot provide for plant transpiration. Plants have some ways to deal with drought stress, to decreases water consumption, increase water uptake, and mitigates impacts of water deficit. First, guard cell stomatal pores are closed upon drought stress and thus the transpirational water loss is minimized. However, this is a relatively quick response. Second, an array of stress-responsive genes is activated.

The products of these genes charge directly or indirectly in drought tolerance. ABA is either required or is involved in all these processes. Under drought stress, plants synthesize ABA, which in turn induces rapid closure of stoma. Stomatal closure takes place to minimize the water loss by transpiration, and ABA plays a basic role in this process by causing stomatal closure through the efflux of K+ anions from guard cells and the withdraw of osmolytes. Thus, stomatal resistance is used as a reference to comparing the intensity of water deficit in different species and growth conditions.

Interestingly, stomatal closure under drought is a response to increasing levels of endogenous ABA synthesized in the roots resulting water loss in soil. Hence, decreases in (Sg) under H2O stress are a wide ranging response in plants. Stomata conductance in kidney beam diminished rapidly 2 days after drought recovered in well-watered plants after 2 days of re-watering as reported by (Kim *et al.*, 2010).

Drought alone can trigger and improves their expression. Drought, high salinity,and cold-inducible gene contains two major elementsABRE and DRE/CRT(C-RepeaT), both of which are involved in stress-inducible gene expression.

LEA protein genes are mediated by both pathways (ie., dependent and independent) uses Ca2+signaling to induces, expression during salinity. Plants expose to salinity is known to increase in ABA concentration. Increase of the endogenous ABA concentration in leaf tissue for salt-stressed plantsas a signal for stomatal closure, induces rapid de-polymerization of cortical actin filaments and slower formation of a new type of actin which is randomly oriented throughout the cell. The stress hormone ABA is up-regulated by salinity and induces genes intricate in salt alleviation. *e.g.*, the tissue distribution and regulation of AtNHX1 expression governed by expression of some transporters involved in salt uptake and compartmentalization.

The protein 2, a serine-threonine protein kinase necessary for Na+ and K+ ion homeostasis, can interact with ABI2, a 2C type phosphatase that negatively regulates ABA signalling. Other 2C type phosphatases, PP2CA, which may act as controller several ABA responses, also interact with K+channels, while K+ channels in guard cells perform vital role in stomatal opening and closing, disturbed K+ homeostasis in roots and other tissues and cell types may contribute to salt sensitivity. Perception of the cold stress signal, transcriptional cascades are the involved in ABA dependent-independent pathways.

In the ABA-dependent pathways, there is a temporary rise in endogenous ABA content, when plants are bare to cold stress. The ABF genes are themselves convince by ABA shows differential regulation by various environmental stresses; ABF1 is induced by cold, ABF2and ABF3 by high salt concentration, and ABF4 by cold, high salt concentration and drought, the differential regulation of ABF expression suggests that separate ABFs are likely to function in these signal transduction cascades through common ABREs. ABA performs a vital role in osmotic stress-regulated gene expression.

Ethylene Signaling

Ethylene signaling perceived by mainly five receptors namely; ETR1, ETR2, ERS1, ERS2, and EIN4. However, the ET receptors may be separated in to two subgroups based on phylogenetic analysis and some structural features, subgroup 1 being composed of ETR1 and ERS1 and subgroup 2 being composed of ETR2, ERS2, and EIN4. The initial discovery that the receptors contain an HK domain and receiver domains, motifs known to participate in His-Asp phosphorlayse, might be relevant in ET signalling. The next immediate elementin the signaling pathway appears CTR1.

CTR1 is regulator of ET signaling shows similar to Raf- group of serine threonine protein kinases in its C-terminal half,CTR1 interacts with the HK domain and receiver domains of the receptor. The response pathway ET, the receptor maintains CTR1 in an active conformation so that domain kinases actively binding of ET induces change in receptor, possibly involving a change in the receptor's kinase activity, and this is transmitted to CTR1. The CTR1 reduces its own ET repression response pathway relives

ET reflexes involve alterations in gene expression by genes in the dehydration responsive element (DRE) binding (DREB) subfamily plays a vital role in plants to abiotic stresses by recognizing DRE.ERF and DREB subgroup in various plant species. Response of Ethylene to Abiotic Stresses ET and oxidative stress, signaling in ozone-exposed tomato leaves and H2O2 or ET treated stomatal guard cells. Although, correlations between ET synthesis, ROS damages tissue the functional exchange between ET biosynthesis and ROS detoxification Ozone had been previously noted as an air pollutant and now is recognized as an abiotic elicitor for activation of an oxidative burst, which provoke a local cell death response similar to that caused by the hypersensitive response. O3 result from its reactions with H2O and other components, while it enters from the stomata. ET synthesis early responses to O3 stress.

An ozone sensitive mutant rcd1 burst. Compared with wild-type plants, rcd1 is more susceptible to o2- than H2 O2 and shows prolonged lesions on leaves even after ozone is removed, suggesting a defect in restraining the toxicity of ROS. Ethylene and Salt Stress A type II ET receptor homolog gene, transgenic plants, with NTHK1 mRNA and protein expression, are salt-sensitive, as would be seen from the severe features of plagiotropic, high electrolyte leakage, and reduces root growth under salt stress NTHK1 improves salt AtERF4, indicating its role in the salt-stress response. AtERF4 stated that transcriptional repressor conferring ET insensitivity in its transgenic plants, over expression to NaCl.

Moreover, these suggest that ET signaling can regulate the salt-stress response by controlling expression of multiple genes. Ethylene and Flood Stress Flood stress plants to oxygen depletion that consists of hypoxia, deficiency of oxygen and anoxia. Oxygen loss depends upon no. of morphological and metabolic adaptations. Yemelyanov and Shishova (2012) stated that an increase in ET production under the lack of oxygen has been observed in a wide variety of cultivated plants. The reaction of ACC oxidation by ACC oxidase (ACO) requires O2 and is blocked by oxygen depletion.

Down-regulation of some ACS genes, such as OsACS5, in flooded rice revealed by expression analysis and hypoxia-induced stimulation demonstrated for expression of ACS2, ACS6, ACS7, and ACS9 in Arabidopsis as well as RpACS1 and RpACO1 in Rumex plants. ET promotes fast apoplastic acidification in flooded plants, which is important for growth promotion. The first protein TIR1, is an auxin receptor. The second protein, AUXIN-BINDING PROTEIN1 (ABP1), contains C-terminus upon binding auxin (Scherer 2011). TIR1 and ABP1 are enhanced by ABP1 apo-plastic auxin concentration.

The receptor TIR1 and its paralogs, auxin signaling F-Box 1(AFB1), AFB2, and AFB3 are the F-box subunits of SCF (Skp1-Cul1-F-box) E3-ubiquitin ligase complex (Iglesias *et al.*, 2010). Auxin binding to SCF TIR1/AFB results in the targeted UQ proteins. Aux/IAA degradation promotes activation of ARF transcription factors and the consequent expression of auxin-responsive genes. TIR1 pre-determines the transcriptional core-pressors to proteolysis by ubiquitination through its E3 ligase activity. Thus, the receptor TIR1 is close to proteolytic regulation of central negative transcriptional co-regulators in the nucleus.

Auxin-Responsive Gene Expression Auxin plays a role to some extent by regulating a group of primary responsive genes areAux/IAA, Gretchen Hagen 3 (GH3), and small auxin- up RNAs (SAURs).Members of the gene family have been studied in regulation of auxin responses Although, none of the SAUR genes are as yet, functionally characterized, the SAUR proteins bind to calcium/ calmodulin suggesting the role the involvement of CAMTA1 (CAMTA-binding motifs encompass the recently identified Ca2+responsive cis-elements) in auxin signaling (Galon *et al.*, 2010). Auxin-responsive genes are involved in a stress. Their expression was investigated by microarray analysis under desiccation, cold, and salt stress Molecular pathways auxin and ABA responses pathway provide evidence for auxin-ABA interaction. The role of IBR5, phosphatases like protein, supported the link between ABA and Auxin sign pathways. Promoters of the auxin-responsive genes expressed under abiotic stress conditions were analyzed to identify cis-acting regulatory elements linked to specific abiotic stress conditions. Even though, no specific to be linked to a specific stress condition analyzed, several ABA and other stress-responsive elements were identified.

Cold stress selectively inhibits the intracellular trafficking of a subset of proteins that include auxin efflux carriers. Cold stress also blocks the asymmetric redistribution and intracellular cycling of PIN3 that facilitates the response to gravity. Molecular and cellular findings reported that polar that deployment of PIN2 from protein to endosome The reduced intracellular cycling affects the functionality of PINs, resulting in reduced shoot ward transport of auxin and diminishing the root's capability to form an auxin gradient. Cold stress-induces change in plant development. Polar intracellular trafficking of auxin carriers (Rahman2013). Some components of the cold signaling pathway are linked to auxin.

SIZ1, a central regulatory component of the cold signaling pathway, reported negatively regulate phosphate-starvation-induced governing the ABA pattern. Downstream regulated cold signaling pathway in Arabidopsis is AtNUP160, plays a vital role in nucleo-cytoplasmic transport under. These results suggest that auxin transport is regulated or governed by water stress in plants. Auxin and salinity stress regulation responded to water stress in plants. Auxin and salinity stress first developmental process is seed germination. NAC transcription factor NTM2 mediates sign crosstalk between Auxin and slat stress via, IAA30gene.

Germination of the ntm^2-deficient, ntm^2- mutant seeds. However, ntm^2-1 mutant over expressing the IAA 30 gene induced by more salinity in NTM 2- dependent manner and molecular link that incorporate auxin sign into salt stress providing a role of seed germination. Cytokinin Signaling Cytokinins plays a vital role in various processes that regulate plant development via complex signaling CKs had been accepted as an main sign that govern the responses in plants to abiotic stress.

In some leguminous crops family like CR are mainly essential for nodulation However, new two genes CRs, MsHK1 from Medicagosativa, and LaHK1 from Lupinusalbus, were involved in the stress. Expression of the MsHK1 gene increased under osmotic stress, were induces the dark-stress that indicates CRs are more closely related to stress responses. Typically in Arabidopsis, transferred via histidine-containing phosphor transfer factors, AHPs, via the C-terminal receiver domains of the sensor histidine kinases CKs and Drought Stress

Physiological and biochemical responses at cellular level toward drought stress, in whichphyto-hormones such as CKs, SA, auxins, GA, and ABA modulate those responses. Signal class travelling from roots to shoots. There the CK content of xylem sap and changes under drought conditions. In grapevines, a decreases in enzymes combined Z and ZR and combined isopentenyl-adenine and isopentenyl adenosine concentrations in xylem sap decreased under drought-stress conditions.

In maize, decrease in Z and ZR concentrations in xylem sap from roots of drought stressed plants, as compared to well- watered controls. Unexpectedly, high concentration of the aromatic CK 6-benzylaminopurine (BAP) was reported in sap of xylem. Increased CK contents following over-expression of the CK biosynthetic gene isopentenyl transferase(IPT) positively correlated with elevated flooding tolerance. The expression of IPT under control of the (SARK) a drought/maturation-induced promoter. In tobacco this together with the stated that plays a vital main role in CKs source and sink responses in drought stress.

In addition, during water stress, IPT gene in rice and increases the expression of BRs genes and repression of jasmonate genes. These genes may mainly make role in mediating the input of CKs into the salt stress response pathway CKs and Cold Stress appears to rapidly up-regulate and the expression of multiple type A. ARRs and contrary to down-regulate the expression of all three CK receptors. Multiple mutant analyses have suggested a complex function for the type-A CK component. Individual mutations in arr5, arr6,or arr7 reported in enhanced cold tolerance AHK2 and AHK3 were found to be primarily involved in mediating cold to express A-type ARRs despite CK deficiency (decreased CK levels).

(COI1) is involved in jasmonate perception and signaling. F-box protein is a CO-1 acts as a substrate- recruiting module of SCF protein UQ E3 ligase. At first, COI1 was suggested to bind directly to JA-Ile and COR, serves as receptor for JA. COI1 degradation of transcriptional repressor JAZ (Jasmonate ZIM domain) proteins, and the complex of both COI1 and JAZ was reported as the true JA receptor In this complex, COI1 might act as part of an SCF E3 ubiquitin ligase to mediate JA signaling, whereas JAZ proteins bind and repress the transcription factors that modulate transcription of JA-responsive genes. (Sheard *et al.*, 2010). The reduction in salt tolerance of ADC2 mutant was loss in chlorophyll from leaves.

The expression of ADC2 were also elevated in JA treatments in barley. Moreover, in maize developing kernels, pattern of expression some genes are in responses to associated with ABA and JA pathways were examined the new genes were positively responsive to drought stress (Luo *et al.*, 2010). In asparagus spear tips increases due to JA application officinalis and in Carica papaya exposed to drought. In alternate to MeJA increases in Cistusalbidusalso subjected to drought.

Application JA or MeJA Exogenously may increases anti- oxidative ability of plants under water stress. Jas Create a role in signallling in drought-induced antioxidant responses, including ascorbate metabolism. Certainly, MeJA promoted increased production of several antioxidative enzymes, including GR, GPX, and APX it has been suggested that this increase may be due Regulator genes for controlling these enzymes or constitutive genes JA, Salinity and Heavy Metal Stress concentrate on the role of JA and its metabolites in the defense response against abiotic stresses, like salinity, and heavy metal toxicity.

Salt stress increased the JA levels in rice roots and in leaves of Iris hexagona. Endogenous application of JA generally increased in response to salinity. Therefore, high levels of JA in salt-tolerant plants accumulated after salt treatments may acts as effective against more salt concentration. Investigating the changes of endogenous levels of JA in rice under various salt stresses, JA concentration in a salt-sensitive cultivars were lower than those in a salt-tolerant cultivar.

Especially, considering the concentration of JA in the shoots of salt-sensitive plants, endogenous JA was rapidly decreased following the increase of NaCl concentration in treatment in salt tolerant cultivar JA treated one is higher compared to the salt-sensitive cultivar. A wounding-JA salinity interaction in tissues, where salt stress induced wound-related genes through activation of the octadecanoid pathway, was reported in tomato. Exogenous application of JA to rice salt stressed seeds recovery, suggesting a role for JA during the response to salinity stress.

In barley, induction JA synthesis genes involved is known as JA-responsive genes was reported as response to salinity. For example, Cu, Cd that induces the rapid accumulation of JA in Phaseoluscoccineus, and Cu have this effect in Arabidopsis plants. The first phase was a rapid increase of JA level occurring after exposure to Cu or Cd for several hours, followed by its rapid decrease. In the next phase, again, increase-but slow-of JA level occurred.

Additionally, the main interruptions indicates that Cu, Cd induces JA responsive events like the vegetative - storage protein 2 (VSP2) transcripts in A.thaliana and MAPK in rice. In a global transcriptome analysis of the response to boron (B) toxicity using microarrays, it was found that high concentrations of boric acid treatment effected in up-regulation of JA-biosynthetic and JA-induced genes in barley leaves. Salicylic Acid Signaling Perception and Transduction SA is synthesized via two distinct and compartmentalized pathways that employ different precursors, the phenyl-propanoid pathway in the cytoplasm starting from phenylalanine, and the iso-chorismate pathway that takes place in the chloroplast. The scientific community has made an important effort to find the SA receptor. NPR1 is the only known gene that, when mutated, renders plants insensitive to SA (Canet *et al.*, 2010).

NPR1has been shown to accumulate in the cytosol and migrate to the nucleus upon SA perception perform a main role in SA perception. SA seems to be crucial in the induction of different signal-transduction pathways. ROS elevation and external Ca2+ influx to be associated with the SA activation mechanism. Unlike tobacco, in the aluminum (Al)-responsive oxidative burst in Arabidopsis cell suspension culture, SA signaling is activated downstream of ROS. The increased expression of ICS1 (isochorismate synthase 1) was involved in SA biosynthesis and NPR1 (nonexpressor of PR-1) for SA perception and transduction. Abiotic Stress-responsive Proteins and Genes Induced by SA.

The abiotic stress tolerance induced by SA might have various causes. Several other genes may be induced by abiotic stress factors, while several "stress-related" compounds, such as ABA and SA, may also induce their expression. Many evidences support the important but not the only involvement of heat shock proteins (HSPs) in thermotolerance in plants. Osmotin is a stress-responsive multifunctional 24-kDa protein and provides osmo-tolerance to plants and there is a direct correlation between overexpression of osmotin and the physiological parameters associated with tolerance against salt or water deficit stress Osmotin could be concerned in osmotic adjustment of solutes. Osmotinis known to protect the proteins with well developed structure during the stress and repair of denatured protein.

Overexpression of the osmotingene in potato provides tolerance to salinity stress. The expression of an osmotingene isolated from a cDNA library constructed from petal protoplast cultures of Petunia hybrid was strongly induced in leaves that were exposed to certain pathogens, or upon wounding in the damaged leaves. SA and Drought Stress The endogenous SA content in leaves of Phillyreaplants increased progressively during drought. During recovery, the SA levels decreased but continued slightly higher than those observed before drought.

Water deficit increased the SA content in the roots of barley plants, whereas the SA content in the leaves did not change. In addition, application of SA exogenously effect in resistance to plants against the more water stress in cell suspensions from turgidity. Application of GB enhanced the yield of sunflower hybrids under different degrees of water stress. Under stress, diameter of the head (inflorescence), number of achenes oil content were reduced. At the molecular level, the improves expression of the activated disease resistance 1 (ADR1) gene.

However, the the dehydration- responsive element (DRE) B2A SA-dependent. In adr1/ADR1 nahG (naphthalene hydroxylase G), adr1/ADR1 eds(enhanced diseases susceptibility)1 and adr1/stress-signaling pathways may explain why the adr1 mutants exhibit increased sensitivity to dehydration stress. SA and High/Low Temperature Stress Not only exogenous SA application modulates stress effects, but also abiotic stress factors may alter the endogenous SA levels in plant cells.

This induction of thermo tolerance was related to changes in the antioxidant enzyme activities. Plants may also tolerate temperature elevation without heat acclimation. This is called as basal thermo tolerance. Endogenous SA correlated with basal thermo tolerance, but SA only partially induces expression of HSP genes. In other plants high temperatures (*i.e.*, heat acclimatization or acquired thermo tolerance) and SA is essential for acquired thermo tolerance. Screening Arabidopsis mutants and NahG transgenic plants for their basal and acquired thermotolerance showed that SA perform a key role in development of acquired thermo tolerance.

Heat caused increased levels of thio-barbituric acid reactive substance (TBARS—an indicator of oxidative damage to membranes) and reduced survival. SA, together with Ca 2+ messenger, ABA, and ET, protect plants against heat stress-induced oxidative damage. In chilling-resistant SA accumulate during cold treatment at (5°C). Plants have restricted growth atthis temperature, and a higher level of SA might mediate this growth inhibition because SA-deficient NahG plants more growth rate rather than wild-type plants. Application of SA increases chilling tolerance of top portion maize, cucumber, and rice seedlings, but not roots.

The SA-induces chilling tolerance in the aerial portions of maize and cucumber plants appeared to be associated to increases the activity of anti-oxidative enzymes such as GR and GPX. SA Stress Plants show a complex molecular response to salt stress, and SA could provide protection against this stress. The salt stress leads oxidative stress and plant survival impairment. Evidence for a role of SA in the oxidative damage generated by NaCl and osmotic stress.

Not only was the endogenous level of SA increased but also the activity of the SA biosynthesis enzyme benzoic acid 2-hydroxylase (BA2H) was induced under salt. The effect of endogenous application of various phytohormones, levels of ABA and JA were generally found to increase, although, interestingly enough, SA declined in response to salinity. The accumulation of heavy metals in soils toxic to plants, and at toxic concentrations they interfere with numerous physiological processes. Exogenous application of SA level in plants increases.

SA could ameliorate the adverse effects of heavy metal toxicity on plants. In one of the first works demonstrating the against an abiotic stress, factor like SA treatment induced tolerance against Cu toxicity in sunflower plants. Pre-treatment of barley seedlings with SA prevented the LP caused by Cd and increased shoot and root. This protection was not, however, the consequence of up-regulation of antioxidant activity. On the other hand, antioxidant enzyme activities were found to increase in Cd-stressed seedlings, but pre-treatment with SA suppressed this effect. Cd treatment also increased the endogenous-free SA content in maize.

Al was also shown to increase the SA concentration of the roots. SA was predictor of nickel (Ni) hyper-accumulation in the six diverse species. Elevation of free SA levels enhances the specific activity of serine acetyl-transferase, leading to elevated glutathione and increased Ni resistance. Gibberellins Gibberellins are a group of plant tetracyclic diterpenes that play roles in growth and development. Over 136 naturally occurring GAs are known, but (GA3),most commonly called as gibberellic acid, is the most important GA in plants. This group of phytohormones is produced in the roots and younger leaves and in seeds of higher plants.

GAs have also been implicated in physiological responses in plants and alter plant metabolism under stress. combination of biochemical and genetic approaches. The exogenous GA3 was able to reverse effect of salt, oxidative, and heat stresses through increasing content of endogenous SA. Moreover, priming with GA3 was very effective in enhancing SA concentration in wheat salt stress (Iqbal *et al.*, 2011). Interestingly, GAs and SA may plays an main role in abiotic stress. The role of GA3 in salinity alleviation of soybean. They found that GA3 application significantly promoted biomass, while they NaCl-induced salt stress.

In contrast, they showed that endogenous SA content decreased under the influence of elevated GA3, whereas it increased in NaCl-treated plants. indicated that GA3 application could decrease excess accumulation of ROS under suboptimal temperature in cucumber hypocotyls. These effects were correlated with the increasing activities of antioxidant enzymes, *e.g.*, Superoxide dismutase (SOD), catalase (CAT), APX, and GPX.

In addition, it has been suggested Brassinosteroids are plant growth regulators of steroidal nature that are synthesized by plants and affect many aspects of plant growth and development, including cell elongation, photo-morphogenesis, xylem differentiation, and seed germination, as well as adaptation to abiotic and biotic environmental stresses BRs is their potential to increase resistance spectrum of stresses, such as low and high temperatures, drought, high salt, and pathogen attack. It is known from the results reported by a few research groups that BRs are involved in tolerance of plants to abiotic stress, *e.g.* (heat, cold, drought and salinity (Divi and Krishna 2010).

Conclusion

Thus, it can be concluded that the multiple stress responses through different phytohormones are important by which plants cope in adverse environmental conditions. Perception, signal transduction, and changes in gene expression are the main aspects in the process of phytohormonalsignaling pathways. Many unique lines of evidence have implicated the key roles of ABA, ET, CKs, IAA, JA, defense against abiotic stresses. However, phytohormones, such as GA, BRsthrough which plants integrate adaptive responses, are not well known. Sign pathway act independently and also have a significant cross-talk with each other in response or tolerance of one or more abiotic stresses. Further integration of molecular, biochemical, and physiological studies need to address how phytohormonesignaling are integrated into phenotype and specific traits. Phytoharmone signalling will improve agricultural practices and production.

REFERENCES

Canet, J.V., Dobón, A., Ibáñez, F., Perales, L., Tornero, P. (2010) Resistance and biomass in *Arabidopsis* : a new model for salicylic acid perception. Plant Biotechnol J 8: 126–141.

Divi UK, Rahman T, Krishna P (2010) Brassinosteroid-mediated stress tolerance in *Arabidopsis* shows interactions with abscisic acid, ethylene and salicylic acid pathways. BMC Plant Biol10: 151 expression of AXR3/IAA17 gene in the cross-talk point with auxin in root development. Galon, Y., Aloni, R., Nachmias, D., Snir, O., Feldmesser, E., Scrase-Field S, Boyce JM, Bouche, Biosci Biotechnol Biochem 70: 768–773.

Hubbard, K.E., Nishimura, N., Hitomi, K., Getzoff, E.D., Schroeder, J.I. (2010) Earlyabscisic acid signal transduction mechanisms: newly discovered components and newly emerging questions. Genes Dev 24: 1695–1708.

Iglesias-Fernández, R., Rodríguez-Gacio, M.C., Matilla, A.J. (2011) Progress in research on dry after ripening. Seed Sci Res 21: 69–80.

Iqbal, N., Nazar, R., Khan, MIR., Masood, A., Khan, N.A. (2011) Role of gibberellins in regulation of source-sink relations under optimal and limiting environmental conditions. Curr Sci 100: 998–1007.

Kim, H., Park, P.J., Hwang H.J., Lee SY, Oh MH, Kim SG (2007) Brassinosteroid signals control Knight MR, Fromm H (2010) Calmodulin-binding transcription activator 1 mediates auxin signaling and responds to stresses in *Arabidopsis.* Planta 232: 165–178.

Luo, M., Liu, J., Lee, R.D., Scully, B.T., Guo, B. (2010) Monitoring the expression of maize genes in developing kernels under drought stress using oligo-microarray. J Integr Plant Biol 52(12): 1059–1074.

País, S.M., Téllez-Iñón, M.T., Capiati D.A. (2009) Serine/threonine protein phosphatases type 2A and their roles in stress signaling. Plant Signal Behav 4: 1013–1015.

Peleg, Z., Blumwald, E. (2011) Hormone balance and abiotic stress tolerance in crop plants. CurrOpin Plant Biol 14: 290–295.

Scherer, GFE. (2011) AUXIN-BINDING-PROTEIN 1, the second auxin receptor: what is the significance of a two-receptor concept in plant signal transduction? J Exp Bot 62: 3339–3357.

Sheard, L.B., Tan, X., Mao, H., Withers, J., Ben-Nissan G, Hinds, TR, Kobayashi Y, Hsu FF, Sharon M, Browse J, He SY, Rizo J, Howe GA, Zheng N (2010) Jasmonate perception by inositol phosphate- potentiated COI1-JAZ co-receptor. Nature 468: 400–405.

Yemelyanov, V.V., Shishova, M.F (2012) The role of phytohormones in the control of plant adaptation to oxygen depletion. In: Khan NA, Nazar R, Iqbal N, Anjum NA (eds) Phytohormones and abiotic stress tolerance in plants. Springer, Berlin, pp 229–248.

Chapter 5

Genome Editing: An Emerging Tool for Plant Breeders

Anand Kumar and Lokendra Singh

Department of Genetics and Plant Breeding, Chandra Shekhar Azad University of Agriculture and Technology, Kanpur – 208002, Uttar Pradesh

Conventional plant breeding has contributed enormously towards feeding the world and has played crucial roles in the development of modern society. The conventional method creates variation by transferring genes between or within the species. In general, these methods are more expensive and take more time, to overcome these limitations, new technology is required. Genome editing is a powerful tool for biotechnology applications, with the capacity to alter the function of any gene. With the availability of gene information for the majority of the traits, genome editing emerged as a potential to create a new variation with the introduction of any transgene. The important genome editing tools used nowadays are ZFNs, TALEN, Pentatricopeptide repeats protein, adenine base editor, RNA interference, and CRISPR/Cas9. These tools have opened a new era for crop improvement. Due to the complex genetic architecture of most traits, it is challenging to edit genes controlling them. To overcome these challenges, genome editing provides a broader perspective. Among the above-mentioned tools, CRISPR/Cas9 is the most powerful tool for gene editing. These technologies are being used to create abiotic and biotic resistance crop varieties.

Keywords: *Conventional method, Cluster regularly interspaced short palindromic sequence, Genome editing, RNA interference, and TALEN.*

Introduction

Genome editing involves the editing of targeted traits in organisms for the creation of new genetic variation. There are two methods to know the function of the gene namely forward and reverse genetics. In forward genetics, the main target is to change a particular genotype for obtaining desired phenotypes while in reverse genetics we target the phenotype and then move towards its genotype. With access to map-based cloning and the T-DNA tag, it is possible to detect and screen mutations in the gene of interest to aid in forward genetics (Page and Grossniklaus, 2002). In reverse genetics, we can study a particular phenotype and then try to relate that with particular genotype effect and detect the candidate gene that has been expressed in the population. This approach also reveals some cells genes expressed in multi-environment with the help of bioinformatics tools or gene chip aiding in genetic transform technology involving the knockout gene or overexpression of the gene of interest (Takahashi*et al.*, 1994). The conventional breeding method creates the variation in the population using physical or chemical mutagens in addition to natural variation (Stadler,1928; Li *et al.*, 2001; Wu *et al.*, 2005; Wang *et al.*, 2008; Talamè *et al.*, 2008; Suzuki *et al.*, 2008), These variations can be exploited in the form of a new variety. These variations involve point mutation, chromosomal mutation, translocation, and deletion in the genome of organism (Kao and Michayluk, 1974) and have been a more widely efficient tool for hybridization. This method enforces the trait to come from both parents that have been utilized in the crossing program. However, this technology can alter the sufficient degree of the genome but still we can't find the target modification easily with this method (Zhang *et al.*, 2017).

Why did We Need it?

Conventional plant breeding has contributed enormously towards feeding the world and has played crucial roles in the development of any new cultivar for human consumption. Conventional breeding methods depend on using old tools such as natural selection, but it takes much more time to develop a new variety. This method cannot edit any gene in the genome, so it needs new technology to create new variation and for knockout of any unwanted gene.

Tools for Genome Editing

Pre-genomic breeding programs led to the development of stress-tolerant and high-yielding crop varieties. In the past, breeding programs were based on natural, hybridization, and mutant-induced genetic variations to select for favorable genetic combinations. The traditional breeding program was based on mutagenesis using chemical compounds or irradiation, followed by screening for desired mutations, and has several drawbacks. Mutagenesis methods, intergeneric crosses, and translocation breeding are non-specific; and sometimes large parts of the genome are transferred instead of a single gene, or sometimes thousands of nucleotides are mutated instead of a single nucleotide. To overcome these problems a sophisticated tool is required for the creation of new variations. There are many tools for genome editing as described below.

TALEN

Transcription activator-like effectors nuclease is an emerged tool for genome editing in crop plants (Ma *et al.*, 2015). TALEN uses a double-strand break (DSB) in a similar type to the use of zinc figure nucleases (ZFNs). Both TALEN and ZNFs contain nonspecific FOK1 endonuclease. However, the FokI domains of ZFNs fuse with specific DNA-binding domains that are highly conserved repeats derived from transcription activator-like effectors [Ma *et al.*, 2015]. Approximately 50 genes have been targeted to induce mutation by TALEN in different crops, for example, Arabidopsis, soybean, barley, tomato, wheat, rice, tobacco, and maize, *etc.* (Ray *et al.*, 2013). TALEN scaffolds were optimized for high activity in plants [Zhang *et al.*, 2013]. Talen scaffolds are demonstrated by the mutagenesis for targeting a trait in Arabidopsis (Christian *et al.*, 2013), brachy podium (Shan *et al.*, 2013), tomato (Lor *et al.*, 2014) and wheat (Wang *et al.*, 2014). In recent times, TALEN created a heritable mutation in rice (Zhang *et al.*, 2014) and which will ultimately aid in plant breeding. In another study, TALEN has been used to engineer disease resistance for *Xanthomonos oryzae pv oryzae* by destroying the target sequence TALE effectors in rice (Li *et al.*, 2012). The FAD2 gene has been used to improve quality in soybean (Haun *et al.*, 2014). In wheat 3 homologous of MLO genes have been successfully knockout simultaneously for conferring heritable disease resistance for powdery mildew (Wang *et al.*, 2014). TALEN has been used to improve the fragrance (Shan *et al.*, 2015) and storage tolerance (Ma *et al.*, 2015) in rice. In potatoes, cold tolerance and processing traits are engineered by the TALEN. These studies are based on protein coding genes for mutagenesis but also have studied for non-coding genes and regulatory elements (Li *et al.*, 2012). There are some studies that have been conducted to generate chromosomal deletion of any crops plant (Chriatian *et al.*, 2013;Malzahn *et al.*, 2017).

Zinc Finger Nuclease

Zinc finger nuclease (ZFNs) is a genome-editing tool involving the cleavage of target DNA that is designed to cut at specific sites of DNA sequence (Carroll 2011). ZFNs create a DSBs through the replacement of homologous recombination genes. Each ZFN consists of a DNA binding domain that is having two finger chain modules. The DNA cleaving domain has a fok1 DNA nuclease domain, that recognized a unique pattern of 6 -bp hexamer in the DNA sequence (Carlson *et al.*, 2012; Gupta *et al.*, 2012). These domains are joined together to form a Zinc finger protein. DNA cleaving and DNA binding domains were fused together, to make one unit work together as specific genomic scissors (Carlson *et al.*, 2012; Gupta *et al.*, 2012). ZFN originates the site-specific cut in double-stranded and breaks the DNA permanently to edit the genome (Carlson *et al.*, 2012; Gupta *et al.*, 2012). ZFN is only dependent on ZFP that can target a specific DNA sequence in the genome. The best possible structure provided by the Cys2His2 ZFP to develop suitable ZFNs that have a required sequence specificity (Pabo *et al.*, 2001). ZFP stabilizes the chelation of zinc ions to conserve the Cys2His2 amino acids that are having approximately 30 amino acids (Thakore and Gersbach 2015; Pavletich and Pabo 1991). The ZF motifs cover and bind the DNA sequence in its genome by incorporating a major groove in the double helix DNA (Pavletich and Pabo 1991). The ZF interacts with ZFN that have specific sequence interaction with specific the DNA sequence (Pavletich and Pabo 1991; Sahi and Berg 1995). In this triplet, the sequence can be bound by each finger. Linking multiples zinc finger motifs to form ZFPs is possible to bind longer DNA strands (Liu *et al.*, 1997; Kim and Pabo 1998). The methylase domain (M),

FokI-cleavage domain (N), transcription activator domain (A), and transcription repressor domain (R) are fused with ZFP to form a ZFN [Kim and Cha 1998; Alwin *et al.*, 2005].

Pentatricopeptide Repeat Proteins

The organellar genome has a low number of promoters. The organellar RNA number is more than that of nuclear RNA. However, the regulation of RNA cleavage, RNA splicing, and RNA editing is insufficient to control the expression of the organellar genome. RNA binding protein is developed by the organelles to regulate the gene at the post-transcriptional level (Small *et al.*, 2013; Schmitz-Linneweber and small 2008). The organellar genome has tetratricopeptide repeats protein

(PPR), that regulates post-transcriptional regulation. Pentatricopeptide repeat protein (PPR) is characterized by the tandem repeats motifs that consist of 35 amino acids (Schmitz-Linneweber and small 2008). PPRs have classes such as P-class (35 amino acids), L-class (35–36 amino acids), S-class (short, approximately 31 amino acids), and E-class (extended domain) [Barken and Small 2014; Shikanai 2006]. DYW tripeptide motif (Aspartate-Tyrosine-Tryptophan) is highly conserved by the PLS type PPRs at the C terminal ends that is the most likely show behavior of the editing domain (Lurin *et al.*, 2004). The DYW domain is having most of the zinc-binding affinity that is necessary to catalytic activity for the editing of any genome (Boussardon *et al.*, 2014). The l- class PPRs have amino acid residue that is involved in RNA binding for the targeted nucleotides. But the E domain is not having proper catalytic activity hence it helps to interact with protein to protein with editing enzymes (Okuda *et al.*, 2007). The helices form a solenoid antiparallel helix-turn-helix structure that provides the basis for a specific sequence for RNA binding. Position 6 and 1 are responsible for the recognition of nucleotide and for RNA binding, as reported by Barkan *et al.* (2012). PPRs recognize the transcription in a modular fashion at positions 1 and 6. The amino acid position at 1 and 6 recognize each PPRs to bind each nucleotide. PPRs bounds the 50 ends of RNA in a parallel fashion. The interaction between nucleotide and PPRs motifs occurred by the Van der Waals interaction and combination with asparagine position 1 and threonine position 6 to recognize adenine nucleotide, whereas asparagine position at 6 and aspartic acid position at 1 recognize uracil nucleotides (Barken *et al.*, 2012; Yin *et al.*, 2013; Yagi *et al.*, 2013). PPRs bounds the RNA sequence as a monomeric by employing the amino acid residues at position 6 and 1 position. The interaction of PPRs with t RNA helped by the amino acid at position 3. This position was created by the hydrophobic amino acid(Fujii *et al.*, 2011). PPRs motifs provide greater versatility for editing due to presence of higher diversity in amino acid composition. PPRs are having the LAGLIDADG that is involved in splicing events (Toole *et al.*, 2008; De Longevialle *et al.*, 2008).

Adenine Base Editor

Spontaneous Conversion of 5-methylcytosine and cytosine to thymine and uracil respectively originate by the hydrolytic deamination in cells and resulting in the mutation of C-G and T-A (Nishida *et al.*, 2016; Zong *et al.*, 2017). This event occurs 100- 500 times per cell in 24 hours (Gaudelli *et al.*, 2017). Therefore, these editing processes are restricted to the conversion of C-G to T-A. Hence the conversion of A-T base pair into G-C base pair in target gene loci is based on SNPs (small nucleotide polymorphism). This hurdle is overcome by the Adenine base editors (ABEs). The ABEs convert the A-T to G-C in bacteria and humans, but nowadays it is also involved in plants (Gaudelli *et al.*, 2017). ABEs are high conversion accuracy of A-T to G-C targeted genomic, and purity (Gaudelli *et al.*, 2017). Adenine base editors generate point mutation than the CRISPR/Cas9 nuclease-based genome editing that is having low rates of INDELS (0.1 per cent) and also having high product purity (99.9 per cent) with less off-target mutations (Gaudelli *et al.*, 2017). In this process, the deoxyadenosine deaminase TadA is fused with catalytically impaired Cas9 nickase with a corresponding single guide RNA (sgRNA) to perform the task [Gaudelli *et al.*, 2017]. Thus, the targeted DNA sequence is bound by the fused TDA and CAS9 that is guided by the program of RNA, thus exhibiting a small single-stranded DNA bubble. The conversion of adenine to inosine within the bubble is with the help of TadA Deoxyadenosine deaminase. The C terminal end of Cas9-nickase fused with TadA instead of the N- terminal ends and starts the editing activity (Gaudelli *et al.*, 2017). Doubling the length of the linker between TadA and Cas9 nickase by using the 32-amino-acid-long (SGGS)2-STEN-(SGGS)2 linker leads to higher editing efficiencies [Gaudelli *et al.*, 2017]. TadA is a homodimer, so one monomer catalyze the activity of deamination process and other acts as a docking station of tRNA substrate (Gaudelli *et al.*, 2017). But this technology exhibits the limited editing of targeted sequences when any genome has multiple adenine nucleotides. But these limitations have been overcome by the use of Tad Cas9 as it targeted two separate sites in which one is TAT and other is TAA in the kanamycin resistance genes.

RNA Interference

This is the post-transcriptional mechanism that regulates the gene expression in a different direction (Zong *et al.*, 2017; Gaudelli *et al.*, 2017; Small 2007). This mechanism can be originated due to siRNA (small interfering RNA) (Axtell *et al.*, 2006; Agrawal *et al.*, 2003), piRNA (PIWI- interacting RNA) (Seto *et al.*, 2004; Klattenhoff and Theurkauf 2008), miRNA (microRNAs) (Carrington and Ambros 2008). siRNA is produced by the double-stranded RNA that encompasses approximately 20-30 bp noncoding single-stranded nucleotides and its guides to DNA cleavage and for translation repression. siRNA is exhibiting a complementary sequence and leads to the knockdown effects of gene function (Jackson *et al.*, 2006; Birmingham *et al.*, 2007). RNAi was discovered by Andrew Fire and Craig Mello; they received the Nobel prize in 2006 and reveals the importance of RNAi. The siRNA acts as a form of RISC (RNA induced silencing complex) that acts as a silencing or degradation target gene in mRNA (Pratt and Macrae 2009; Kawamata and Tomari 2010). In siRNA methylation of RNA binding sites can alter the chromatin configuration (Zhou *et al.*, 2014; Kawasaki *et al.*, 2004). miRNA is directly derived from the genome whereas siRNA is derived from exogenous sources. Mostly it has been observed that miRNA exhibits the mismatched pair due to the presence of extended terminal D-loop, whereas siRNA exhibits complete and correct base pairing (Zhu *et al.*, 2013; Lau *et al.*, 2001). The origin of siRNA and miRNA is different but both require RISC. A few siRNA is produced from the plant cells while the majority of siRNA are from viruses. But the presence of cell walls in plant cells makes it difficult to transfer siRNA into cells. Transfer of siRNA into plant cells is achieved by the hairpin loop hairpin RNA that folds back to generate a double-stranded region. DCL/DICER like enzymes recognized by double-stranded the region which later makes complexes with Argonaute for the regulation of the silencing process (Kurzynska-Kokorniak *et al.,* 2015; Kurihara and Watanabe 2004). Most of the modified virus is used to induce the RNAi and knockdown genes this process called the virus-induced gene silencing (VIGS) (Lu *et al.,* 2003; Burch-Smith *et al.,* 2006). VIGS is very helpful to transfer transgene into the target gene. This process is very helpful to those plants that are recalcitrant to genetic transformation. While miRNA and siRNA are produced from the endogenous genome that is exhibiting a natural origin. Dicer enzymes cleave the pre-miRNA that is having a hairpin precursor with specificity (Kurihara and Watanabe 2004). The expression of several genes is controlled by the miRNA that is conserved evolutionary small RNA (Warnefors *et al.,* 2014; Li and Mao 2007). The target transcript is cleaved by generating artificial miRNA (Tiwari *et al.,* 2014; Gasparis *et al.,* 2017). Inhibit expression of genes with the help of a small 7-nucleotide complementarity of the siRNA. However, forecasting off targeted effects using siRNA is more difficult. Limitation of sequence specificity of siRNA, difficult to separate screening of expressed genes. Mutation in gene changes the stable effects and functional effects of mutation can be predicted easily. However, RNA interference shows the effects on genes. Hence, two plants that have been interfered with by RNA to show different effects.

Cisgenesis and Intragenesis

This is the main technique used to induce genome editing, target transgene generation, and genetic mutation. Recent time's somatic hybridization used to transfer genes from two different species separated by the genetic barrier. The development of new technology led to the isolation of genes from cross-species, this gene called the cisgenesis. Cisgenesis or intragenesis refers to the transfer of genes from between the same species or genetically cross-compatible species without linkage drag, this may be helped by the DNA sequencing (Hartung *et al.,* 2014; Schouten *et al.,* 2006). Genes remain unchanged in cisgenesis, but in intragenesis only a part of a gene such as a promoter or a regulator is transferred. Cisgenesis may result in a new organism that might not be distinguishable by one obtained from conventional crossbreeding, whereas intragenesis results in an organism that cannot be obtained by conventional crossbreeding [Hou *et al.,* 2014]. A large no of cisgenesis gene has been isolated with the help of DNA sequencing technology; these genes do not have any marker genes with them. It leads to genome modification while remaining the gene pool. Cisgenesis plants are highly demanded by the public than transgenic plants. Cisgenes has been successfully utilized in cereal plants where the phytase gene was introduced to confer bioavailability of phosphate (Holme *et al.,* 2014). In wheat durum species The 1D × 5 and 1Dy10 glutenin subunits with native endosperm promoter and terminator are using a transformation method. The bread quality is enhanced by the using transformation method in wheat (162). In poplar for gibberellin signaled five cisgenesis genes such as PtGA20ox7, PtGA2ox2, Pt RGL11, PtRGL12, and PtGAI1were transferred to Populustremula x alba (Han *et al.*, 2011). PtGA20ox7is responsible for the early growth and rate of shoot generation (Han *et al.*, 2011). PtRGL12 is responsible for longer xylem fibers while PtGAI1 cisgenesis is responsible for increasing regeneration (Han *et al.*, 2011). Cisgenesis and transgenesis is the same approach except that in cisgenesis transfer of genes from the same plants or crossable species by breeding technique as well. (Mohanta *et al.*, 2017).

7 CRISPR/CAS9

In the era of high throughput genotyping, precise and effective genome editing technologies are needed to implement reverse genetics. Genome editing is of great importance for functional characterization and genetic improvement of agricultural plants. There are various technologies that are being available including transcription activator-like effectors nucleases (TALENs), zinc fingers, and homing meganucleases for target genome modifications(Stoddard, 2005; Doyon *et al.*, 2011; Xie *et al.*, 2011). Although with the availability of such tools, there is a need for a new technology for genome editing which could be affordable, robust, scalable, and easy to engineer. This has led to the emergence of type II clustered regularly interspaced short palindromic sequences repeat (CRISPR) technology using RNA-guided nucleases (Cas9) for precise insertions, deletions, and sequence changes in a broader range of cell types and organisms. CRISPR allows precise genome editing with being comparatively easy, inexpensive, user-friendly, and rapidly adopted editing technology(Kim *et al.*, 2011). Traditionally, crops were being improved by using conventional breeding but with the constrained of ever declining existing genetic variation in plants, hampering the production for future feeding. The development of genetically modified crops with CRISPR is similar to those by mutation or conventional breeding, but CRISPR isan extremely versatile tool for providing sustainable agriculture for the rapidly increasing population. In the last decade, CRISPR has been successfully performed for targeted mutagenesis and regulation of transcriptional control in the model and agriculturally important crop plants, demonstrating the revolutionary aspect of this novel system. CRISPR has been extensively used for gene replacement, multiplex editing, gene knockouts, and targeted gene editing for improvement or designing a new crop plant.

The functioning of CRISPR Cas9 contains combinations of two Cas9 genes. The sequence from the non-coding RNA element is known as crRNA while the other sequence is from trans-encoded CRISPR RNA termed as trans acting crRNA (tracrRNA). Guide RNA is formed by complex of these two RNA sequences (crRNA + tracrRNA) ultimately determining the specificity of the target sequence in the genome along with the Protospacer Adjacent Motif (PAM) (Jinek*et al.*, 2014). The cleavage of DNA took place within this protospacer boundaries when targeted by guide RNA. CAS9 protein is an endonuclease responsible for double-strand break (DSB) when targeted by guide RNA and facilitates endogenous DNA repair mechanism resulting in edited DNA(Mali *et al.*, 2013). There are various variants available for Cas9 nucleases, guide RNA, and their combination for genome editing (Doudna and Charpentier, 2014). The desired combination or type depends upon the crop species you are working and the gene of interest.

Till now, genome editing in plants is restricted to changes in enzymatic functions for a single purpose. CRISPR-Cas9 provides the most robust and reliable system for genome editing for multipurpose plant systems. There is an ongoing study at John Innes center for characterizing the pathways for nitrogen fixation in plants using bacteria (Xie *et al.*, 2012). Some studies are working to incorporate these pathways into wheat (*Triticum aestivum*). This will ultimately result in a reduction in dependency on inorganic fertilizer as plants will be able to fix atmospheric nitrogen(Cook Martin and Bastow, 2014). The other major focus in the present world is to convert C3 rice (*Oryzae sativa*) to C4 using CRISPR, as engineered C4 rice has scope for increasing yield. This conversion requires the conversion of a single cell C3 cycle to a two-celled C4 cycle (Baltes *et al.*, 2014). There are various successful studies using CRISPR to remove unwanted cells from different plants, but there is scope for new genic regulatory sequences insertion to plant genome. The transfer of new DNA segments into plants is of special interest for conferring new biological functions. With CRISPR-Cas9 there is tremendous potential for multiple gene stacking with very little variability in gene expression. The gene stacking approach has been successfully demonstrated in maize (*Zea mays*) where 5 per cent of transgenic progeny witnessed proper integration (Alagoz *et al.*, 2016). Similar gene stacking was performed in cotton (*Gossypium hirsutum*) (D'Halluin *et al.*, 2013). There is no doubt that plant breeders are working hard to engineer disease resistance, high yielding, drought tolerance and nutritional varieties taking genes from other varieties or species. But this effort, the introgression of desired traits, takes several years in conventional breeding approaches. CRISPR-Cas9 is now widely used to improve agricultural important crop plants. A wheat line is recently developed resistant to powdery mildew using a gene-editing approach. Jointly with other gene-editing tools, CRISPR-Cas 9 is emerging as a game-changer for transformation in plants.

Since the last decade, genome editing has been adopted from mammals to plants to achieve a wide range of modifications from single nucleotide alterations to deletions of megabases in the DNA. Plant genome editing has great potential for the ambiguous project for the plant improvements. Most of the CRISPR-Cas9 information is obtained from mammals and it is assessed that most of these findings are universal which could be executed in diverse crop species. Future research will focus on optimizing the strategies for designing the guide RNA variants, as it is critical for inducing breaks in the DNA. There is a huge scope for designing guide RNA for polyploid species like wheat, potato (*Solanum tuberosum*), and sugarcane (*Saccharum officinarum*) as it requires information from variation in diverse allelic forms. Another major challenge in gene editing is the lack of a high throughput platform for identifying edited gene events. Keeping in consideration for all the upcoming challenges and the number of researchers involved in CRISPR-Cas9 there are potentials form swift improvements with a new batch of genome editing tools.

Applications of Genome Editing

Genome editing technology has several advantageous resulting in a wide application in crop improvement. This is one of the biggest tools for modern biotechnology, crop varieties are continuously improving with the help of genome editing, such as biotic, and abiotic resistances and also improved nutritional values in crops (Osakabe *et al.*, 2016; Xiong *et al.*, 2015; Kissoudis *et al.*, 2014; Chantrenongpiur *et al.*, 2016). Genome editing tools are used in plant breeding for the improvement of new desirable characters to create the point mutation that is similar to natural SNPs (Jacobs *et al.*, 2015; Xu *et al.*, 2017). There have been many modifications with the help of genome editing such as alteration of genes, knockout of genes, gene pyramiding, repression or activation of genes expression, epigenetic changes, *etc.* For instance, in *Arabidopsis thaliana* and *Zea mays* has been edited with the help of ZFN tools for herbicide resistance (Osakabe *et al.*, 2010; Townsend *et al.*, 2009; Shukla *et al.*, 2009). In the last few years, genome editing is used to produce new varieties for insect resistance, disease resistance, increased yield, and improved quality, *etc.*

Improved Yield

Yields of the plant mainly depend on grain number, grain weight and grain size, these characters are contributed by the quantitative trait loci (Xing and Zhang, 2010; Bai *et al.*, 2012). For improvement of yield, need knockout of genes those who have negative effects such as DEP1, GW2, TGW6, GS5, Gn1a, and GS3 in rice, this is the direct way to the improvement of crops. Gn1a, Gs3 and DEP1 have been utilized to improve traits through the use of CRISPR/Cas9 (Li *et al.*, 2016b; Shen *et al.*, 2016). TGW6, GW2 and GW5 have been knocked out, resulting in approximately 29.8 per cent increase in thousand-grain weight in the triple mutant (Xu *et al.*, 2016). Thousand kernel weight may be increased if these genes knockouts, effects of three homo-alleles of GASR7 who have the negative effects of kernel width and weight in bread wheat are improved with the help of CRISPR/Cas9 (Zhang *et al.*, 2016b). It is nevertheless important to remark that increased grain yield per plant and higher the thousand-grain weight does not necessarily translate into improved crop yield because large-scale field trials are necessary to verify the potential agronomic improvements.

Disease Resistance

Severe loss in yield due to infection by the disease. The disease affects consumption, freshness, and food processing safety in many crop plants (Savary *et al.*, 2012). Disease resistance may be increased through the use of genome editing of targeted traits in any crop. Bacterial pathogen *Xanthomonus oryzaepv. Oryzae* that is causing bacterial blight, this secret type IInd effector protein activated to OsSWEET14 is a host disease- susceptibility gene (Li *et al.*, 2012b). One more gene, CsLOB1 host susceptibility plays a pivotal role in both activities such as enhance the erumpent pustule formation and for pathogen growth ((Hu *et al.*, 2014). CsLOB1 promoter has been modified with the help of CRISPR/Cas9 in which two groups have been generated in citrus canker (Jia *et al.*, 2017; Peng *et al.*, 2017). In rice, ERF transcription factor OsERF922, a negative regulator of blast susceptible has been knockout, resulting in an enhancement in yield (Wang *et al.*, 2016a). Another one more example in wheat TaMLO that is a contribution to disease susceptibility has been edited, resulting in disease resistance (Wang *et al.*, 2014). MLO allele has been knocked out in Arabidopsis, tomato and barley that causes the powdery mildew disease to produce durable resistance and broad-spectrum *Blumeriagraminis f. sp.*

Tritici (Bgt)s (Wang *et al.*, 2014). This gene creates the mutation in all three TaMLO homologous genes induced by the TALEN, these genes produced heritable variation for powdery mildew resistance in bread wheat (Wang *et al.*, 2014).In tomato, gene SIMLO1 produces the powdery mildew has been disrupted with the help of using CRISPR/Cas9 (Nekrasov *et al.*, 2017). There are many bacteria, eukaryotic and other viruses that have been edited through using the genome editing approach. Tobacco and tomato are most affected by the virus, so these plants have been generated for targeted viral genome sequences through the use of the CRISPR/Cas9 editing approach (Ali *et al.*, 2015; Baltes *et al.*, 2015; Ji *et al.*, 2015). The eukaryotic translation initiation factor eIF4E is a host factor required by plant RNA viruses to maintain their life cycle and mutations in this gene have produced broad virus resistance in T3 non-transgenic cucumbers (Chandrasekaran *et al.*, 2016) (Ali *et al.*, 2015; Baltes *et al.*, 2015; Ji *et al.*, 2015).

Herbicide Tolerance

In the past, the herbicide resistance crops were produced by using transgenic technology (Lombardo *et al.*, 2016). Nowadays genomic editing approach has created a new approach for herbicide resistance plants to edit endogenous plant genes such as ALS and EPSPS, resulting in herbicide tolerance (Lombardo *et al.*, 2016). The acetolactate synthase enzymes encoded by the ALS gene, that is being participated in the biosynthesis of branched-chain amino acids likes isoleucine, valine, and leucine (Lee *et al.*, 1988; Chipman *et al.*, 1998). Inhibitors of ALS are used as herbicides that slowly starve affected plants of these amino acids, eventually leading to inhibition of DNA synthesis, but specific point mutations within the conserved region of ALS can confer resistance to these herbicides (Lee *et al.*, 1988; Chipman *et al.*, 1998; Svitashev *et al.*, 2015). Precise mutation in the ALS gene, there are many plants that have been developed through the replacement of genes with the help of genome editing approach, the first example of tobacco, use of donor template and ZNF has been reported in 2009 (Townsend *et al.*, 2009). Site-directed DNA base changes in the ALS gene have been created with the help of CRISPR/Cas9 for herbicide resistance in tomato, soybean, and maize (Li *et al.*, 2015; Svitashev *et al.*, 2015; Li *et al.*, 2016c). A 5-enolpyruvylshikimate-3- phosphate synthase is encoded by the EPSPS that is necessary for biosynthetic aromatic amino acid that is essential for plant survival (Kishore and Shah, 1988). EPSPS has widely used herbicide in plants that is responsible for glyphosate that binds to EPSPS functional sites and prevents its activity (Kishore and Shah, 1988). The usual method to introduce glyphosate tolerance in plants is to modify the EPSPS protein structure in order to disrupt herbicide binding site while maintaining its catalytic activity (Sammons and Gaines, 2014).In *LinumUsitatissimum* (Flax) substitution of two nucleotides through HDR-based genomes editing that is responsible for EPSPS binding sites that have been used with the help of single- stranded DNA repair templates and CRISPR/Cas9 (Sauer *et al.*, 2016).The EPSPS modified genes show a high level of herbicide tolerance than control. There is one more similar approach that has been used a modification of EPSPS gene in rice in which substitution of two nucleotides has been edited for glyphosate resistance (Li *et al.*, 2016a).

Improved Oil Composition

In polyunsaturated fatty acids particularly, linoleic acid has poor availability, poor frying and oxidative nature, these criteria put restrictions on oil use. Improvement of quality and fatty acid composition requirement to change the FAD genes (fatty acid desaturase). Conversion of oleic acid into linoleic acid is responsible with the help of the family FAD2 genes, while the FAD3 genes family encodes enzymes to catalyze the production of linoleic acid into linolenic acid (Demorest *et al.*, 2016). Two soybean FAD2 genes such as FAD2-1A and FAD2-1B has been used for the improvement of oil quality by the TALEN editing approach, in which linoleic acid reduced from 50 per cent to <4 per cent but oleic acid increased from 20 per cent to 80 per cent (Haun *et al.*, 2014).TALENs were used for the improvement of oil in soybean in which FAD3A introduced into fad2-1a/fad2-1b in soybean plants, resulting in a decrease of linolenic acid and increased linoleic acid (Demorest *et al.*, 2016). Three FAD2 homoeologous genes in the allohexaploid, camelina sativa have been simultaneously knockout using two independent groups CRISPR/Cas9, resulting in producing a composition oil (Morineau *et al.*, 2016; Jiang *et al.*, 2017).

Healthy Potatoes

Potato is a good source of carbohydrates, minerals, vitamins, *etc.* The requirement for protection of potatoes from infection, diseases, and pests. We need to conserve potatoes in cold storage. Cold storage provides a sufficient quantity of potato, saves from sprouting and also gives the continuous supply to the public domain but there is one more problem created with a potato that reduces the accumulation of sugars. This sugar produces free amino acid that originates during high temperatures and produces browning, bitter test and increases the level of acrylamides, to overcome this problem, the requirement of new technology for human safety (Clasen *et al.*, 2015). Conversion of sucrose into glucose and fructose is catalyzed by the vacuolar invertase that is encoded by the VINV, VINV plays an indispensable role to reduce sucrose in storage potato tubers. TALEN used to mutate VINV in a russet potato variety that nowadays is used for commercial use in which potato having undetectable levels of reducing sugars. The heating of potatoes in the cold storage is responsible for reduced levels of acrylamide and producing high levels of lightly colored chips (Clasen *et al.*, 2015).

Other Examples

Mutate of TMS gene in rice with the help of CRISPR/Cas9 leads to the development of a temperature-sensitive line for use of hybrid seed production (Zhou *et al.*, 2016). Granule-bound starch synthase is encoded by the maize waxy gene (Wx) that is responsible for the synthesis of amylose in the kernels (Nelson and Rines, 1962). Wild type maize kernels contain 25 per cent amylase and 75 per cent amylopectin whereas wx/wx lines contain approximate 100 per cent amylopectin so-called the waxy maize. Waxy gene knockout from maize to use of CRISPR/Cas that is having more economic value (Chilcoat *et al.*, 2017). Knockout of the starch branching enzymes genes such asSBEI and SBEIIb that creates high amylose rice with potential health advantages has been generated to the use of CRISPR/CAS9 (Sun *et al.*, 2017; Zhang *et al.*, 2017).

For Functional Genomics

Nowadays, there are many tools used for genome editing such as ZFN, TALEN, ODM, and CRISPR/Cas9. Such types of new tools create the point mutation, deletion and insertion of new nucleotide or substitution of a nucleotide in the targeted genome [Gaj *et al.*, 2013; Kumar and Jian 2015; Chain *et al.*, 2016; Zheng *et al.*, 2016; Gao *et al.*, 2015; Lowder *et al.*, 2016; Perez-Pinera *et al.*, 2012]. Introducing a new gene into the genome is to detect the function of foreign genes. Researchers have proved which one individual gene or cell has a function. For example, the unique ability of the CRISPR/Cas9 system to selectively bind to specific DNA sites helps to regulate gene activity [Cong *et al.*, 2011; Zhang *et al.*, 2011; Lowder *et al.*, 2016]. Promoter activity for protein activation or repressing that is controlled by the function of a gene can be attached to the catalytic inactive Cas9 protein. It has been proved that the binding of targeted DNA at binding sites can be stimulated or inhibited the function of target trait (Lowder *et al.*, 2016). In addition, targeted to different genome sites has been simultaneously introduced in the cells with the help of CRISPR/Cas9 system [Li *et al.*, 2016; Cong *et al.*, 2013; Wang *et al.*, 2013]. In crop domestication, there are many genes that have been identified using this CRISPR/Cas9 technology. CRISPER/Cas9 helps to decide the intergenic interaction of any genes (Lowder *et al.*, 2016).

REFERENCES

Agrawal, N.; Dasaradhi, P.V.N.; Mohmmed, A.; Malhotra, P.; Bhatnagar, R.K.; Mukherjee, S.K., 2003. RNAInterference: Biology, mechanism, and applications. Microbiol. Mol. Biol. Rev., 67, 657–685.

Alagoz, Y. *et al.* (2016) 'Manipulating the biosynthesis of bioactive compound alkaloids for next- generation metabolic engineering in opium poppy Using CRISPR-Cas 9 genome editing technology', *Scientific Reports*. Nature Publishing Group, 6(May), pp. 1–9. doi: 10.1038/srep30910.

Ali, Z., Abulfaraj, A., Idris, A., Ali, S., Tashkandi, M., and Mahfouz, M. M. 2015. CRISPR/Cas9- mediated viral interference in plants. Genome Biol. 16: 238.

Alwin, S.; Gere, M.B.; Guhl, E.; Effertz, K.; Barbas, C.F.; Segal, D.J.;Weitzman, M.D.; Cathomen, T., 2005. Custom zinc-finger nucleases for use in human cells. Mol. Ther., 12, 610–617.

Axtell, M.J.; Jan, C.; Rajagopalan, R.; Bartel, D.P.A., 2006 two-hit trigger for siRNA biogenesis in plants. Cell, 127, 565–577.

Bai, X., Wu, B., and Xing, Y. 2012.Yield_related QTLs and their applications in rice genetic improvement.J. Integr. Plant Biol. 54: 300–311.

Baltes, N. J. *et al.* (2014) 'DNA replicons for plant genome engineering', *Plant Cell*, 26(1), pp. 151–163. doi: 10.1105/tpc.113.119792.

Baltes, N. J., Hummel, A. W., Konecna, E., Cegan, R., Bruns, A. N., Bisaro, D. M., and Voytas,

D. F. 2015. Conferringresistance to geminiviruses with the CRISPR–Cas prokaryotic immune system.Nature.1: 15145.

Barkan, A.; Rojas, M.; Fujii, S.; Yap, A.; Chong, Y.S.; Bond, C.S.; Small, I.A., 2012 combinatorial amino acid code for RNA recognition by pentatricopeptide repeat proteins.PLoS Genet., 8, e1002910.

Birmingham, A.; Anderson, E.; Sullivan, K.; Reynolds, A.; Boese, Q.; Leake, D.; Karpilow, J.; Khvorova, A.A, 2007. protocol for designing siRNAs with high functionality and specificity. Nat. Protoc., 2, 2068–2078.

Burch-Smith, T.M.; Schiff, M.; Liu, Y.; Dinesh-Kumar, S.P., 2006. Efficient virus-induced gene silencing in Arabidopsis.Plant Physiol., 142, 21–27.

Carlson, D.F.; Fahrenkrug, S.C.; Hackett, P.B., 2012.Targeting DNA with fingers and TALENs.Mol.Ther.Nucleic Acids 1.

Carrington, J.C.; Ambros, V., 2003. Role of microRNAs in plant and animal development. Science, 301, 336–338.

Carroll, D., 2011 Genome engineering with zinc-finger nucleases. Genetics, 188, 773–782. Chandrasekaran, J., Brumin, M., Wolf, D., Leibman, D., Klap, C., Pearlsman, M., Sherman, A., Arazi, T., and Galon, A. 2016.Development of broad virus resistance in non-transgenic cucumber using CRISPR/Cas9 technology.Mol. Plant Pathol. 17: 1140–1153.

ChantreNongpiur, R., LataSingla-Pareek, S. and Pareek, A., 2016. Genomics approaches for improving salinity stress tolerance in crop plants. *Current genomics*, *17*(4), pp.343-357.

Chen, B., Hu, J., Almeida, R., Liu, H., Balakrishnan, S., Covill-Cooke, C., Lim, W.A. and Huang, B., 2016.Expanding the CRISPR imaging toolset with Staphylococcus aureus Cas9 for simultaneous imaging of multiple genomic loci. *Nucleic acids research*, *44*(8), pp.e75-e75.

Chilcoat, D., Liu, Z. B., and Sander, J. 2017. Use of CRISPR/Cas9 for crop improvement in maize and soybean.Prog.Mol. Biol. Transl. Sci. 149: 27–46.

Chipman, D., Barak, Z., and Schloss, J. V. 1998. Biosynthesis of 2-aceto-2-hydroxy acids: Acetolactate synthases and acetohydroxyacidsynthases. Biochim. Biophys.Acta. 1385: 401– 419. Christian, M., Qi, Y., Zhang, Y. and Voytas, D.F., 2013. Targeted mutagenesis of Arabidopsis thaliana using engineered TAL effector nucleases. *G3: Genes, Genomes, Genetics*, *3*(10), pp.1697- 1705.

Clasen, B. M., Stoddard, T. J., Luo, S., Demorest, Z. L., Li, J., Cedrone, F., Tibebu, R., Davison, S., Ray, E. E., Daulhac, A., Coffman, A., Yabandith, A., Retterath, A., Haun, W.,Baltes, N. J., Mathis, L., Voytas, D. F., and Zhang, F. 2015. Improving cold storage and processing traits in potato through targeted gene knockout. Plant Biotechnol. J. 14: 169–176.

Cong, L., Ran, F.A., Cox, D., Lin, S., Barretto, R., Habib, N., Hsu, P.D., Wu, X., Jiang, W., Marraffini, L.A. and Zhang, F., 2013. Multiplex genome engineering using CRISPR/Cas systems. *Science*, *339*(6121), pp.819-823.

Cong, L., Ran, F.A., Cox, D., Lin, S., Barretto, R., Habib, N., Hsu, P.D., Wu, X., Jiang, W., Marraffini, L.A. and Zhang, F., 2013. Multiplex genome engineering using CRISPR/Cas systems. *Science*, *339*(6121), pp.819-823.

Cook, C., Martin, L. and Bastow, R. (2014) 'Opportunities in plant synthetic biology', *Journal of Experimental Botany*, 65(8), pp. 1921–1926. doi: 10.1093/jxb/eru013.

D'Halluin, K. *et al.* (2013) 'Targeted molecular trait stacking in cotton through targeted double- strand break induction', *Plant Biotechnology Journal*, 11(8), pp. 933–941. doi: 10.1111/pbi.12085.

De Longevialle, A.F.; Hendrickson, L.; Taylor, N.L.; Delannoy, E.; Lurin, C.; Badger, M.; Millar, A.H.; Small, I., 2008. Thepentatricopeptide repeat gene OTP51 with two LAGLIDADG motifs is required for the cis-splicing of plastid ycf3 intron 2 in *Arabidopsis thaliana*. Plant J., 56, 157–168.

Demorest, Z. L., Coffman, A., Baltes, N. J., Stoddard, T. J., Clasen, B. M., Luo, S., Retterath, A., Yabandith, A., Gamo, M. E., Bissen, J., Mathis, L.,Voytas, D. F., and F, Z. 2016. Direct stacking of sequence-specific nuclease-induced mutations to produce high oleic and low linolenic soybean oil. BMC Plant Biol. 16: 225.

Doudna, J. A. and Charpentier, E. (2014) 'The new frontier of genome engineering with CRISPR- Cas9', *Science*, 346(6213). doi: 10.1126/science.1258096.

Doyon, Y. *et al.* (2011) 'Enhancing zinc-finger-nuclease activity with improved obligate heterodimeric architectures', *Nature Methods*, 8(1), pp. 74–79. doi: 10.1038/nmeth.1539. factor gene OsERF922. PloS One. 11: e0154027.

Fujii, S.; Bond, C.S.; Small, I.D., 2011. Selection patterns on restorer-like genes reveal a conflict between nuclear and mitochondrial genomes throughout angiosperm evolution. Proc. Natl. Acad. Sci. USA, 108, 1723–1728.

Gaj, T., Gersbach, C.A. and Barbas III, C.F., 2013.ZFN, TALEN, and CRISPR/Cas-based methods for genome engineering. *Trends in biotechnology, 31*(7), pp.397-405.

Gao, J., Wang, G., Ma, S., Xie, X., Wu, X., Zhang, X., Wu, Y., Zhao, P. and Xia, Q., 2015. CRISPR/Cas9-mediated targeted mutagenesis in Nicotianatabacum. *Plant molecular biology, 87*(1-2), pp.99-110.

Gasparis, S.; Ka³a, M.; Przyborowski, M.; Orczyk, W.; Nadolska-Orczyk, A., 2017. Artificial MicroRNA-based specific gene silencing of grain hardness genes in polyploid cereals appeared to be not stable over transgenic plant generations. Front. Plant Sci., 7, 2017.

Gaudelli, N.M., Komor, A.C., Rees, H.A., Packer, M.S., Badran, A.H., Bryson, D.I. and Liu, D.R., 2017. Programmable base editing of A • T to G • C in genomic DNA without DNA cleavage. *Nature, 551*(7681), pp.464-471.

Gaudelli, N.M.; Komor, A.C.; Rees, H.A.; Packer, M.S.; Badran, A.H.; Bryson, D.I.; Liu, D.R., 2017. Programmable base editing of A_T to G_C in genomic DNA without DNA cleavage. Nature, 1–27.

Gupta, A.; Christensen, R.G.; Rayla, A.L.; Lakshmanan, A.; Stormo, G.D.; Wolfe, S.A., 2012. An optimizedtwo-finger archive for ZFN-mediated gene targeting. Nat. Methods 9, 588–590.

Han, K.M.; Dharmawardhana, P.; Arias, R.S.; Ma, C.; Busov, V.; Strauss, S.H., 2011. Gibberellin- associated cisgenesmodify growth, stature and wood properties in Populus. Plant Biotechnol. J., 9, 162–178.

Hartung, F.; Schiemann, J.; Quedlinburg, D., 2014. Precise plant breeding using new genome editing techniques: Opportunities, safety and regulation in the EU. Plant J., 78, 742–752.

Haun, W., Coffman, A., Clasen, B. M., Demorest, Z. L., Lowy, A., Ray, E., Retterath, A., Stoddard, T., Juillerat, A., Cedrone, F., Mathis, L., Voytas, D. F., and Zhang, F. 2014.Improved soybean oil quality by targeted mutagenesis of the fatty acid desaturase 2 gene family.PlantBiotechnol. J. 12: 934–940.

Haun, W., Coffman, A., Clasen, B.M., Demorest, Z.L., Lowy, A., Ray, E., Retterath, A., Stoddard, T., Juillerat, A., Cedrone, F. and Mathis, L., 2014. Improved soybean oil quality by targeted mutagenesis of the fatty acid desaturase 2 gene family. *Plant biotechnology journal*, *12*(7), pp.934- 940.

Holme, I.B.;Wendt, T.; Holm, P.B., 2013. Intragenesis and cisgenesis as alternatives to transgenic crop development. Plant Biotechnol. J., 11, 395–407.

Hou, H.; Atlihan, N.; Lu, Z.-X., 2014. New biotechnology enhances the application of cisgenesis in plant breeding. Front. Plant Sci., 5, 389.

Hu, Y., Zhang, J., Jia, H., Sosso, D., Li, T., Frommer, W. B., Yang, B., White, F. F., Wang, N., and Jones, J. B. 2014. Lateral organ boundaries 1 is a disease susceptibility gene for citrus bacterial canker disease. Proc. Natl. Acad. Sci. USA.111: 521–529.

Jackson, A.L.; Burchard, J.; Schelter, J.; Chau, B.N.; Cleary, M.; Lim, L.; Linsley, P.S., 2006. WidspreadsiRNA"off-target" transcript silencing mediated by seed region sequence complementarity. RNA, 12, 1179–1187.

Jacobs, T.B., LaFayette, P.R., Schmitz, R.J. and Parrott, W.A., 2015. Targeted genome modifications in soybean with CRISPR/Cas9. *BMC biotechnology*, *15*(1), p.16.

Ji, X., Zhang, H., Zhang, Y., Wang, Y., and Gao, C. 2015. Establishing a CRISPR-Cas-like immune system conferring DNA virus resistance in plants.Nat. Plants.1: 15144.

Jia, H., Zhang, Y., Orbovic, V., Xu, J., White, F., Jones, J., and Wang, N. 2017. Genome editing of the disease susceptibility gene CsLOB1 in citrus confers resistance to citrus canker. Plant Biotechnol. J. 15: 817–823.

Jiang, W. Z., Henry, I. M., Lynagh, P. G., Comai, L., Cahoon, E. B., and Weeks, D. P. 2017. Significant enhancement of fatty acid composition in seeds of the allohexaploid,Camelinasativa, usingCRISPR/Cas9 gene editing. Plant Biotechnol. J. 15: 648–657.

Jinek, M. *et al.* (2014) 'Structures of Cas9 endonucleases reveal RNA-mediated conformational activation', *Science*, 343(6176). doi: 10.1126/science.1247997.

Kao, K.N., andMichayluk,M. R. (1974).Amethod for high-frequency intergeneric fusion of plant protoplasts. Planta 115, 355–367. doi: 10.1007/BF00388618

Kawamata, T.; Tomari, Y. 2010. Making RISC.TrendsBiochem. Sci., 35, 368–376

Kawasaki, H.; Taira, K., 2004. Induction of DNA methylation and gene silencing by short interfering RNAs in human cells. Nature, 431, 211–217.

Kim, Hyojin*et al.* (2011) 'Surrogate reporters for enrichment of cells with nuclease-induced mutations', *Nature Methods*, 8(11), pp. 941–943. doi: 10.1038/nmeth.1733.

Kim, J.-S.; Pabo, C.O., 1998. Getting a handhold on DNA: Design of poly-zinc finger proteins with femtomolardissociation constants. Proc. Natl. Acad. Sci. USA, 95, 2812–2817.

Kim, Y.G.; Cha, J.; Chandrasegaran, S., 1996. Hybrid restriction enzymes: Zinc finger fusions to Fok I cleavage domain. Proc. Natl. Acad. Sci. USA, 93, 1156–1160.

Kishore, G. M., and Shah, D. M. 1988. Amino acid biosynthesis inhibitors as herbicides.Annu. Rev. Biochem. 57: 627–663.

Kissoudis, C., van de Wiel, C., Visser, R.G. and van der Linden, G., 2014. Enhancing crop resilience to combined abiotic and biotic stress through the dissection of physiological and molecular crosstalk. *Frontiers in Plant Science, 5*, p.207.

Klattenhoff, C.; Theurkauf, W., 2008. Biogenesis and germline functions of piRNAs. Development, 135, 3–9.

Kumar, V. and Jain, M., 2015. The CRISPR–Cas system for plant genome editing: advances and opportunities. *Journal of experimental botany*, *66*(1), pp.47-57.

Kurihara, Y.; Watanabe, Y., 2004. Arabidopsis micro-RNA biogenesis through Dicer-like 1 protein functions.Proc. Natl. Acad. Sci. USA, 101, 12753–12758.

Kurzynska-Kokorniak, A.; Koralewska, N.; Pokornowska, M.; Urbanowicz, A.; Tworak, A.; Mickiewicz, A.; Figlerowicz, M., 2015. The many faces of Dicer: The complexity of the mechanisms regulating Dicer gene expression and enzyme activities. Nucleic Acids Res., 43, 4365–4380.

Lau, N.C.; Lim, L.P.; Weinstein, *e.g.*;Bartel, D.P., 2001. An Abundant class of tiny RNAs with probable regulatory roles in Caenorhabditiselegans. Science, 294, 858–862.

Lee, K. Y., Townsend, J., Tepperman, J., Black, M., Chui, C. F., Mazur, B., Dunsmuir, P., and Bedbrook, J. 1988.The molecular basis of sulfonylurea herbicide resistance in tobacco.EMBO J. 7: 1241–1248.

Li, A.; Mao, L. 2007. Evolution of plant microRNA gene families. Cell Res., 17, 212–218.

Li, J., Meng, X., Zong, Y., Chen, K., Zhang, H., Liu, J., Li, J., and Gao, C. 2016a. Gene replacements and insertions in rice byintron targeting using CRISPR-Cas9.Nat. Plants.2: 16139.

Li, J.F., Norville, J.E., Aach, J., McCormack, M., Zhang, D., Bush, J., Church, G.M. and Sheen, J., 2013. Multiplex and homologous recombination–mediated genome editing in Arabidopsis and Nicotianabenthamiana using guide RNA and Cas9. *Nature biotechnology*, *31*(8), pp.688-691.

Li, M., Li, X., Zhou, Z., Wu, P., Fang, M., Pan, X., Lin, Q., Luo, W., Wu, G., and Li, H. 2016b. Reassessment of the four yield-related genes Gn1a, DEP1, GS3, and IPA1 in rice using a CRISPR/Cas9 system. Front Plant Sci. 7: 377.

Li, T., Liu, B., Chen, C. Y., and Yang, B. 2016c. TALEN-mediated homologous recombination produces site-directedDNA base change and herbicide-resistant rice. J. Genet. Genomics. 43: 297– 305.

Li, T., Liu, B., Spalding, M.H., Weeks, D.P. and Yang, B., 2012. High-efficiency TALEN-based gene editing produces disease-resistant rice. *Nature biotechnology, 30*(5), p.390.

Li, X., Song, Y., Century, K., Straight, S., Ronald, P., and Dong, X. (2001). A fast neutron deletion mutagenesis-based reverse genetics system for plants. Plant J. 27, 235–242. doi: 10.1046/j.1365- 313x.2001.01084.x

Li, Z., Liu, Z. B., Xing, A., Moon, B. P., Koellhoffer, J. P., Huang, L., Ward, R. T., Clifton, E., Falco, S. C., and Cigan, A. M. 2015. Cas9-guide RNA directed genome editing in soybean. Plant Physiol. 169: 960–970.

Liu, Q.; Segal, D.J.; Ghiara, J.B.; Barbas, C.F., 1997. Design of polydactyl zinc-finger proteins for unique addressing within complex genomes. Proc. Natl. Acad. Sci. USA, 94, 5525–5530.

Lombardo, L., Coppola, G., and Zelasco, S. 2016. New technologies for insect-resistant and herbicide-tolerant plants.TrendsBiotechnol. 34: 49–57.

Lor, V.S., Starker, C.G., Voytas, D.F., Weiss, D. and Olszewski, N.E., 2014. Targeted mutagenesis of the tomato PROCERA gene using transcription activator-like effector nucleases. *Plant physiology, 166*(3), pp.1288-1291.

Lowder, L., Malzahn, A. and Qi, Y., 2016. Rapid evolution of manifold CRISPR systems for plant genome editing. *Frontiers in plant science, 7*, p.1683.

Lu, R.; Martin-Hernandez, A.M.; Peart, J.R.; Malcuit, I.; Baulcombe, D.C., 2003. Virus-induced gene silencing in plants. Methods, 30, 296–303.

Lurin, C.; Andrés, C.; Aubourg, S.; Bellaoui, M.; Bitton, F.; Bruyère, C.; Caboche, M.; Debast, C.; Gualberto, J.; Hoffmann, B., 2004. Genome-Wide Analysis of Arabidopsis pentatricopeptide repeat proteins reveals their essential role in organelle biogenesis. Plant Cell, 16, 2089–2103.

Ma, L., Zhu, F., Li, Z., Zhang, J., Li, X., Dong, J. and Wang, T., 2015. TALEN-based mutagenesis of lipoxygenase LOX3 enhances the storage tolerance of rice (Oryza sativa) seeds. *PLoS One, 10*(12).

Mali, P. *et al.* (2013) 'CAS9 transcriptional activators for target specificity screening and paired nickases for cooperative genome engineering', *Nature Biotechnology*, 31(9), pp. 833–838. doi: 10.1038/nbt.2675.

Morineau, C., Bellec, Y., Tellier, F., Gissot, L., Kelemen, Z., Nogu_e, F., and Faure, J. D. 2016.Selective gene dosage by CRISPR-Cas9 genome editing in hexaploidCamelinasativa.PlantBiotechnol. J. 15: 729–739

Nekrasov, V., Wang, C., Win, J., Lanz, C., Weigel, D., and Kamoun, S. 2017.Rapid generation of a transgene-free powdery mildew resistant tomato by genome deletion.Sci. Rep. 7: 482.

Nelson, O. E., and Rines, H. W. 1962.The enzymatic deficiency in the waxy mutant of maize.Biochem.Biophys. Res. Commun. 9: 297–300.

Nishida, K., Arazoe, T., Yachie, N., Banno, S., Kakimoto, M., Tabata, M., Mochizuki, M., Miyabe, A., Araki, M., Hara, K.Y. and Shimatani, Z., 2016. Targeted nucleotide editing using hybrid prokaryotic and vertebrate adaptive immune systems. *Science, 353*(6305), p.aaf8729.

O'Toole, N.; Hattori, M.; Andres, C.; Iida, K.; Lurin, C.; Schmitz-Linneweber, C.; Sugita, M.; Small, I., 2008. On the expansion of the pentatricopeptide repeat gene family in plants. Mol. Biol. Evol., 25, 1120–1128.

Okuda, K.; Myouga, F.; Motohashi, R.; Shinozaki, K.; Shikanai, T., 2007Conserved domain structure of pentatricopeptide repeat proteins involved in chloroplast RNA editing. Proc. Natl. Acad. Sci. USA, 104, 8178–8183.

Osakabe, K., Osakabe, Y. and Toki, S., 2010. Site-directed mutagenesis in Arabidopsis using custom-designed zinc finger nucleases. *Proceedings of the National Academy of Sciences, 107*(26), pp.12034-12039.

Osakabe, Y., Watanabe, T., Sugano, S.S., Ueta, R., Ishihara, R., Shinozaki, K. and Osakabe, K., 2016. Optimization of CRISPR/Cas9 genome editing to modify abiotic stress responses in plants. *Scientific reports, 6*, p.26685.

Pabo, C.O.; Peisach, E.; Grant, R.A., 2001. Design and selection of novel Cys2His2 zinc finger proteins.Annu. Rev. Biochem, 70, 313–340.

Page, D.R. and Grossniklaus, U., 2002. The art and design of genetic screens: *Arabidopsis thaliana. Nature Reviews Genetics, 3*(2), pp.124-136.

Pavletich, N.P.; Pabo, C.O., 1991. Zinc finger-DNA recognition: Crystal structure of a Zif268- DNA complex at 2.1 A. Science, 252, 809–817.

Peng, A., Chen, S., Lei, T., Xu, L., He, Y., Wu, L., Yao, L., and Zou, X. 2017. Engineering canker- resistant plants through CRISPR/Cas9-targeted editing of the susceptibility gene CsLOB1 promoter in citrus.PlantBiotechnol.J. doi: 10.1111/pbi.12733.

Perez-Pinera, P., Ousterout, D.G. and Gersbach, C.A., 2012.Advances in targeted genome editing. *Current opinion in chemical biology, 16*(3-4), pp.268-277.

Pratt, A.J.; MacRae, I.J., 2009. The RNA-induced silencing complex: A versatile gene-silencing machine. J. Biol. Chem., 284, 17897–17901.

Ray, D.K., Mueller, N.D., West, P.C. and Foley, J.A., 2013. Yield trends are insufficient to double global crop production by 2050. *PloS one, 8*(6), p.e66428.

Sauer, N. J., Narv_aez-V_asquez, J., Mozoruk, J., Miller, R. B., Warburg, Z. J., Woodward, M. J., Mihiret, Y. A., Lincoln, T. A., Segami, R. E., and Sanders, S. L. 2016. Oligonucleotidemediatedgenome editing provides precision and function to engineered nucleases and antibiotics in plants. Plant Physiol. 170: 1917–1928.

Savary, S., Ficke, A., Aubertot, J. N., and Hollier, C. 2012.Crop losses due to diseases and their implications for global food production losses and food security.Food Security.4: 519– 537.

Schmitz-Linneweber, C.; Small, I., 2008. Pentatricopeptide repeat proteins: A socket set for organelle gene expression. Trends Plant Sci., 13, 663–670.

Schouten, H.J.; Krens, F.A.; Jacobsen, E., 2006. Do cisgenic plants warrant less stringent oversight? Nat. Biotechnol., 24, 753.

Seto, A.G.; Kingston, R.E.; Lau, N.C., 2007. The coming of age for Piwi proteins. Mol. Cell, 26, 603–609.

Shan, Q., Wang, Y., Chen, K., Liang, Z., Li, J., Zhang, Y., Zhang, K., Liu, J., Voytas, D.F., Zheng, X. and Zhang, Y., 2013. Rapid and efficient gene modification in rice and Brachypodium using TALENs. *Molecular plant*, *6*(4), pp.1365-1368.

Shan, Q., Zhang, Y., Chen, K., Zhang, K. and Gao, C., 2015. Creation of fragrant rice by targeted knockout of the Os BADH 2 gene using TALEN technology. *Plant biotechnology journal*, *13*(6), pp.791-800.

Shen, L., Wang, C., Fu, Y., Wang, J., Liu, Q., Zhang, X., Yan, C., Qian, Q., and Wang, K. 2016. QTL editing confers opposing yield performance in different rice varieties. J. Integr. Plant Biol. doi: 10.1111/jipb.12501

Shi, Y.; Berg, J.M.A., 1995. direct comparison of the properties of natural and designed zinc-finger proteins.Chem. Biol., 2, 83–89.

Shikanai, T., 2006. RNA editing in plant organelles: Machinery, physiological function and evolution. Cell. Mol. Life Sci. C., 63, 698–708.

Shukla, V.K., Doyon, Y., Miller, J.C., DeKelver, R.C., Moehle, E.A., Worden, S.E., Mitchell, J.C., Arnold, N.L., Gopalan, S., Meng, X. and Choi, V.M., 2009. Precise genome modification in the crop species Zea mays using zinc-finger nucleases. *Nature*, *459*(7245), pp.437-441.

Small, I., 2007. RNAi for revealing and engineering plant gene functions. Curr.Opin. Biotechnol., 18, 148–153.

Small, I.D.; Rackham, O.; Filipovska, A., 2013. Organelle transcriptomes: Products of a deconstructed genome. Curr.Opin.Microbiol., 16, 652–658.

Stadler, L. J. (1928a). Mutations in barley induced by X-rays and radium. Sciencen 68, 186–187. Stoddard, B. L. (2005) 'Homing endonuclease structure and function', *Quarterly Reviews of Biophysics*, 38(1), pp. 49–95. doi: 10.1017/S0033583505004063.

Sun, Y., Jiao, G., Liu, Z., Zhang, X., Li, J., Guo, X., Du, W., Du, J., Francis, F., and Zhao, Y. 2017. Generation of high-amylose rice through CRISPR/Cas9-mediated targeted mutagenesis of starch branching enzymes. Front Plant Sci. 8: 298.

Suzuki, T., Eiguchi, M., Kumamaru, T., Satoh, H., Matsusaka, H., and Moriguchi, K. (2008).MNU-induced mutant pools and high performance TILLING enable finding of any gene mutation in rice. Mol. Genet. Genomics 279, 213–223. doi: 10.1007/s00438-007-0293-2.

Svitashev, S., Young, J. K., Schwartz, C., Gao, H., Falco, S. C., and Cigan, A. M. 2015. Targeted mutagenesis, precise gene editing and site-specific gene insertion in maize using Cas9 and guide RNA. Plant Physiol. 169: 931–945.

Takahashi, J. S., Pinto, L. H., and Vitaterna, M. H. (1994). Forward and reverse genetic approaches to behavior in the mouse. Science 264, 1724–1733. doi: 10.1126/science.8209253

Talamè, V., Bovina, R., Sanguineti,M. C., Tuberosa, R., Lundqvist, U., and Salvi, S. (2008). TILLMore, a resource for the discovery of chemically induced mutants in barley. Plant Biotechnol. J. 6, 477–485. doi: 10.1111/j.1467-7652.2008.00341.x

Thakore, P.I.; Gersbach, C.A., 2015. Genome engineering for therapeutic applications. In Translating GeneTherapy to the Clinic; Laurence, J., Franklin, N., Eds.; Academic Press: Boston, MA, USA, pp. 27–43,ISBN 978-0-12-800563-7.

Tiwari, M.; Sharma, D.; Trivedi, P.K., 2014. Artificial microRNA mediated gene silencing in plants: Progress and perspectives. Plant Mol. Biol., 86, 1–18.

Townsend, J. A., Wright, D. A., Winfrey, R. J., Fu, F., Maeder, M. L., Joung, J. K., and Voytas, D. F. 2009. High-frequency modification of plant genes using engineered zinc-finger nucleases. Nature. 459: 442–445.

Wang, F., Wang, C., Liu, P., Lei, C., Hao, W., Gao, Y., Liu, Y. G., and Zhao, K. 2016a. Enhanced rice blast resistance by CRISPR/Cas9-targeted mutagenesis of the ERF transcription

Wang, H., Yang, H., Shivalila, C.S., Dawlaty, M.M., Cheng, A.W., Zhang, F. and Jaenisch, R., 2013.One-step generation of mice carrying mutations in multiple genes by CRISPR/Cas-mediated genome engineering. *Cell, 153*(4), pp.910-918.

Wang, N., Wang, Y., Tian, F., King, G. J., Zhang, C., and Long, Y. (2008). A functional genomics resource for Brassica napus: development of an EMS mutagenized population and discovery of FAE1 point mutations by TILLING. New Phytol. 180, 751–765. doi: 10.1111/j.1469- 8137.2008.02619.x

Wang, Y., Cheng, X., Shan, Q., Zhang, Y., Liu, J., Gao, C., and Qiu, J. L. 2014. Simultaneous editing of three homoeoallelesin hexaploid bread wheat confers heritable resistance to powdery mildew. Nat. Biotechnol. 32: 947–951.

Warnefors, M.; Liechti, A.; Halbert, J.; Valloton, D.; Kaessmann, H., 2014. Conserved microRNA editing in mammalian evolution, development and disease.Genome Biol., 15, R83.

Wu, J. L.,Wu, C., Lei, C., Baraoidan,M., Bordeos, A., andMadamba,M. R. (2005). Chemical- and irradiation-induced mutants of indica rice IR64 for forward and reverse genetics. Plant Mol. Biol. 59, 85–97. doi: 10.1007/s11103-004-5112-0

Xie, F. *et al.* (2012) 'Legume pectate lyase required for root infection by rhizobia', *Proceedings of the National Academy of Sciences of the United States of America*, 109(2), pp. 633–638. doi: 10.1073/pnas.1113992109.

Xing, Y., and Zhang, Q. 2010. Genetic and molecular bases of rice yield. Annu. Rev. Plant Biol. 61: 421–442.

Xiong, J.S., Ding, J. and Li, Y., 2015. Genome-editing technologies and their potential application in horticultural crop breeding. *Horticulture research*, 2(1), pp.1-10.

Xu, R., Qin, R., Li, H., Li, D., Li, L., Wei, P. and Yang, J., 2017. Generation of targeted mutant rice using a CRISPR Cpf1 system. *Plant biotechnology journal, 15*(6), pp.713-717.

Xu, R., Yang, Y., Qin, R., Hao, L., Qiu, C., Li, L., Wei, P., and Yang, J. 2016. Rapid improvement of grain weight via highly efficient CRISPR/Cas9-mediated multiplex genome editing in rice. J. Genet. Genomics. 43: 529–532.

Yagi, Y.; Hayashi, S.; Kobayashi, K.; Hirayama, T.; Nakamura, T., 2013. Elucidation of the RNA recognition code for pentatricopeptide repeat proteins involved in organelle RNA editing in plants. PLoS ONE, 8, e57286.

Yin, K., Gao, C. and Qiu, J.L., 2017. Progress and prospects in plant genome editing. *Nature plants, 3*(8), pp.1-6.

Zhang, F., Wen, Y. and Guo, X., 2014. CRISPR/Cas9 for genome editing: progress, implications and challenges. *Human molecular genetics, 23*(R1), pp.R40-R46.

Zhang, H., Gou, F., Zhang, J., Liu, W., Li, Q., Mao, Y., Botella, J.R. and Zhu, J.K., 2016. TALEN mediated targeted mutagenesis produces a large variety of heritable mutations in rice. *Plant biotechnology journal, 14*(1), pp.186-194.

Zhang, K., Raboanatahiry, N., Zhu, B. and Li, M., 2017. Progress in genome editing technology and its application in plants. *Frontiers in Plant science, 8*, p.177.

Zhang, Y., Liang, Z., Zong, Y., Wang, Y., Liu, J., Chen, K., Qiu, J. L., and Gao, C. 2016b. Efficient and transgenefreegenome editing in wheat through transient expression of CRISPR/Cas9 DNA or RNA. Nat. Commun. 7: 12617.

Zhang, Y., Zhang, F., Li, X., Baller, J.A., Qi, Y., Starker, C.G., Bogdanove, A.J. and Voytas, D.F., 2013. Transcription activator-like effector nucleases enable efficient plant genome engineering. *Plant physiology, 161*(1), pp.20-27.

Zhen Xie, Liliana Wroblewska, Laura Prochazka, Ron Weiss, Y. B. (2011) 'Multi-Input RNAi- Based Logic Circuit for Identification of Specific Cancer Cells', *Science*, 333(September), pp. 1307–1312. doi: 10.1126/science.1205527.

Zhou, H., He, M., Li, J., Chen, L., Huang, Z., Zheng, S., Zhu, L., Ni, E., Jiang, D., and Zhao, B. 2016. Development of commercial thermo-sensitive genic male sterile rice accelerates hybrid rice breeding using the CRISPR/Cas9-mediated TMS5 editing system. Sci. Rep. 6: 37395.

Zhou, H.-L.; Luo, G.;Wise, J.A.; Lou, H., 2014. Regulation of alternative splicing by local histone modifications: Potential roles for RNA-guided mechanisms. Nucleic Acids Res., 42, 701–713.

Zhou, Q., Liu, W., Zhang, Y., and Liu, K. K. 2007. Action mechanisms of acetolactatesynthase- inhibiting herbicides.PestBiochem. Physiol. 89: 89–96.

Zhu, H.; Zhou, Y.; Castillo-González, C.; Lu, A.; Ge, C.; Zhao, Y.-T.; Duan, L.; Li, Z.; Axtell, M.J.; Wang, X.-J.; Zhang, X., 2013. Bi-directional processing of pri-miRNAs with branched terminal loops by Arabidopsis Dicer-like1. Nat. Struct. Mol. Biol., 20, 1106–1115.

Zong, Y., Wang, Y., Li, C., Zhang, R., Chen, K., Ran, Y., Qiu, J.L., Wang, D. and Gao, C., 2017. Precise base editing in rice, wheat and maize with a Cas9-cytidine deaminase fusion. *Nature biotechnology*, *35*(5), p.438.

Zong, Y.;Wang, Y.; Li, C.; Zhang, R.; Chen, K.; Ran, Y.; Qiu, J.-L.;Wang, D.; Gao, C., 2017. Precise base editing in rice, wheat and maize with a Cas9-cytidine deaminase fusion. Nat. Biotechnol., 35, 438–440.

Chapter 6

Plant Probiotics and its Role in Crop Improvement

***K. Greeshma*[1], *A. Akhil Reddy*[2], *K. Anvesh*[3] *and Sonal Athnere*[4]**

[1]Department of Plant Pathology, [2]Department of Seed Science and Technology, [3]Department of Entomology, College of Agriculture, PJTSAU, Hyderabad, Telangana

[4]Department of Agronomy, Maharana Pratap University of Agriculture and Technology, Udaipur, Rajasthan

With the demand of augmenting population and the need of producing abundant edibles with in the same land resulted in intensification of agriculture and therefore resulting the intensive using of pesticides and fertilizers. Unfortunately, the high and indiscriminate usage of the chemicals resulted in depletion effects on soil, water, human and animal health. Some terrestrial bacteria harbor some of the contrivances to enhance plant growth, which includes the biocontrol of plant pathogens, nutrient mobilization, Phyto stimulation and abiotic stress protection. These microbes also have the capacity to promote food quality of vegetable. Hence, these are also said to be Plant Probiotic Bacteria. These plant probiotic microorganisms (PPM), are also known as biofertilizers, biostimulants, biocontrollers or bioprotectants, are beneficial microorganisms that offer a promising alternative and reduce the health problems and constitute an ecofriendly manner to contribute the production of food needed for world population to sustain.

Keywords: *Bacteria, Mechanisms, Sustainable, Biofertilizers, Protection.*

Introduction

Now a days with the increasing population globally, there is an increase in food demand with the limited resources the agricultural production need to be increased. The intensification of agriculture resulted in increase in use fertilizers and manures which increased occurrence of pests and disease. This situation inevitably implies a search for more efficient procedures to produce animal feed and food for humans, with higher yields and more resistance to biotic and abiotic stresses, while protecting animal and human health and being environment friendly.

Plants and microbes occur in close association interwined functionally effecting the overall metabolism of the plant some microbes are harmful pathogens deteriorating plant health and become problematic for food production and ecosystem stability globally. Other class of microbes serves a beneficial role by enhancing the plant health acting as biocontrol agents, due to the intensified agricultural practices from the last few years, producers are highly dependent on the agrochemicals as a comparatively consistent way of crop protection which helps in financial constancy of their practices. However, using of more synthetic inputs, causes numerous deleterious effects, causing worldwide contamination of the water and food chain by pesticide residues. With the increasing cost of pesticides, especially in low economic regions of the world, similarly there are numerous diseases which are difficult to control by chemical means, few or non-existent.

Also, the demand for high quality food is also gaining importance at present and in future. In this sense, the application of useful microbes especially bacteria, with plant growth promoting features, the so-called plant probiotics used in the crop protection may be a possible solution. We have been using probiotics from many years in the fermentation of bakery products. Though, the possible use of probiotics in non-dairy edible food stuffs and farming have not yet gained formal identification. In current times, there has been an increasing awareness to agricultural and food applications of using the probiotics avoiding the hazardous complications related to the application of different synthetic pesticides and moreover, allowing the better quality food (Flores-Felix *et al.*, 2015).

Probiotics

Probiotic is resultant of Latin and Greek word, it means 'for life', firstly coined by Kollath (1953).According to Lilly and Stillwell (1965), probiotics defined as substances released by one microorganism that stimulate the growth of another. The joint panel of FAO/WHO defined the term probiotics as living micro-organisms, which when administered in sufficient amounts, leads to a health benefits on the host (FAO/WHO, 2002). These are orally administered or given along with the food supplements and are present inside the gastrointestinal tract of human beings. These are also been used for the health benefits of animals, fishes and birds. Most of the probiotics are bacteria, among them, the most mutual one is lactic acid bacteria (LAB) however few yeasts and molds can also be used as probiotics. The different genera usually considered as probiotics contains the bacteria like- *Bifidobacterium,*

Lactobacillus, Streptococcus, Lactococcus, Propionibacterium, Carnobacterium,Leuconostoc, Enterococcus, Pediococcus and *Bacillus* species, *Saccharomyces* yeasts, (Amara and Shibl, 2015).

As a whole, mainly probiotics are gram-positive, generally catalase-negative rods with curved edges and find in pairs, small or long chains. They are non-motile, non-flagellated and do not form spores and are sensitive to salt. Optimum temperature for probiotic growth is 37°C but some species such as *L. casei* favor 30°C and the suitable pH for primary growth is 6.5-7.0.Probiotics also occur naturally in the fermented foodstuff product such as yoghurt, sauerkraut, kefir, cabbage kimchee and soybean-based miso and natto. Many health benefits have been recognized by the use of probiotics. Some strains of probiotics may also be used as delivery systems for immunoglobulins, vaccines and other protein-based therapy. A good probiotic must possess the following characters such as:

- The organism must have the capacity of exerting a positive or advantageous effect on the hosts,
- Higher growth/resistance to the disease,
- It has to be non-pathogenic and non-toxic,
- It should be stable under storage and field conditions (Fuller, 1989).

Plant Probiotics

Immense work has been done on the human probiotics and the application of probiotics is expanding. Plant probiotics has been gaining the attention recently. Plant probiotics are microbial culture which has biocontrol potential and promote plant growth by virtue of their typical mechanisms comprising nitrogen fixation, phosphate solubilization, siderophore production and upgraded plant invulnerability in contradiction of various diseases.

These also enhances the soil structure by amalgamating the soil units together by releasing some extracellular metabolites, promoting the cessation of composite materials and insoluble nutrients into simpler molecules so that it can be easily available for plant growth and also provide resistance against stress and diseases (Maheshwari *et al.*, 2012). It is assumed that ample populations of useful soil microbes are mandatory for good soil structure and also provide better plant development. Microbes mainly bacteria that facilitate phosphate solubilization, nitrification and bio-control of plant diseases are the ample examples of plant probiotics (Prakash *et al.*, 2015). Apart from these rhizobacteria, the findings on endophytic bacteria and fungi showed that bacteria existing in the core background helps the plants in nutrient attainment, overall growth and stress tolerance. Thus, the introduction of plant probiotics for promoting plant health, development and better potentiality is an eco-friendly substitute to inorganic fertilizers which sustain soil health as well as encourage organic farming.

Results achieved through broad studies have shown that microbes are an important constituent for plant growth and can also increase resistance, tolerance and resilience of plants in biotic and abiotic stress conditions.

Mixtures of one or more of these bacteria are framed into goods and used as biofertilizers, improving the crop yields by making the availability of nutrients to the plants, without fully reliable on inorganic fertilizers and thus, keeping the surroundings safe. It is the prime time to initiate a New Green Revolution that can use the benefits of certain microorganisms, called plant probiotics, in crop improvement, production and productivity.

Probiotic bacteria have plant growth promoting ability through a wide range of activities, which can be categorized based on their way of action in:

(i) The direct assimilation of affluences by plants,

(ii) The nutrient mobilization,

(iii) The plant stress resistance induction,

(iv) The disease prevention.

The microbial interfaces with plants are divided into different classes based on the site and mode of interaction.

Types of Probiotic Plant Growth-Promoting Rhizobacteria

The plant useful rhizobacteria serves as probiotic for plant as well as to soil because it not only improves plant health but also maintains soil fertility and texture. Plant growth-promoting rhizobacteria having different types of associations with plant roots. Based on the relationship of the bacteria with plant,the PGPR classified into several types:

- ☆ Endophytic
- ☆ Symbiotic
- ☆ Associative
- ☆ Free living

Plant Probiotics in Crop Production

The rhizosphere has a significant effect on plant growth, not because of the nutrients concentration it contains, but also owing to other organisms that populate the rhizosphere zone. Rhizobacteria, mentioning to bacterial population obtain in soil surrounding the roots, can have different influences on the plant growth.

Nitrogen Fixation

Although nitrogen is quite abundant on the earth, it is the utmost limiting nutrient factor for plants, which is essential for the synthesis of amino acids and proteins. Some prokaryotic bacteria have the ability of conversion of freely available atmospheric nitrogen into organic forms, by a symbiotic mechanism known as Biological nitrogen fixation, which is directly assimilated by the plants. Amongest free-living rhizobacteria, the genera *Azospirillum, Bacillus, Paenibacillus, Beijerinckia, Azotobacter, Burkholderia, Glucono acetobacter* and *Herbaspirillum* were reported as important nitrogen-fixing microorganisms. The genus *Azospirillum* commonly associated with cereals in temperate zones, has shown increasing crop yields in cereals as well as some legumes and sugarcane. Members of the genus *Azotobacter* which are non-symbiotic bacteria are able to fix nitrogen in rice crops (Sahoo *et al.*, 2014). Moreover, some species of this genus are tested as biofertilizers for several cereals, such as wheat, barley, oat, rice or maize; oil plants such as linseeds and sunflowers; and other variety of plants, such as beetroot, coconut,tobacco, tea and coffee

Some species found in the genera like *Gluconacetobacter, Herbaspirillum* and *Azospirillum* are frequent sugarcane endophytes and act as nitrogen fixers. The genera *Bacillus* and *Paenibacillus* are free-living nitrogen fixers and have other PGP traits, which make them suitable candidates for application. Some strains of the genera *Rhizobium* and *Bradyrhizobium,* which were found in association with rice and wheat roots, increase this nutrient concentration in those plant yields.

Table 6.1: Nitrogen Fixing Probiotic Bacteria

Nitrogen fixation	*Azoarcus*	Rice	Reinhold-Hurek, B. *et al.*, 1998
	Azorhizobium	Wheat	Sabry, SRS *et al.*, 1997
	Azospirillum	Several cereals, sugarcane, bean, soybean	Tejera, N. *et al.*, 2005
	Azotobacter	Several cereals, lineseed, tobacco, sunflower, tea, coffee, beetroot, tomato	Ramakrishnan, K. *et al.*, 2012
	Bacillus	Rice	Beneduzi, A. *et al.*, 2008
	Brevundimonas	Wheat	Rana, A. *et al.*, 2011
	Burkholderia	Rice	Kao, C.M. *et al.*, 2003
	Enterobacteriales	Maize, wheat, sugarcane	Berge, O. *et al.*, 1991
	Rhizobium and related genera	Leguminous plants	Taule, C. *et al.*, 2012

Phytohormone Biosynthesis

Many bacterial probiotics are able to synthetize phytohormones, which are defined as organic molecules participated in some processes at different stages of plant growth and development A wide spectrum of beneficial bacteria are capable of synthesizing them and are involved in plant growth and development. Amongst which more focus is for PGP rhizobacteria producing auxins, cytokinins, gibberellins and ethylene which are intricate in key processes of plant development.

Auxin-producing *Bacillus* spp. have been testified to have a progressive effect in the development of several crops, such as *Solanum tuberosum* (potato) or *Oryza sativa* (rice). The genus *Paenibacillus* was also reported as a good phytohormone producer. *Enterobacter* and related Enterobacteriales are also good phytohormone producers and have PGP effects in sugarcane, wheat, pepper and soybean. Some bacteria generates ACC deaminase (1-aminocyclopropane-1-carboxylate deaminase), to hydrolyze the ethylene precursor in plants (ACC), to get α-ketobutyrate and ammonia, which can be used as a source carbon and nitrogen. Thus, these bacteria modulate ethylene levels in plants and therefore reduces the negative effects formed by high ethylene concentration. Moreover, these molecules have an essential role in the nodulation process between a rhizobial strain and a legume.

Table 6.2: Probiotic Bacteria in Phytohormone Synthesis

Phytohormone biosynthesis (auxins, gibberelings, cytokynins, ethylene and ACC desaminase synthesis)	*Azobacter*	Cucumber	Aloni R., 2006
	Bacillus	potato, cucumber, oriental thuja, pepper	Rana, A. *et al.*, 2011; Ahmed, A. *et al.*, 2010.
	Enterobacteriales	sugarcane, wheat, pepper, soybean	Rana, A. *et al.*, 2011; Taule, C. *et al.*, 2012
	Hartmannibacter	summer barley	Suarez, C. *et al.*, 2015
	Phyllobacterium	*Arabidopsis*	Galland, M. *et al.*, 2012
	Pseudomonas	wheat, mung bean	Shaharoona, B. *et al.*, 2008

Nutrient Mobilization

Phosphorous Solubilizers (PSB)

Phosphorous (P) is the second major nutrient for plants, after nitrogen (N) and the major part of the reservoirs are not available for them as it is insoluble in soils is applied exogenously as chemical P fertilizers. However, when given as fertilizer to crop fields, P translocates quickly to convert in insoluble and thus, unavailable to plants. Therefore, the application of P-solubilizing bacteria (PSB) might represent a green alternative for these environment-damaging chemical P fertilizers. Soil bacteria such as the genera *Micrococcus, Pseudomonas, Bacillus, Paenibacillus, Deftia, Azotobacter, Klebsiella, Pantoea* and *Flavobacterium*, have been reported as efficient phosphate solubilizers. Moreover, there are many phosphate-solubilizing rhizobial strains, which promote the growth of several crops of economic interest. The genus *Mesorhizobium* has also strains that are good P-solubilizers, promoting the growth of chickpea and barley.

Table 6.3: Phosphorous Solubulising Probiotic Bacteria

Phosphate solubilization	*Bacillus*	Rice	Beneduzi, A. *et al.*, 2008
	Burkholderia	Rice	Estrada, G.A. *et al.*, 2013
	Enterobacteriales	Wheat	Rana, A. *et al.*, 2011
	Herbaspirillum	Rice	Estrada, G.A. *et al.*, 2013
	Paenibacillus	Rice	Beneduzi, A. *et al.*, 2008
	Phyllobacterium	Strawberries	Flores-Felix, J.D. *et al.*, 2005
	Rhizobium and related genera	Pepper, tomato, lettuce, carrot, strawberries, carnation, chickpea	Menendez, E., 2016
	Streptomyces	Wheat	Jog, R. *et al.*, 2014

Potassium Solubilizers (KSB)

After nitrogen and phosphorous, potassium (K) is the third most important nutrient vital for plant growth. Few rhizobacteria have ability to make available the insoluble potassium forms. A huge diversity of K-solubilizing bacterial genera (KSB) are present, among firmicutes, there are various K-solubilizing bacteria.

Most of the K-solubilizing bacteria are reported in *Bacillus* and *Paenibacillus* genera. *Bacillus edaphicus* has been shown increased K utilization in wheat and *Paenibacillus glucanolyticus* was attributed to improves the dry matter of black pepper. Sudan grass injected with the potassium-solubilizing bacterium *Bacillus mucilaginosus* had more biomass yields. Also, *Bacillus mucilaginosus* with the phosphate-solubilizing *Bacillus megaterium* induced the growth of groundnut, eggplant, pepper and cucumber. Amongst Proteobacteria, the genus *Pseudomonas* was described as KSB, promoting growth and development of tea plants (*Camellia sinensis*) and tobacco (Zhang and Kong, 2014).

Table 6.4: Potassium Solubulising Probiotic Bacteria

Potassium solubilization	*Bacillus*	Wheat, maize Sudan grass, eggplants, pepper, cucumber, cotton, rape, groundnut	Sheng, X.F., 2005
	Enterobacteriales	Tobacco	Velazquez, E., 2016
	Frauteria	Tobacco	Subhashini, D.V., 2015
	Microbacterium	Tobacco	Zhang, C., 2014
	Paenibacillus	Black pepper	Velazquez, E., 2016
	Pseudomonas	Tobacco, tea	Bagyalakshmi, B., 2012

Plant Probiotics in Crop Protection

Plant Probiotics which are the Plant Rhizobacteria with their role in increasing the crop productivity are also known to promote plant health. Apart from growth stimulation, these plant probiotics also act as biocontrol agents competing with the plant pathogens in the root zone. The probiotic rhizobacteria act as biological control agents mainly by indirect mechanisms. Probiotics involve different modes for biocontrol activity in plant which includes production of antibiotics and antifungal metabolites, siderophores production, competition for nutrients, exclusion of niche and induced systemic resistance (ISR). They also act on cell wall and cause structural, physiological and biochemical changes involved in plant defense mechanisms.

Siderophore Production

Iron is an essential element for the growth and development of all living organisms. In soil iron occurs as ferric iron (Fe III) and is unavailable for direct assimilation by microorganisms. The shortage of bioavailable iron on plant surfaces and in soil habitats results a competition. Under such iron-limiting or scarcity conditions the probiotic PGPR produce siderophores which are low-molecular-weight iron binding compounds that acquire ferric ion from soil competitively, and result in death of the pathogenic fungi due to scarcity of the essential element. The siderophores produced by microbes differ in overall structure but most of them contain catechol and hydroxamate groups which are mainly involved in chelating the ferric ion.

Biosynthesis of Siderophore is mainly regulated by GacS and GacA, the global regulators the iron-sensitive Fur proteins, RpoS and FpvI, the sigma factors, *N*-acyl homoserine lactone which is a quorum sensing auto inducer and some site specific recombinases. The factors such as level and form of iron, pH, the presence or absence of trace elements, availability of nutrients like carbon, nitrogen and phosphorus and other environmental factors influence siderophores synthesis (Radzki*et al.*, 2013). Rhizosphere microbial community structure is greatly influenced by Iron nutrition of the plant. *Bacillus* and *Pseudomonas* produce siderophores that inhibit several fungal pathogens, the role of pyoverdine siderophore produced by the bacteria *Pseudomonas* has been clearly demonstrated, in the control of soil borne *Pythium* and *Fusarium* species.

Table 6.5: Siderophore Producing Probiotic Bacteria

Siderophore production	*Bacillus*	Maize, pepper, rice	Beneduzi, A. *et al.*, 2012
	Chryseobacterium	Tomato	Radzki, W. *et al.*, 2013
	Enterobacteriales	Wheat	Rana, A. *et al.*, 2011
	Micrococcus	Maize, canola	Ghavami,N. *et al.*, 2016
	Phyllobacterium	Strawberries	Flores, J.D. *et al.*, 2015
	Pseudomonas	Potato, maize	Beneduzi, A. *et al.*, 2012
	Rhizobium and related genera	Pepper, tomato, lettuce, carrot, strawberries, carnation, chickpea	Menéndez, E. *et al.*, 2016

Antibiotic Production

Production of antibiotics is the most effective mechanisms employed by the probiotic PGPR to prevent proliferation of pathogenic microorganisms. Bacteria that produce antibiotics are suitable for biocontrol. They must not also compete successfully with other microorganisms for nutrients released from the root. Antibiotics are the organic compounds with low-molecular weight that are undesirable for the growth and metabolic activities of other microorganisms even in the low concentration. A number of antibiotics produced by *Pseudomonas, Bacillus, Streptomyces* have been discovered in recent times, which include phenazine, 2,4-diacetylphloroglucinol (DAPG), pyoluteorin, amphisin, pyrrolnitrin and cyclic lipopeptides produced by *pseudomonas* and oligomycin A, kanosamine, produced by *Streptomyces* and *Stenotrophomonas* spp.

Regulatory aspects of these antibiotics involve GacA, RpoS, *N*-acetyl homoserine lactone derivatives and a positive auto regulation. Antibiotic synthesis is firmly linked to the nutrient availability of the major and important minor minerals, the type source and supply of carbon, which in turn governs overall metabolic activity of the cell, the factors such as pH, temperature along with other parameters influence the overall genetic stability of bacteria, in turn affecting the ability of bacteria to produce antibiotics and other secondary metabolites. A number of strains belonging to *Bacillus* spp., are able to produce antibiotics that are active against many phyto pathogenic bacteria and fungi.

Enzyme Production

Most biocontrol agents employ hyperparasitic effect by using hydrolytic enzymes which includes cellulases, chitinases and glucanases which are involved in degrading the cell wall and acting as the biocontrol agents of pathogens, chitin β-glucan and cellulose are the major components of cellwall for most of the fungi. Chitinase and 1, 3-glucanase degrade chitin, an insoluble linear polymer of 1,4-N-acetylglucoseamine, bacteria that produce these enzymes such inhibit the fungal growth.

Similar to the antibiotic and siderophore production the regulation of enzyme production involves GacA/GacS or GrrA regulatory system. The growth and infection caused by *Rhizoctonia solani* and *Phytophthora capsici*, is inhibited by the enzymes chitinases and β-glucanases produced by the *Pseudomonas* spp. *Pseudomonas fluorescens*LPK2 produce chitinases and β-glucanases and resulted in control the disease caused by *Fusarium udum* (Kumar *et al.*, 2010).

Induced Systemic Resistance

A defensive mechanism developed by a plant on inoculation of selected PGPR strains is Induced systemic resistance (ISR) when stimulated appropriately, PGPR strains has shown to trigger a plant mediated resistance in above ground plant parts. This type of resistance is often referred to as induced systemic resistance (ISR).

Probiotic bacteria has shown to provide resistance against a number of plant pathogens causing diseases by triggering a phenomenon called induced systemic resistance (ISR) which is phenotypically almost similar to systemic acquired resistance (SAR). The SAR mainly develops when plants are infected by a pathogen and successfully activate their defense mechanism by responding to certain molecules referred as elicitors which are secreted by microorganism. whereas ISR induces a phenomenon called as hypersensitive reaction (HR) by which the pathogen becomes limited to a local necrotic lesion of desiccated tissue. SAR differs from ISR in not causing any visible symptoms on the host and SAR is induced systemically after inoculation with necrotic plant pathogens but ISR is induced by non-pathogenic rhizobacteria. salicyclic acid is the main signaling molecule in plants that activates SAR and is accompanied by the induction of pathogenesis related proteins. Moreover, ISR involves jasmonic acid and ethylene signaling and is independent of salicyclic acid. Both the mechanisms ISR and SAR are highly effective against a broad range of plant pathogenic micro organisms.

ISR elicited by probiotic bacteria was first observed on carnation and cucumber showing a low level of susceptibility to wilt caused by fungus *Fusarium* spp., and foliar disease caused by *Colletotrichum orbiculare*, respectively after inoculation with a probiotic bacterial strain. The inoculation has resulted in the structural changes like formation of barriers, thick cell wall, synthesis and accumulation of phenolic compounds at site of pathogen attack. The Biochemical or physiological changes occurring in plants include induced accumulation of pathogenesis-related proteins (PR proteins) and some peroxidases. However, certain probiotic bacteria do not induce synthesis of PR proteins but rather increase accumulation of enzymes such as phenylalanine ammonia lyase, polyphenol oxidase or chalcone synthase and phytoalexins.

Many probiotic bacteria with biocontrol potential have been reported to produce antifungal metabolites like hydrogen cyanide, pyrrolnitrin, DAPG(2, 4-diacetyl phloroglucinol), pyoluteorin, and tensin (Bhattacharyya and Jha, 2012). Microbial antagonism includes the inhibiting the growth of microbes, competition for nutrients, minerals and degradation of pathogenicity factors. Thus, plant growth-promoting rhizobacteria (PGPR) function as probiotics for plants as they improve the plant growth and productivity by combination of various direct and indirect mechanisms.

Table 6.6: Probiotic Bacteria in Disease Control

Probiotic Bacteria	*Pathogen*	*Mode of Action*	*Reference*
Pseudomonas fluorescens	*Sclerospora graminicola*	The probiotic isolares offered protection ranging from 20 to 75 per cent against downy mildew to pearl millet.	Niranjan *et al.*, 2004
P. putida	*Fusarium oxysporum f.sp. melonis*	Seed treatment of *P. putida* strain 30 and strain 180 resulted in Control on muskmelon up to 63 per cent and 40 per cent respectively	Bora *et al.*, 2004
Pseudomonas fluorescens	*X. Oryzae* pv.*oryzae*	Showed resistance to the rice bacterial blight pathogen.	Vidhyasekaran *et al.*, 2001
Bacillus cereus, Bacillus lentimorbus, Bacillus pumilus	*X. Compestris* pv.*Compestris*	When the antagonists were applied the Incidence and severity of black rot of cabbage were reduced.	Massomo *et al.*, 2004
Bacillus amyloliquefaciens, B. subtilis, B. pumilus	Cucumber mosaicvirus (CMV)	PGPR mediated ISR occurred against CMV following mechanical inoculation on tomato.	Zehnder *et al.*, 2000
Fusarium oxysporum f. sp. Ciceri	*Paenibacillus lentimorbus* NRRL B 30488	Seed treatment with strain B 30488 caused greater mortality in non-inoculated seedlings compared to inoculated seedlings of chickpea.	Dasgupta *et al.*, 2006.

Inoculation of Probiotics

Microbial biocontrol agents popularly known as "biopesticides, biofertilizers, orbio-inoculants" are formulations that contain active ingredient which is are living microorganisms. These formulations when applied to host parts such as seed, leaves, branches other parts of plant or to soil, colonize the host rhizosphere, surface of plant or the inner parts of the plant and thereby increase availability of primary nutrients and promote growth of the host plant. Biofertilizers supply nutrients to plants and there by promote synthesis of growth-promoting substances. The micro organisms present as active ingredient of biofertilizers helps to increase the natural soil nutrient cycle and there by help in building soil organic matter thus maintain the soil fertility. Of all types of formulations of probiotics. Probiotic inoculants as seed treatment is the most commonly used. In a report, sowing of treated chickpea seeds with talc-based formulation of *P. fluorescens* strain-Pf1 @ 4 g/kg (107cfu/g) for the management of chickpea wilt resulted in establishment of rhizobacteria in the rhizosphere. PGPR also called plant probiotics are the most preferred to be used as microbial inoculants and are gaining the acceptance worldwide because they are safe to handle and are cheaper than chemical pesticides. In order to have beneficial effects it is important to introduce these bacteria into the soil. PGPR strains when inoculated to soil improves the soil health and positively influence the growth of plant. PGPR act as probiotic by acting as biofertilizer for crop productivity and biocontrol agent for plant health.

Probiotics Application Challenges

Plant Probiotics industry faces a number of challenges equally when claiming the benefits, several challenges are identified in the development and application of probiotics including: selection of strain, growth and survival during processing, inoculation, viability during storage, the viable counts, particularly when multiple probiotic strains are used. When there are also starter cultures added changes and the effects on sensory properties (Herrmann *et al.*, 2013).Other challenges such as: maintaining of probiotics, diversity and survival to being active.

The challenges faced by plant probiotics, also include the performance under field conditions the PGPR has to withstand adverse environmental conditions that influence their activity PGPR need to be propagated to regain their viability and biological activity. The natural variation is an issue because it is difficult to predict how bacteria will act in laboratory and when placed in field. The effectiveness of plant probiotics in field conditions can also be affected by climatic conditions, specifically fluctuating environmental conditions and the sheer volume of more competitive, already naturalized microbes already present in the soil. "Regardless of delivery method, insufficient storage during transportation and on the farm are an even greater barrier. As with some other biologicals (*e.g. rhizobium* bacteria inoculants used in soybeans), improper storage commonly means growers are applying dead or low-activity products.

Conclusion

The probiotics usage and their applications have shown an enormous increase in the recent past. Probiotics use has shown many health benefits to plants by increasing the productivity and acting as a biocontrol agent against phytopathogens. The probiotic applications faces many challenges, in addition to the viability and environmental barriers it must be kept in mind for selection of suitable strain, inoculation and processing of starter cultures. Indeed, the formulations must show the viability throughout the storage period.

REFERENCES

Ahmed, A. and Hasnain, S. 2010. Auxin producing *Bacillus* spp: Auxin quantification and effect on the growth of *Solanum tuberosum*. *Pure Appl Chem.*82: 313–319.

Amara, A.A. and Shibl, A. 2015.Role of probiotics in health improvement, infection control and disease treatment and management – *Saudi pharmaceutical Journal.* 23(2): 107-114.

Ahmed, A. and Hasnain, S. 2010. Auxin producing *Bacillus* spp: Auxin quantification and effect on the growth *Solanum tuberosum*. *Pure Appl Chem*. 82: 313–319.

Asaf, S., Khan, M. A. and Khan, A.L. 2017. Bacterial endophytes from arid land plants regulate endogenous hormone content and promote growth in crop plants: an example of *Sphingomonas* sp. and *Serratiama rcescens*. *J Plant Interact*. 12: 31–38.

Bagyalakshmi, B., Ponmurugan, P. and Marimuthu, S. 2012. Influence of potassium solubilizing bacteria on crop productivity and quality of tea (*Camellia sinensis*). *Afr J Agric Res*. 7: 4250- 4259.

Bakker, P. A. H. M. 2014. Induced systemic resistance by beneficial microbes. *Annu. Rev. Phytopathol*. 52: 347-375.

Beneduzi, A., Ambrosini, A. and Passaglia, L.M. 2012. Plant growth-promoting rhizobacteria (PGPR): Their potential as antagonists and biocontrol agents. *Genet Mol Biol*. 35: 1044–1051.

Bermudez, B. M., Plaza, D. J., Munoz, Q. S and Gomez, L.C. 2012.Probiotic Mechanisms of Action. *Ann Nutr Metab*. 61: 160–174.

Bhattacharyya, P.N. and Jha, D. K.2012. Plant growth-promoting rhizobacteria (PGPR): emergence in agriculture. *World J Microbiol Biotechnol*. 28: 1327–1350.

Borriss, R. 2015. *Bacillus*, a plant-beneficial bacterium, In: *Principles of Plant-Microbe Interactions.Springer International Publishing*.379–391.

Brigido, C., Nascimento, F.X. and Duan, J. 2013. Expression of an exogenous 1- aminocyclopropane-1-carboxylate deaminase gene in *Mesorhizobium* spp. reduces the negative effects of salt stress in chickpea. *FEMS Microbiol Lett*. 349: 46–53.

Dastager, S.G., Deepa, C.K and Pandey, A. 2010. Isolation and characterization of novel plant growth promoting Micrococcus sp. NII-0909 and its interaction with cowpea. *Plant Physiol Biochem*. 48: 987–992.

Estrada, G.A., Baldani, V.L.D and de Oliveira, D.M. Selection of phosphate-solubilizing diazotrophic *Herbaspirillum* and *Burkholderia* strains and their effect on rice crop yield and nutrient uptake. *Plant Soil*. 2013;369: 115–129.

Flores-Felix, J.D, Marcos-García, M. and Silva, L.R. 2015. *Rhizobium* as plant probiotic for strawberry production under microcosm conditions. *Symbiosis*. 67: 25–32.

Flores-Felix, J.D, Menendez, E. and Rivera, L.P. 2013. Use of *Rhizobium leguminosarum* as a potential biofertilizer for *Lactuca sativa* and *Daucuscarota* crops. *J Plant Nutr Soil Sci*. 176: 876–882.

Flores-Felix, J.D., Silva, L.R. and Rivera, L.P.2015. Plants probiotics as a tool to produce highly functional fruits: the case of *Phyllobacterium* and vitamin C in strawberries. *PLoS One* 10.

Galland, M., Gamet, L. and Varoquaux, F. 2012. The ethylene pathway contributes to root hair elongation induced by the beneficial bacteria *phyllobacterium brassicacearum*STM196. *Plant Sci*.190: 74–81.

Gamalero, E. and Glick, B.R. 2015.Bacterial modulation of plant ethylene levels. *Plant Physiol.* 169: 13–22.

Ghavami N., Alikhani, H.A. and Pourbabaei, A.A. 2016. Effects of two new siderophore producing rhizobacteria on growth and iron content of maize and canola plants. *J Plant Nutr.*40: 736–746.

Ghosh, P.K., Sen, S.K. and Maiti, T.K. 2015. Production and metabolism of IAA by *Enterobacter* spp. (Gammaproteo bacteria) isolated from root nodules of a legume *Abrus precatorius* L. *Biocatal Agric Biotechnol.* 4: 296–303.

Grady, E.N., MacDonald, J. and Liu, L. 2016. Current knowledge and perspectives of *Paenibacillus*: a review. *Microb Cell Fact.*15: 203.

Herrmann, L., Atieno, M., Brau, L. and Lesueur, D. 2013.Microbial quality of commercial inoculants to increase BNF and nutrient use efficiency. In: de Bruijn FJ (ed) Molecular microbial ecology of the rhizosphere. Wiley-Blackwell, Hoboken.

Hungria, M., Nogueira, M.A. and Araujo, R.S. 2013. Co-inoculation of soybeans and common beans with rhizobia and azospirilla: strategies to improve sustainability. *BiolFert Soils.* 49: 791–801.

Isawa, T., Yasuda, M. and Awazaki, H. 2010. *Azospirillum*sp. strain B510 enhances rice growth and yield. *Microbes Environ.*25: 58–61.

Jaiswal, D.K., Verma, J.P. and Prakash, S. 2016. Potassium as an important plant nutrient in sustainable agriculture: a state of the art, In: *Potassium Solubilizing Microorganisms for Sustainable Agriculture,* Springer India.21–29.

Jog, R., Pandya, M. and Naresh kumar, G. Mechanism of phosphate solubilization and antifungal activity of *Streptomyces* spp. isolated from wheat roots and rhizosphere and their application in improving plant growth. *Microbiology.* 2014;160: 778–788.

Kang, S.M., Khan, A.L. and Waqas, M. 2015. Gibberellin-producing *Serratia nematodiphila* PEJ1011 ameliorates low temperature stress in *Capsicum annuum* L. *Eur J Soil Biol.* 68: 85–93.

Kanmani, P., Kumar, S. R., Yuvaraj, N., Paari, A. K., Pattukumar, V and Arul, V. 2013.Probiotics and its functionally valuable products-A review. *Critical Reviews in Food Science and Nutrition.* 53: 641 658.

Kumar,H., Bajpai, V.K. and Dubey, R.C. 2010. Wilt disease management and enhancement of growth and yield of *Cajanus cajan* (L) var. Manak by bacterial combinations amended with chemical fertilizer. *Crop Protect.* 29: 591–598.

Liu F, Xing, S. and Ma, H. 2013.Cytokinin-producing, plant growth-promoting rhizobacteria that confer resistance to drought stress in *Platycladus orientalis* container seedlings.*ApplMicrobiolBiotechnol.*97: 9155–9164.

Maheshwari, D. K., Saraf, M. and Aeron, A. 2012.Bacteria in agrobiology: crop productivity. Springer, Berlin.127–165.

Menendez, E., Escribano-Viana, R. and Flores-Felix, J.D. Biological Nitrogen Fixation and Beneficial Plant-Microbe Interaction. Springer International Publishing; 2016.Rhizobialbiofertilizers for ornamental plants.13–21.

Menendez, E., Escribano-Viana, R and Flores-Felix, JD.2016.Rhizobialbiofertilizers for ornamental plants, In: *Biological Nitrogen Fixation* and *Beneficial Plant-Microbe Interaction,Springer International Publishing*.13–21.

Nascimento, F. X., Brigido, C. and Glick, B.R. 2016.The role of rhizobial ACC deaminase in the nodulation process of leguminous plants. *Int J Agron.*

Radzki W, Gutierrez Manero F.J., Algar, E., Lucas García, J.A., García-Villaraco, A. and Ramos Solano, B. 2013. Bacterial siderophores efficiently provide iron to iron-starved tomato plants in hydroponics culture. *Antonie Van Leeuwenhoek.* 104(3): 321-330.

Ramakrishnan, K. and Selvakumar, G. Effect of biofertilizers on enhancement of growth and yield on Tomato (*Lycopersicum esculentum* Mill.) *Int. J Res Bot.* 2012; 2: 20–23.

Rana, A., Saharan, B. and Joshi, M. 2011. Identification of multi-trait PGPR isolates and evaluating their potential as inoculants for wheat. *Ann Microbiol.* 61: 893–900.

Sahoo, R.K, Ansari, M.W. and Pradhan, M. 2014. Phenotypic and molecular characterization of native *Azospirillum* strains from rice fields to improve crop productivity. *Protoplasma.* 251: 943–953.

Sahoo, R. K., Ansari, M.W. and Dangar, T.K. 2014.Phenotypic and molecular characterization of efficient nitrogen-fixing *Azotobacter* strains from rice fields for crop improvement. *Protoplasma*. 251: 511–523.

Sangeeth, K.P., Bhai,R.S and Srinivasan, V. 2012.*Paenibacillus glucanolyticus*, a promising potassium solubilizing bacterium isolated from black pepper (*Piper nigrum* L.) rhizosphere. *JSpices Arom Crops*.21.

Sharma, S.B., Sayyed, R. Z. and Trivedi, M.H. 2013. Phosphate solubilizing microbes: sustainable approach for managing phosphorus deficiency in agricultural soils. *Springerplus*.2: 587.

Sokolova, M.G., Akimova, G. P. and Vaishlya, O.B. 2011. Effect of phytohormones synthesized by rhizosphere bacteria on plants. *App Biochem Microbiol.* 47: 274–278.

Suarez, C., Cardinale, M. and Ratering, S. 2015. Plant growth-promoting effects of *Hartmannibacter diazotrophicus* on summer barley (*Hordeum vulgare* L.) under salt stress. *Appl Soil Ecol*.95: 23–30.

Subhashini, D.V. 2015. Growth promotion and increased potassium uptake of tobacco by potassium-mobilizing bacterium *Frateuria aurantia* grown at different potassium levels in vertisols. *Commun Soil Sci Plant Anal.* 46: 210–220.

Taule, C., Mareque, C. and Barlocco, C. 2012.The contribution of nitrogen fixation to sugarcane (*Saccharum officinarum* L.), and the identification and characterization of part of the associated diazotrophic bacterial community. *Plant Soil*.356: 35–49.

Velazquez, E., Silva, L.R. and Ramirez-Bahena, M.H. 2016. Potassium Solubilizing Microorganisms for Sustainable Agriculture. *Springer India*; Diversity of potassium-solubilizing microorganisms and their interactions with plants; 99–110.

Wan, L. Y. M., Chen, Z. J., Shah, N. P. and El-Nezami, H. 2016. Modulation of intestinal epithelial defense responses by probiotic bacteria. *Critical Reviews in Food Science and Nutrition*. 56: 2628–2641.

Yanni, Y.G., Dazzo, F.B. and Squartini, A. 2016. Assessment of the natural endophytic association between *Rhizobium* and wheat and its ability to increase wheat production in the Nile delta. *Plant Soil*. 407: 367–383.

Zhang, C. and Kong, F. 2014.Isolation and identification of potassium-solubilizing bacteria from tobacco rhizospheric soil and their effect on tobacco plants. *Appl Soil Ecol*. 82: 18–25.

Chapter 7

Species Composition and Richness in different Sized Agroecosystems of Indian Central Himalaya, Along an Altitudinal Gradient

Charu Shahi, Surendra Singh Bargali

Department of Botany, D.S.B. Campus, Kumaun University, Nainital – 263001, Uttarakhand

The present study has been performed to analyze the plant species composition, species richness (SR), Family Importance Value (FIV), and similarity indices (SI) among different-sized Central Himalayan agroecosystems (AGEs). The AGEs have been divided into three size classes i.e. small, medium, and large size class based on the available landholdings among four different altitudes i.e. very low (VLA), low (LA), mid (MA) and high altitude (HA). The results revealed that among all the sites a total of 58 plant species belonging to 25 families were recorded within AGEs, out of these, 17 were tree species, 21 herb species were cultivated, 15 herb species were wild and 5 herb species were categorized as wild- cultivated due to the cultivation of wild species for forage purpose. SR was reported as VLA (18) > LA (17)> MA (16) HA> (16) and depicted a positively significant relation with altitudes (r^2= 0.607) whereas it was not affected by AGE sizes. It was observed that the higher altitudes (MA and HA) were proliferated with vegetable-based agroecosystems (Brassica oleracea var. capitata, Phaseolus vulgaris, Pisum sativum, Solanum tuberosum, etc.) whereas the grain (Triticum aestivum, Oryza sativa, and Zea mays) and pulse (Glycine max) based agroecosystems were common in lower altitudes. The highest FIV was reported for family Poaceae followed by family Fabaceae among *all altitudes. Similarity indices among agroecosystem size classes and along altitudes depicted that the highest similarity (48 per cent) of plant species composition was recorded between VLA- LA, followed by HA-MA (43.90 per cent).*

Keywords: *Agroecosystems, Altitude variation, Size variation, Species composition, FIV.*

Introduction

The ecosystem saga begins with the human, as a rational being, transforms the natural environment in a new living space anthroposphere, which is a component of the biosphere. To maintain the anthroposphere, people receive materials, energy, and food from the surrounding environment including the biosphere. In the process of evolution, people have transformed a huge part of materials and energy resources in different forms, human use mainly processed animal and plant raw materials. The plant's raw material comes from the agroecosystems, which is created by people and requires appropriate human care and intervention for continuing its existence (Stoev, 2018).

Agricultural objectives generally have been to achieve maximum yields, operate with a maximum profit, minimize year-to-year instability in production, and prevent long-term degradation of the productive capacity of the agricultural system. Due to the interfaces between human society and natural ecosystem agroecosystems are considered as ecological, socioeconomic systems and communities of plants or animals (biological component) interacting with their physical and chemical environments that have been modified by people with space and time for the production, consumption, and processing (Caporali, 2015).

Biodiversity refers to all existing plant species, animals, and microorganisms that interact within an ecosystem and provides the basis for all living things. In traditional agroecosystems the expansion of complex and diversified cropping systems is of key importance to the stability of peasant farming systems, allowing crops to reach acceptable productivity levels in the midst of environmentally stressful conditions. Generally, the traditional agroecosystems are less exposed to the catastrophic accidents because they grow a wide variety of crops, trees, animals, and associated biota in various spatial and temporal arrangements (Shahi, 2020).

The degree of biodiversity in agroecosystems depends on four main features of agroecosystem: one is the diversity of plant vegetation within and around the agroecosystem second one is the permanence of the various crops within the agroecosystem, next is the intensity of management or land use patterns and the last one is the extent of the isolation of the agroecosystem from natural vegetation (El-Ramady *et al.*, 2012).

Agrobiodiversity through agriculture, manage the interactions between crops, domestic animals, and their associated biodiversity and the environment, which provides (directly or indirectly from plants) most of our food with less than 5 per cent coming from the wild. It has been estimated that more than 80 per cent of our calories and edible dry weight derives from crop plants. So it is of prime importance to know about the regional agrobiodiversity to sustain the global socio-economy in a long run.

Materials and Methods

This study has been performed in agroecosystems of Indian Central Himalaya. Four different altitudes *i.e.* very low (VLA), low (LA), mid (MA) and high altitude (HA) have been selected and the AGEs were categorized into three size classes *i.e.* small, medium, and large size based on the available landholdings. The AGEs were evaluated for species composition and richness (the number of plant species present in AGEs), family importance value, and similarity index. Species richness is the number of species represented by a community or area.

Family Importance Value (FIV)

FIV index was used to evaluate floristic composition at the species family level, and it combines richness, density, and dominance, according to the given formulae

$$\text{Family relative density (\%)} = \frac{\text{Number of individuals in a family}}{\text{Total number of individuals}} \times 100$$

$$\text{Family relative diversity (\%)} = \frac{\text{Number of species in a family}}{\text{Total number of species}} \times 100$$

FIV is the sum of family relative diversity and relative density.

Index of Similarity

Index of similarity (IS) was calculated as:

$$IS = \frac{2C}{A+B} \times 100$$

where,

A the number of species at site A, B the number of species at site B and C common species in both the sites.

Results

Across all the sites a total of 58 plant species belonging to 25 families were recorded within agroecosystems, out of these, 21 herb species were cultivated, 15 herb species were wild and 5 herb species were categorized as wild- cultivated due to the cultivation of wild species for forage purpose. Among all the sites a total of 17 tree species were reported in agroecosystems mainly grown at bunds and boundaries of the cultivated areas.

Table 7.1: Species composition and richness in agroecosystems across all sized agroecosystems along the altitudes

Sl.No.	Species	Altitudes			
	Cultivated crops	VLA	LA	MA	HA
1	Abelmoschus esculentus	-	*	-	-
2	Amaranthus caudatus	-	-	-	**
3	*Brassica oleracea var. botrytis*	-	-	***	-
4	*Brassica oleracea var. capitata*	-	-	**	****
5	*Cannabis sativa*	-	-	-	*
6	*Eleusine coracana*	*	-	****	******
7	*Glycine max*	****	*	-	-
8	*Glycine soja*	-	*****	-	-
9	*Oryza sativa*	****	****	**	-
10	*Parilla frutescence*	*	-	-	-
11	*Phaseolus lunatus*	-	-	-	******
12	*Phaseolus vulgaris*	-	-	****	-
13	*Pisum sativum*	***	-	****	****
14	*Raphanus sativus*	-	-	*	-
15	*Solanum lycopersicum*	-	-	****	-
16	*Solanum tuberosum*	-	-	****	****
17	*Tagetes erecta*	*	-	-	**
18	*Triticum aestivum*	****	****	***	****
19	*Vicia faba*	**	-	-	-
20	*Vigna mungo*	-	***	-	-
21	*Zea mays*	****	****	-	****
	Wild species				
22	Ageratum conyzoides	****	****	****	-
23	*Anagallis arvensis*	-	-	-	****
24	*Avena lanata*	-	****	-	
25	*Avena sativa*		-	-	-
26	*Commelina benghalensis*	*	-	-	-
27	*Cynodon dactyon*	****	****	-	*****
28	*Cyperus rotundus*	****	-	***	-
29	*Fragaria vesca*	-	-	***	*****
30	*Galinsoga parviflora*	-	****	****	****
31	*Oxalis corniculata*	-	****	****	****
32	Oxalis stricta	***	****	-	-
33	*Phalaris minor*	****	****	-	-
34	*Poa annua*	****	-	-	-
35	*Polygonum plebeium*	***	****	-	-
36	*Sonchus oleraceus*	-	******	-	-

Sl.No.	Species	Altitudes			
	Cultivated crops	VLA	LA	MA	HA
37	*Sorghum bicolor*	-	*	-	-
38	*Stellaria media*	****	-	****	****
39	*Trifolium alexandrinum*		*****	-	-
40	*Trifolium pratense*	-	-	***	-
41	*Urtica dioca*	-	-	-	***
	Trees	-	-	-	-
42	*Annona reticulata*	***	-	-	-
43	*Artocarpus integrifolia*	*****	-	-	-
44	*Cinnamomum tamala*	-	*	-	-
45	*Eucalyptus globulus*	***	-	-	-
46	*Ficus auriculata*	-	-	*	-
47	*Ficus ovata*	-	-		-
48	*Ficus semicordata*	-	***	-	-
49	*Ficus auriculata*	-	**	-	-
50	*Grevia optiva*	-	-	******	-
51	Juglans regia	-			***
52	*Litchi chinensis*	*****	-	-	-
53	*Malus domestica*	-	-	-	****
54	*Mangifera indica*	**	******	-	-
55	*Morus alba*	*****	-	-	-
56	*Prunus domestica*	-	-	-	*
57	*Prunus persica*	-	-	-	*****
58	*Psidium guajawa*	*	*	-	-
	Species richness	27	23	20	21

VLA: Very low altitude; LA: Low altitude; MA: Mid altitude; HA: High altitude.

*: Large; **: Medium+large; ***: Small; ****: Small+medium+large; *****: Medium; ******: Small+medium.

i. Very Low Altitude

Total species richness at this site was 27 out of which 11 species have most commonly occurred in all sized AGEs (Table 7.1). In the small sized AGEs, 17 plant species belonging to 9 families were recorded, of these 5 species were unique for small sized AGEs. Out of the total plant species recorded, 35.29 per cent were cultivated, 47.06 per cent were wild and 17.65 per cent were wild- cultivated species in small sized AGEs (Figure 7.1) of this altitude. The family importance value was recorded highest for Poaceae (FIV= 83), followed by Fabaceae (FIV= 45) and Caryophyllaceae (FIV= 21) whereas lowest FIV (7) was recorded for family Annonaceae, Asteraceae, Lamiaceae, and Myrtaceae (Figure 7.2).

In the medium sized AGEs, from 8 different families, 16 plant species were recorded and out of these, 3 plant species were exclusive (Table 7.1). The plant species were present in descending order as cultivated (56.25 per cent) > wild (31.25 per cent) > wild-cultivated (12.5 per cent) in this size class of AGEs (Figure 7.1). The FIV was observed (Figure 7.2) minimum for Myrtaceae (FIV= 6) and maximum for Poaceae (FIV= 100).

In the large sized AGEs, the species richness was 19 and out of these, 5 species were exclusively present in this size of agroecosystems (Table 7.1). The presence percentile was minimum for wild-cultivated species (15.79 per cent) and maximum for cultivated species (52.63 per cent) (Figure 7.1). Family importance value (Figure 7.2) in this size class was ranged from 7 (Anacardiaceae) to 107 (Poaceae).

ii. Low Altitude

The species richness for this altitude was recorded as 23, and out of them, the small AGEs were supported with 16 plant species belonging to 7 families. 3 plant

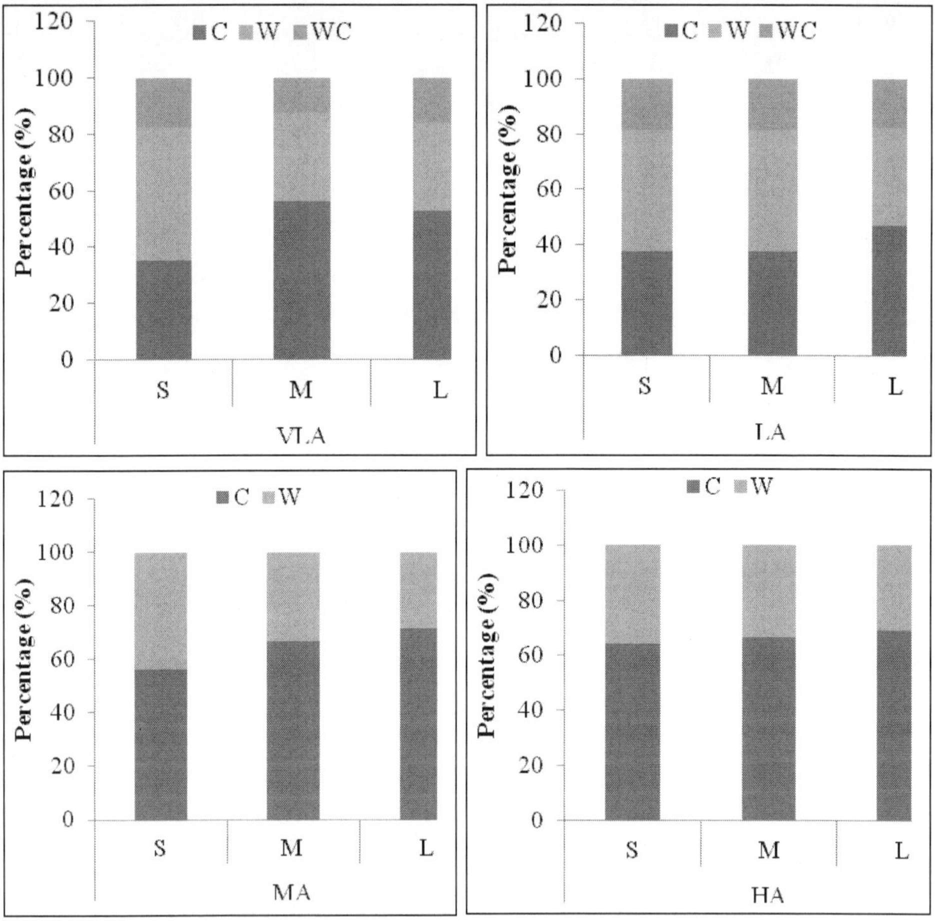

Figure 7.1: Per cent Contribution of Species Composition (Cultivated, Wild, and Wild-cultivated) Across all the Sites.

VLA: Very low altitude; LA: Low altitude; MA: Mid altitude; HA: High altitude; S: Small; M: Medium; L Large; C: Cultivated; W: Wild; WC: Wild cultivated species.

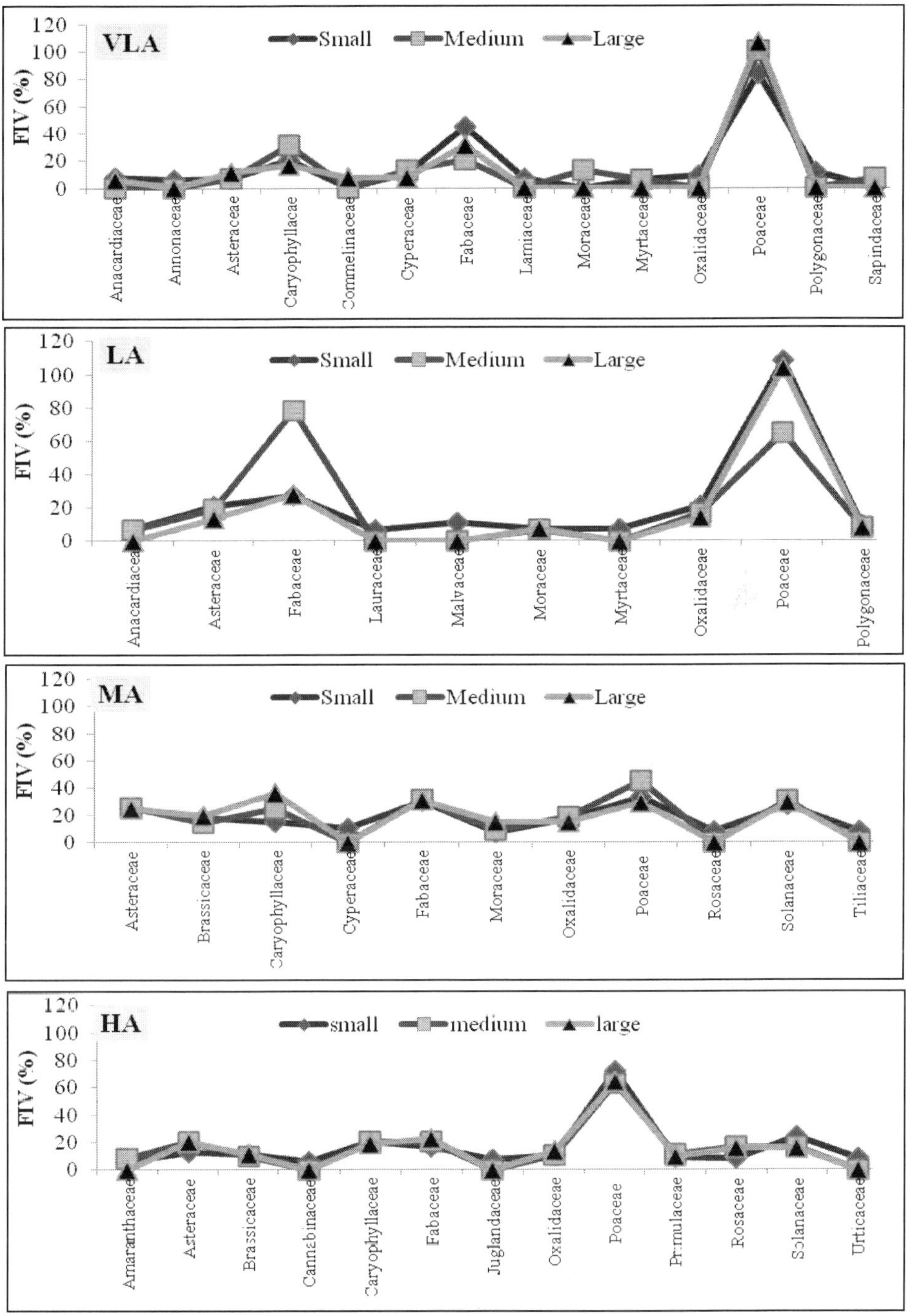

Figure 7.2: Family Importance Value of AGE Species Along Altitudes with Size Classes.

VLA: Very low altitude; LA: Low altitude; MA: Mid altitude; HA: High altitude; S: Small; M: Medium; L Large.

species occurred exclusively in the small sized AGEs while 11 species were most common which occurred in all sized AGEs (Table 7.1). In the small sized AGEs, the highest percentage was recorded for wild species (43.75 per cent), followed by cultivated (37.5 per cent) and wild-cultivated (18.75 per cent) at this altitude (Figure 7.1). The FIV in this size class AGEs (Figure 7.2) was reported highest for Poaceae (FIV= 108) which followed by Fabaceae (FIV= 27), Oxalidaceae (FIV=21), *etc.*

Medium sized AGEs depicted a similar pattern as small sized AGEs where only 2 plant species were unique in this size class of AGEs (Table 7.1). The plant species were present in the same order as reported in small sized AGEs (Figure 7.1). The family importance value was least reported (Figure 7.2) for Moraceae (FIV= 6) while it was reported highest for Fabaceae (FIV= 78) which further followed by family Poaceae (FIV= 65) and Asteraceae (FIV= 19), *etc.*

In the large sized AGEs, a total of 17 species from 9 different families were reported and out of these, 5 species were unique in these agroecosystems (Table 7.1). The presence of species was minimum for wild-cultivated species (17.65 per cent) and maximum for cultivated species (47.06 per cent) (Figure 7.1). The FIV was recorded minimum (Figure 7.2) for family Moraceae (FIV= 7) and maximum for Poaceae (FIV= 104).

iii. Mid Altitude

A total of 20 plant species were observed at this altitude and out of which 9 species were common in all the three categories of AGEs. In the small sized AGEs, 16 plant species (from 11 families) were reported and 5 plant species were unique over them, (Table 7.1). High presence percentile was recorded for cultivated species (56.25 per cent) followed by wild species (43.75 per cent) and wild-cultivated species were not reported in AGEs of this altitude (Figure 7.1). The family importance value (Figure 7.2) was maximum for Poaceae (FIV= 33) (FIV= 27) while the minimum value was reported for Moraceae (FIV= 7).

In the medium sized AGEs, 12 species belonging to 8 families were enumerated and there were no unique species (Table 7.1). Out of these reported species 66.67 per cent were cultivated species and 33.33 per cent were wild species (Figure 7.1). The FIV (Figure 7.2) was recorded highest for Poaceae (FIV= 45) while the lowest for Moraceae (FIV= 9).

In the large sized AGEs, a total of 14 species belonging to 8 families were reported and out of these, 2 species were only present in this size class (Table 7.1). The species percentage was higher for cultivated species (71.43 per cent) and lower for wild species (28.57 per cent) (Figure 7.1). FIV (Figure 7.2) was observed maximum for Caryophyllaceae (FIV= 37) whereas it was minimum for Moraceae and Oxalidaceae (FIV= 15).

iv. High Altitude

The AGEs of this altitude supported 21 plant species, out of which 10 species were co-ordinately found in all sized AGEs. A total of 14 plant species belonging to 10 families was reported and 2 plant species were exclusive in the small sized

AGEs (Table 7.1). Cultivated species were reported higher (64.29 per cent) than wild species (35.71 per cent) and none of the wild-cultivated species were reported in AGEs of high altitude (Figure 7.1). The FIV was reported maximum for Poaceae (FIV= 72) and minimum (FIV= 6) for Amaranthaceae and Cannabinaceae (Figure 7.2).

In the medium sized AGEs, 18 plant species belonging to 11 families were encountered. There were 4 unique species in these AGEs (Table 7.1). Cultivated plant species were recorded with 66.67 per cent and wild species with 33.33 per cent (Figure 7.1). The FIV (Figure 7.2) ranged from 8 (Amaranthceae) to 63 (Poaceae).

In the large sized AGEs, a total of 13 plant species were reported belonging to 10 families, of which only one species was specific to this size class of AGEs (Table 7.1). The species percentage was higher for cultivated species (69.23 per cent) and lower for wild species (30.77 per cent) (Figure 7.1). The family importance value was recorded highest for Poaceae (FIV= 65), and the lowest FIV (10) was recorded for Primulaceae (Figure 7.2).

Discussion

Agrobiodiversity pledge an enormous array of biological estates tied with agriculture such as crops, livestock, soil fauna, agricultural landscape, *etc.* Traditional agroecosystems, which maintain, support, and protect biological diversity, are shifting firmly to modern agroecosystems. Modern agroecosystems are based on intensive cropping (monoculture, plantations, *etc.*), and have evolved principally to increase yield and economic returns, this trend is resulting in considerable loss of biological diversity and depletion of natural resources (Shahi, 2020). There is an obvious weakening of linkages among the components of traditional agroecosystems and consequently the erosion of agrobiodiversity (Shahi *et al.*, 2019).

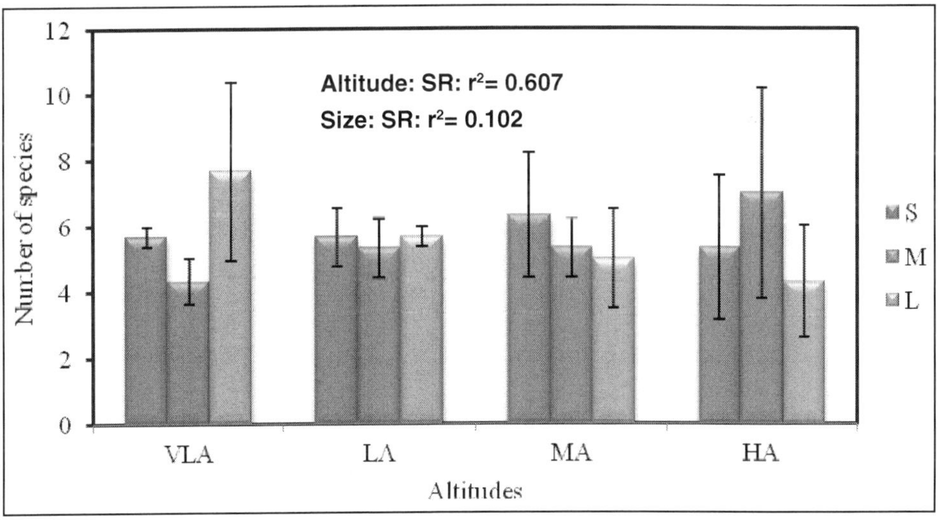

Figure 7.3: Species Richness (SR) Across Agroecosystem Sizes and Along Altitudes. VLA: Very low altitude; LA: Low altitude; MA: Mid altitude; HA: High altitude; S: Small; M: Medium; L Large.

Among studied agroecosystems, the species richness was reported highest at VLA (18) followed by LA (17), MA (16), and HA (16). In this study mean species richness per agroecosystem varied from 4-8 species (Figure 7.3). Species richness, in the present study, was positively significant with altitudes (r^2= 0.607) whereas it was not affected by AGE size (r^2= 0.102). Many earlier researchers stated that the farm size doesn't have any effect on crop species richness after analyzing different farms in Southern Ethiopia and Indonesia, respectively (Shahi, 2020).

The present study reported that the higher altitudes (MA and HA) were proliferated with vegetable-based agroecosystems (*Brassica oleracea* var. *capitata, Phaseolus vulgaris, Pisum sativum, Solanum tuberosum, etc.*) whereas the grain (*Triticum aestivum, Oryza sativa,* and *Zea mays*) and pulse (*Glycine max*) based agroecosystems were common in lower altitudes. The topography, climate, and farm management practices of higher altitudes supporting the cash crop AGEs while in lower altitudes the climatic conditions support grain and pulse-based AGEs.

In this study, the highest FIV was reported for family Poaceae followed by family Fabaceae along all altitudes (in the following order), which depicts the importance of Poaceae and Fabaceae over all other families (Figure 7.4).

VLA- Poaceae (97)> Fabaceae (32)> Caryophyllaceae (23)>.

LA- Poaceae (92)> Fabaceae (44)> Asteraceae (18)>.

MA- Poaceae (36)> Fabaceae (31)> Solanaceae (29)>.

HA- Poaceae (66)> Fabaceae, Caryophyllaceae (20)> Solanaceae (19)>.

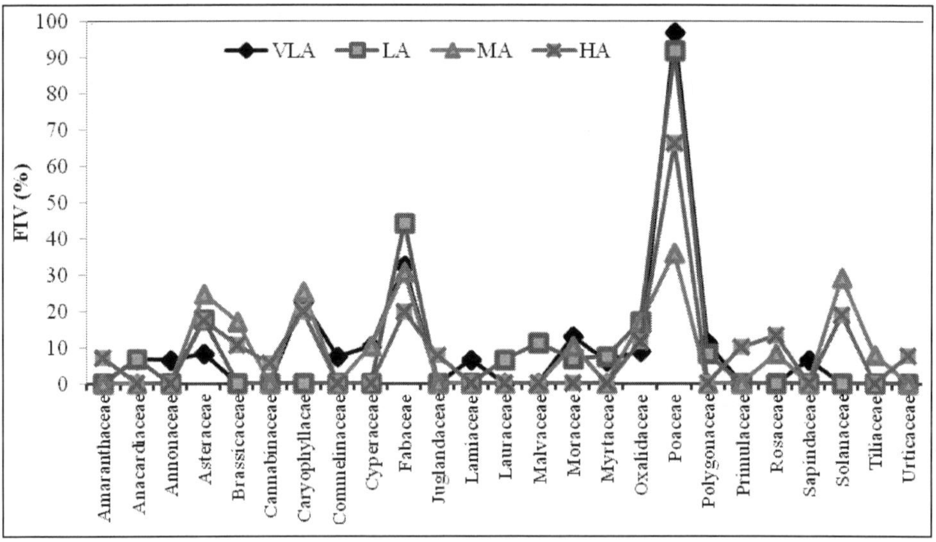

Figure 7.4: Family Importance Value (FIV) of Reported Plant Species in Agroecosystems, along the Altitudinal Gradient.

VLA: Very low altitude; LA: Low altitude; MA: Mid altitude.

The present study supports the statement of many previous researchers that the proliferation of cash crops significantly decreasing the traditional crop diversity among Himalayan agroecosystems, though there are so many functions which govern by vegetation diversity (Ratnadas *et al.*, 2012) still it is declining prominently. This trend was also reported in many other mountain countries such as Nepal, Pakistan, Afghanistan, Bhutan, Tibet, *etc.*, by many researchers (Shahi, 2020).

The proliferation of cash crops at upper altitudes definitely, enhancing the economy of farmers but at a cost of increased vulnerability to market and climatic fluctuations. Many of the cash crops (potato, pea, cabbage, *etc.*) do not have any fodder value for instant or long term as grain residues contain which reflect the lesser fodder production from private farms and indirectly implies pressure on natural linking forest resources. Generally, the agroecosystems of higher altitudes lack the trend of forage and fodder cultivation, which promotes the intensive lopping and removal of forest floor biomass ultimately posing a threat to forest flora. Many researchers pointed out these situations in the context of Himalayan agroecosystems, and present study recognized that those reversions are still downgrading the Himalayan agroecosystems in an alarming rate and enhancing the vulnerability among sustainable agroecosystems (Shahi, 2020).

Similarity indices among agroecosystem size classes and along altitudes depicted that the highest similarity (48 per cent) of plant species composition was recorded between VLA- LA, followed by HA-MA (43.90 per cent) (Table 7.2). Among size classes the similarity indices showed the pattern as; large-medium (75.95 per cent)> small-medium (73.42 per cent)> small-large (69.23 per cent), which indicates that up to some extent, the size of an agricultural landscape may have the determining effect on the number of species composition.

Table 7.2: Similarity Indices (Per cent) between Agroecosystems along Altitudes and Across the Agroecosystem Sizes

Altitudes	*VLA*	*LA*	*MA*
VLA	100		
LA	48	100	
MA	29.79	27.91	100
HA	29.17	22.73	43.90
Size	*Small*	*Medium*	*Large*
Small	100		
Medium	73.42	100	
Large	69.23	75.95	100

VLA: Very low altitude; LA: Low altitude; MA: Mid altitude.

Altieri *et al.* (2012) defined that agroecosystems are profoundly rooted in the ecological rationale of traditional small-scale agriculture, delineating the long-established examples of successful agricultural systems characterized by a tremendous diversity of domesticated crops and animal species maintained and enhanced by candid soil, water, and biodiversity management governance, nourished by intricate folk knowledge systems. Altieri *et al.* (2012) further stated that these "traditional agroecosystems have the potential to bring solutions to many uncertainties in an era of climate change, energy, and financial crisis".

El-Ramady *et al.* (2012) have premised on the verity that agrobiodiversity is irreplaceably important in its own right. Agrobiodiversity is important for providing most of our food and its management will determine our future, both in cities and the countryside. Agrobiodiversity determines the fate of agroecosystems which always provided the essential ecosystem service of food production and can be designed to deliver a further range of ecosystem services as needs and knowledge change (Lenné and Wood, 2011). Present knowledge extends from a greater appreciation of traditional agriculture and the needs of farmers, through classical agricultural research in livestock husbandry, genetics, statistics, plant breeding, agronomy, crop protection, information management, and many more.

Conclusion

The present study revealed that the very low and low altitude AGEs supported somehow traditional crops (cultivation of cereal, pulse crop, *etc.*) whereas the higher altitude agroecosystems proliferated with cash crops (cultivation of cash crops like vegetables). Lack of irrigation facilities is one of the main constraints that have restricted grain production in high altitude farms. Additionally, the hardcore efforts of peasants and relatively cool climate at higher altitudes supported the cash crop agroecosystems in those areas. This study suggests that managing agroecosystems in a sustainable and organized manner (like AGEs of mid and high altitudes in this study) could be beneficial for humanity and economy which combated sufficiently with the agroecosystems of very low and low altitudes even with the lacking of modern mechanized farming machinery.

Acknowledgements

The cooperation and response of villagers during fieldwork is duly acknowledged by authors.

REFERENCES

Altieri MA. Funes-Monzote FR, Petersen P. (2012). Agroecologically efficient agricultural systems for smallholder farmers: contribution to food security. *Agronomy for Sustainable Development*. 32: 1–13.

Caporali F. (2015). History and Development of Agroecology and Theory of Agroecosystems. *Law and Agroecology*. 3-29.

El-Ramady HR, Belal AB, El-Marsafawy SM, Belal E. (2012). Contemporary Environmental Readings: Climate change, a blessing or a curse for agriculture. Lambert Academic Publishing.

Lenné JM, Wood D. (2011). Agrobiodiversity Revisited. *In*: Agrobiodiversity Management for Food Security: A Critical Review. Eds. J. M. Lenné and D. Wood CAB International. CABI Head Office, Nosworthy Way, UK. Chapter 1: 1-11.

Ratnadas A, Fernandes P, Avelino J, Habib R. (2012). Plant species diversity for sustainable management of crop pests and diseases in agroecosystems: A review. *Agronomy for Sustainable Development*. 32(1): 273-303.

Shahi C. (2020). Agroecosystems of Kumaun Himalaya: Ecology, Productivity, Nutrient Dynamics and Energy Budget.Ph. D Thesis, Kumaun University, Nainital

Shahi C, Bargali SS, Bargali K. (2019). Effects of size and altitude on plant diversity in agroecosystems of Kumaun Himalaya, India. Eds. Richa Singh, Sharad Vaish, Vineeta Lal, Ragvendra Pratap Narayan. Shrinkhala Publishing House, Ayodhya (U.P). 38-51 pp.

Stoev A. (2018). Anthropogenic Factors and Plant in Agroecosystems. *Agroecosystems Analysis*. 21(2): 117-124.

Chapter 8

Food Security and Crop Management

Minakshi Chakraborty

DGM-Rural and Agri Economist, Mahindra and Mahindra

Agriculture sector in India has shown remarkable progress since the Green Revolution period. After achieving self-reliance in food production, the focus has now shifted towards improving nutritional security. However, despite the increased focus, India appears amongst the bottom 50 countries concerning Global Food Security. Key factors resulting in low food security are poor protein quality, low dietary diversity and low micro-nutrient availability. Growth of agriculture sector failed to address issues concerning affordability and availability of nutritious food. On the demand side, large section of the Indian population comes under various food safety net programs that give them easy access to food grains, but they can barely afford nutritious food. Similarly, on the supply side, with various measures taken by the government to incentivize production of cereal crops, like, rice and wheat, there is a huge disbalance in agricultural production across geographies. Several researchers and policymakers highlighted the need for 'Right crop in right geography' for a diversified agriculture system. However, farmers can shift to the crops of high nutritious value if they are incentivized with higher returns. Many proven technologies and planting improvement practices have promises to boost returns and reduce cost of production. This paper identifies the expected profitability of farmers by either shifting from rice cultivation to other crops of high nutritious value or inter cropping with other crops.

Keywords: *Agricultural diversification, Food security, Nutritional security, Mechanization.*

Introduction

An agricultural growth is seen to have strong effects on reducing poverty. Growth in agriculture not only help in earning within the sector but also induces growth in other sectors. Cross-country estimates have shown that GDP growth resulting from agriculture is at least twice as effective in reducing poverty as compared to GDP growth outside agriculture (Anjum and Tarique, 2017). Loayza and Raddatz (2010) showed growth in sectors which are highly labour-intensive have strong effect on poverty reduction. In India, agriculture sector engages 65 per cent of the workforce and is the most labour -intensive sector. Therefore, successful poverty elimination in India is possible with growth in agriculture sector.

Agriculture in India has shown rapid strides. Green revolution during the 1960s has taken India to the path of National Food Security and more precisely self-reliance in food production. During, 2001 to 2018, food grain production increased by almost 1.5 times and horticulture production almost doubled. Similarly, poverty reduction in India has also been significant. The Multi-Dimensional Poverty Index for 105 nations, showed that 55 per cent of Indians were classified as 'Poor' in 2005-06 which came down to 28 per cent in the next 10 years. In absolute terms, the number of poor fell from 630 million in 2005-06 to 360 million in 2015-16, *i.e.*, 270 million people were out of poverty in 10 years. However, despite the remarkable performance in agriculture sector and poverty reduction, India is still amongst the bottom 50 countries concerning Global Food Security. Key factors resulting in low food security in India are - poor protein quality, low dietary diversity and low micro-nutrient availability.

The primary focus of agricultural research, policy, and practice so far has been on increasing agricultural productivity. However, very little attention has been given on improving the nutrient output of farming systems. Increasing yields is important but the nutritional quality of crops is of utmost importance to improve food security of the country.

High food security is possible only with a diversified agriculture system. Diversified agriculture system requires proper crop management which can be done in two ways — Firstly, by maximizing returns through increased level of mechanization. Secondly, price protection for producers and consumers. This transition is possible if farmers are assured of higher profitability by shifting to nutritious crops.

The Chapter is organised in the following way. The following section gives an overview of India's position concerning food security in the world map which emphasizes the need for diversifying the diet of the consumers. The subsequent section gives an overview of the agriculture trends in India, and the need for diversifying the agricultural production system. It also briefly explains the crops that can be produced along with the staple crops to support nutritional diet of the population. The section is followed with how farm mechanization can lead to the maximisation of returns and therefore incentivize farmers for crop diversification. In the immediate next section need for price protection to all crops which provides nutritional support. Next section concludes the paper. Finally, the chapter concludes with recommendations.

India's Position in Terms of Food Security

The Global Food Security Index (GFSI) developed by the Economist Intelligence Unit and Cortevea Agri Science provides a comprehensive assessment of the state of food security through a dynamic quantitative and qualitative benchmarking model. The GFSI is based on four components, *i.e.*, affordability, availability, quality and safety in 113 developed and developing countries.

The components used in developing GFSI is illustrated below:

Affordability: This component focuses on ability to purchase in terms of food cost, income, presence of food safety programs, *etc.* The measures included in these are: food consumption as a share of household expenditure, proportion of population under the global poverty line, per capita income, agricultural import tariff, the presence of food safety net programs and access to credit.

Availability: Availability is measured in terms of food supply with respect to production, distribution, food loss, *etc.* Measures included in these are: sufficiency of supply, public expenditure on agriculture research and development, agricultural infrastructure, volatility of agricultural production, political stability risk, corruption, urban absorption capacity and food loss.

Quality and Safety: This component focuses on dietary diversity, nutritional standards, micronutrient availability and protein quality.

Natural Resources and Resilience: This component is measured based on climate exposure and the three natural assets -land, water and ocean. It includes exposure to risks from natural calamities, sensitivity, adaptive capacity and demographic stress.

The Figure 8.1, gives the top 10 countries and bottom 10 countries measured in terms of GFSI – 2019. India's GFSI index is way below the top countries, and just slightly above the bottom 10 countries. India is ranked 72 in a list of 113 countries.

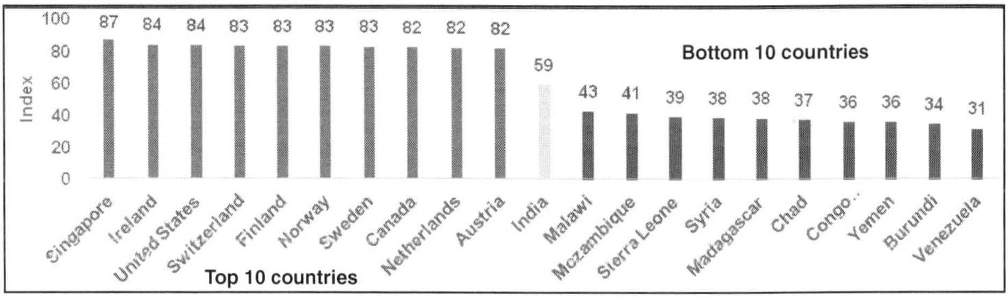

Figure 8.1: Global Food Security Index.
(*Source*: https://foodsecurityindex.eiu.com/Index)

India scores below global average in all the components of GFSI. Evidently from the Table 8.1, India scores 64 as against a global average of 68 regarding affordability measure of food security. However, in terms of presence and quality of food safety net programmes, India tops the chart. There has also been reduction in food cost which makes food more affordable. However, the score on GDP per capita is way below global score. This indicates that with the targeted public distribution system of the government, basic food grains that are distributed, *i.e.*, rice and wheat has become cheaper and affordable to the large segment of the population. However, very low per capita income suggests that the affordability to have nutritious food and dietary diversity is poor.

Table 8.1: GFSI: India vs Global Average

Index Component	*India Score*	*Global Score*
Affordability	64.2	67.5
Availability	58.4	59.4
Quality and Safety	47.0	61.0
Natural resources and resilience	46.7	57.1

Source: https://foodsecurityindex.eiu.com/Index.

India scores just marginally below global average with respect to availability. Increase in food grain production over the years has ensured sufficiency of supply. However, what essentially holds back in concerning food availability is poor agriculture infrastructure, high level of corruption and extremely low public expenditure on agriculture research and development.

Quality and safety of food is the real concern in India and stands significantly below global average. Poor protein quality, low dietary diversity and low micro-nutrient availability are the key factors resulting in the low food security index.

Agriculture is the Foundation of the Food Supply

Agricultural interventions have an important role to play in efforts aimed at improving the nutritional status of population (Masset *et al.*, 2012). Policies designed to increase the production of a wide variety of nutrient-dense foods and make them available at affordable prices can increase intakes of deficient micronutrients and, ultimately, increase food security within the country.

Key factors influencing the nutritional status are the following:

- **Cropping Pattern**: Changes in cropping pattern to include more nutrient dense crops will increase the nutrient crop of farming system. Graham et. al (2007) proposed cropping system improvement and alternative crop combinations as strategies for meeting nutritional goals while maintaining or improving agronomic or societal acceptability within farming system (Miller and Welch, 2013)
- **Soil Fertility**: Soil micro-nutrient status, and soil amendments are important factors that impact the micro-nutrient content of crops. Well-nourished crops grown of fertile soils have higher micro-nutrient density as compared to nutrient stressed crops grown on infertile soils. Graham et. al (2012) showed the associations between zinc deficient soils and zinc deficiency in humans in developing countries.
- **Allied food products**: Animal source foods are good source of micronutrients. Livestock and fish production can make important contributions to the nutritional health of low-income families. For example,

in a large cross-sectional study in Guatemala, Democratic Republic of Congo, Zambia, and Pakistan, Krebs *et al.* (2011) showed that meat consumption by the toddlers was associated with lower rates of stunting.

- ✰ **Biofortification**: Biofortification encompasses the use of plant breeding and agronomic approaches, such as, micronutrient fertilizer applications to increase concentrations of key nutrients in staple food crops (Bouis *et al.*, 2011).

Thus, the key factors that emerge from the literature review on agriculture system for food security is diversification. Diversification from traditional food crops to more nutritious crops, and also allied agri-sectors, like livestock and fisheries, can lead to the next revolution of nutritional security.

Looking at the trend of cropping pattern in the last few decades, it is evident that the staple crops, paddy and wheat dominate the cropped land. Cropping pattern is usually directly proportional to the increase in value share. For instance, area share of fruits and vegetables has expanded significantly from 2 per cent in the 1960s to 7 per cent during 2010 to 2018 signalling the orientation of demand and production for high value crops. At the other hand, area share of nutri-cereals decreased considerably over the period. It has decreased from 26 per cent in the 1960s to 12 per cent during 2010 to 2018. The declining value share during this period explains the shift away from nutri-cereals. However, while change in cropping pattern of nutri-cereals and fruits and vegetables is directly proportional to change in returns, paddy and wheat show a completely opposite trend. Value share of paddy dropped in the last 3 decades, yet, cropped area increased from 31 per cent in the 1960s to 37 per cent during 2010 to 2018. This is because of the returns that the farmers are assured through the Minimum Support Price and procurement policy of the government.

Table 8.2: Area Share to Total Cropped Land vs Value Share to Total Value of Output

Crops	*Area Share (Per cent)*		*Value Share (Per cent)*	
	F61-F69	*F10-F18*	*F61-F69*	*F10-F18*
Paddy and Wheat	**31.3**	**37.3**	*31.3*	*23.8*
Nutri-cereals	*25.5*	*12.7*	*10.7*	*3.7*
Pulses	14.7	12.2	*6.2*	*5.4*
Oilseeds	**9.5**	**13.9**	**7.7**	**8.1**
Sugar	**1.5**	**2.3**	**5.4**	**6.3**
Cotton and Jute	5.8	5.9	4.9	5.4
Condiments and Spices	1	1.5	**2.3**	**4.5**
Fruits and Vegetables	**1.9**	**6.5**	**12.6**	**27.7**
Other crops	8.8	7.8	19.8	15.0

Bold Data: Increasing trend; *Italic* Data: Decreasing trend.

Source: National Accounts Statistics, MoSPI.

Clearly, one of the key factors that guide the cropping pattern is high remuneration. However, remuneration of farm commodity is subject to high market risks. Therefore, assured returns on crops play a significant role in determining the cropping pattern. Effective procurement mechanism of the government has been one of the guiding factors in determining the cropping pattern for the last several years. For instance, in Punjab, 78 per cent of the rice and wheat production is procured which has led to 82 per cent of the cropped area under these crops. Similarly, Haryana, Chhattisgarh, Orissa are the other states where a large part of the cropped are is under rice and wheat which also coincides with high procurement in these states.

Table 8.3: Cropped Area vs Procurement (Rice and Wheat)

States	*Per cent of Cropped Area (Rice and Wheat)*	*Per cent of Production Procured (Rice and Wheat)*
Punjab	82 per cent	78 per cent
Telangana	45 per cent	78 per cent
Haryana	60 per cent	78 per cent
Chhattisgarh	72 per cent	59 per cent
Andhra Pradesh	29 per cent	59 per cent
Odisha	82 per cent	57 per cent
Madhya Pradesh	34 per cent	39 per cent
Uttarakhand	56 per cent	32 per cent
Tamil Nadu	33 per cent	21 per cent
Uttar Pradesh	62 per cent	14 per cent
Rajasthan	13 per cent	13 per cent
Maharashtra	10 per cent	13 per cent
West Bengal	59 per cent	12 per cent
Bihar	71 per cent	8 per cent
Jharkhand	68 per cent	5 per cent
Assam	62 per cent	2 per cent
Karnataka	11 per cent	2 per cent
Gujarat	14 per cent	0 per cent

Source: Ministry of Agriculture, Food Corporation of India.

Continuous adoption of rice-wheat cropping system in the North-Western plains of Punjab, Haryana and West Uttar Pradesh has resulted in depletion of ground water and deterioration of soil quality, posing a serious threat to its sustainability (CACP, 2020). The need for diversification therefore stems from the discriminatory policy leading to disbalance in the agri eco-system.

Rice is a crop of tropical climate and grows well in humid to sub-humid regions. In India, rice is grown under different climatic conditions, including, temperate, tropical and sub-tropical. It is grown in all types of soil though with varying productivity (Pathak *et al.*, 2020). Using the three bio -physical factors-rainfall, temperature and soil, the rice suitability index developed by Pathak *et al.*, 2020, identified 145 districts unsuitable for rice production. These districts cover 19 per cent of the kharif rice area and contribute 23 per cent to rice production. These unsuitable districts are mainly in 5 states- Punjab, Haryana, Gujarat, Rajasthan and Uttar Pradesh. About 93 per cent of the unsuitable rice cultivated areas are located in these states, and they contribute 21 per cent to the total rice production in the country. Crop shift and diversification is crucial to correct the disbalance in agri ecosystems and include more nutritious crops in the production system.

Mechanisation and Food Security

Many proven technologies and planting improvement practices have promises to boost agricultural production and improve nutritional security in developing countries. Globally, mechanization has been one of the key components of high agricultural growth and therefore, high food security. As evident from the Table 8.4, countries rated high in mechanization are also high in Food Security Index. Agricultural technology played a part in lifting the living standards of the developed world, and it is assumed that mechanisation of the developing world could deliver a similar outcome (Piesse,2019).

Table 8.4: Food Security Index and Level of Mechanisation

Countries	*Food Security Index (Score: 0-100; 0- Worst; 100- Best)*	*Level of Mechanisation (Per cent)*
USA	76.5	95
Russia	69.7	80
Brazil	70.1	75
China	71.0	55
India	58.9	40

Source: https://foodsecurityindex.eiu.com/; Indian Council for Food and Agriculture.

Mechanisation can help in improving the level food security index in two ways. Use of improved implements have the potential to increase productivity up to 30 per cent and reduce the cost of cultivation up to 20 per cent (NABARD, 2018). Labour forms a major component of the operational cost of cultivation for all crops (Table 8.5). Agriculture needs time management which requires adequate human resource at each stage of operations. The labour cost required at each stage forms a significant component of the total cost of cultivation. Evident from the table below, maize, jowar and groundnut are the least mechanised crops, and more than 60 per cent of the total cost of cultivation per hectare is incurred on labour. Gram, urad, paddy have a relatively higher level of mechanisation as the cost incurred on labour with respect to machine cost is relatively lower than compared to other crops.

Table 8.5: Crop-wise Labour Cost with Respect to Machine Cost in Operational Cost of Cultivation

Crops	*Per cent of Labour Cost per Hectare*	*Ratio of Machine Labour to Manual Labour (Human Labour + Animal labour)*
Paddy	56 per cent	0.30
Maize	64 per cent	0.27
Jowar	63 per cent	0.28
Groundnut	65 per cent	0.23
Pearl millet	58 per cent	0.39
Arhar	58 per cent	0.29
Gram	48 per cent	0.41
Urad	57 per cent	0.33

Source: Author's own estimates using data from Directorate of Economics and Statistics, Ministry of Agriculture.

The number of field operations and labour demand varies for different crops. Prabhakar *et al.* (2011) reported that nursery, main land preparation and transplanting require 62 person day/ha, first weeding and fertilizer application needed 47 person days/ha, second weeding and irrigation require 47 days/ha and for harvesting 50 person days/ha required for paddy cultivation. The requirement of labour in other crops with low level of mechanisation, like, maize, jowar, groundnut is even higher. Replacing use of labour with machines in different stages of field operation wherever possible can play a decisive role in reducing the cost of cultivation.

Table 8.6: Field Operations: Manual Operations vs Machine

Stage of operations	Manual operations	Machine
Manuring/composting	Loading and deloading Spreading	
Field preparation	By plough	By tractor
Sowing/transplanting	Line sowing Spreading	By transplanter By tractor/cultivator
Irrigation	Flooding Check basin/bunding/furrow	Sprinkler/drip
Weeding	Manual	Mechanical Chemical
Fertilizer application	Basal application by labour	Power driven basal application
Insect pest control	Chemical spray using labour	Sprayer
Harvesting	Manual Bullock	Harvester/combine Thresher
Cleaning of grain/seed	Hand cleaning	Benower

Source: Prasad, Sarda (2017), "Shortages in Agriculture Labour Market and Changes in Cropping Pattern" in: Bathla, Seema and Amaresh Dubey (Ed.), Changing Contours of Indian Agriculture. Springer Press, pp. 181-204.

With the current level of mechanisation, for most of the cereal crops, apart from arhar and groundnut, average returns is much lower in relation to paddy. By replacing machine labour at each stage of operations for different crops, operational cost of cultivation is estimated to decrease by at least 20-35 per cent. Assuming 10 per cent increase in productivity coupled with decline in cost, gross returns are estimated to double for maize, gram and bajra, and increase is almost thrice for jowar and soybean. Average relative returns at a current level of mechanisation is higher than paddy for arhar and groundnut. At an optimal level of mechanisation, average relative returns with respect to paddy is expected to increase for all crops. Relative returns for maize crop is expected to be higher than that of paddy.

Table 8.7: Profitability – Current Level of Mechanisation vs Optimum Level of Mechanisation

Crops	*Current Level of Mechanisation*			*Optimal Level of Mechanisation*		
	CoC (Rs/ha)	*Gross Returns over CoC (Rs/ha)*	*Average Returns Relative to Paddy*	*CoC (Rs/ha)*	*Gross Returns over CoC (Rs/ha)*	*Average Returns Relative to Paddy*
Paddy	45,049	18,481	100	38,292	25,238	100
Maize	38,272	13,640	74	24,877	27,113	107
Jowar	26,758	5,004	27	18,463	13,394	53
Bajra	24,884	5,302	29	17,916	12,360	49
Arhar (Tur)	41,188	29,846	161	30,479	15,448	149
Gram	19,375	6,745	36	13,756	12,442	49
Urad	21,033	15,324	83	14,723	21,743	86
Groundnut	51,586	24,526	133	41,269	35,072	139
Soybean	28,999	4,011	22	21,459	11,650	46

Source: Author's own estimates using data from Directorate of Economics and Statistics, Ministry of Agriculture.

The above table therefore illustrates that with mechanisation, diversification either through inter cropping or crop shift can increase profitability of the farmers significantly.

Price Protection to Farmers and Consumers

The second step for incentivising farmers for crop diversification is to ensure remunerative returns for alternative crops with high nutritious value. Volatility in crop prices leads to high risks both for farmer and the consumer. High prices may incentivise farmers but reduces the affordability of the consumers. On the other hand, low prices may incentivise consumers for a balanced diet but reduces the income of the farmers.

Price volatility of all crops is measured using monthly wholesale price index for the period Jan-2015 to Aug-2020 in the following way:

Price volatility = [Stdev (Crop 'I')/Average (crop 'I')] *100

Evident from above (Figure 8.3), crop prices of paddy are the least volatile while prices of pulses, like, gram and arhar are highly volatile. Low volatility in paddy prices stems from price protection from the government through procurement at MSPs and stable demand. Such measures have been introduced by the Government of India for pulses and oilseeds as well. For instance, a new umbrella scheme, "Pradhan Mantri Annadata Aay Sanrakshan Abhiyan (PM-AASHA)" comprising Price Support Scheme (PSS) for pulses, oilseeds, and copra, Price Deficiency Payment Scheme (PDPS) for oilseeds and Pilot of Private Procurement and Stockist Scheme (PPSS) for oilseeds in selected districts.

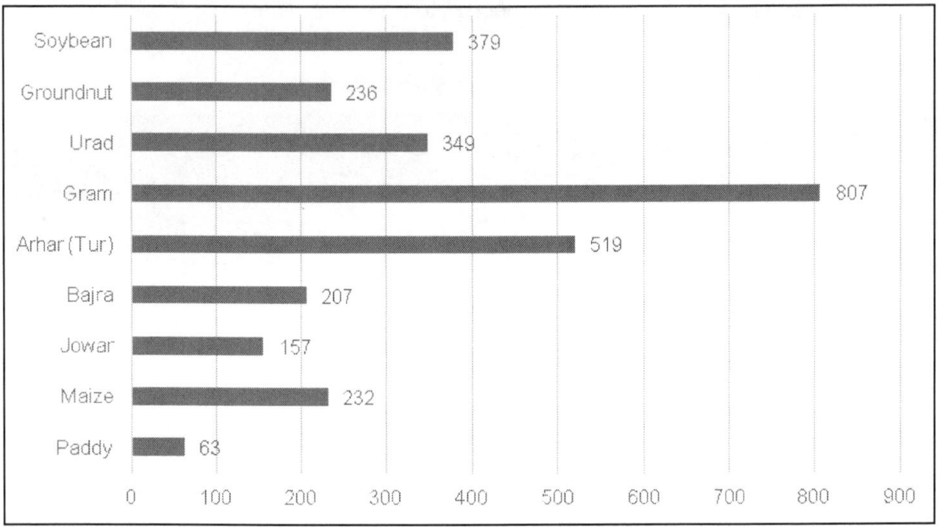

Figure 8.3: Price Volatility.

(*Source*: Author's own estimate using data from National Account Statistics).

Further, for achieving food security it also important that consumers get at an affordable price. PDS is an important tool of Government's policy for the management of food economy. The Dept. of Food and Public Distribution (DFPD) makes allocation of food grains to States/UTs under Targeted Public Distribution System (TPDS) and Other Welfare Schemes (OWS) and maintains their offtake record. Initially the focus of the policymakers was on assuring the availability and price stability of basic foodstuffs at the international and national level. Later, in the early 80s focus was broadened also to ensure food security. During the nineties, issues such as, food safety, nutrition, dietary needs and food preferences were also considered important ingredients of food security. However, proportion of coarse cereals, pulses, and oilseeds procured from the farmers and distributed at a subsidised rate continues to be significantly low.

While 1/3rd of rice and wheat production is procured by the government, approximately 16 per cent of pulses produce is procured and even lower for oilseeds, and coarse cereals. Low procurement leads to low distribution at a subsidized price. Low per capita income, and less coverage of nutritious crops in the TPDS explains the poor food security in the country. Thus, the key to achieving nutritional security would be to expand the food distribution system to include more of nutritious crops in an effective manner.

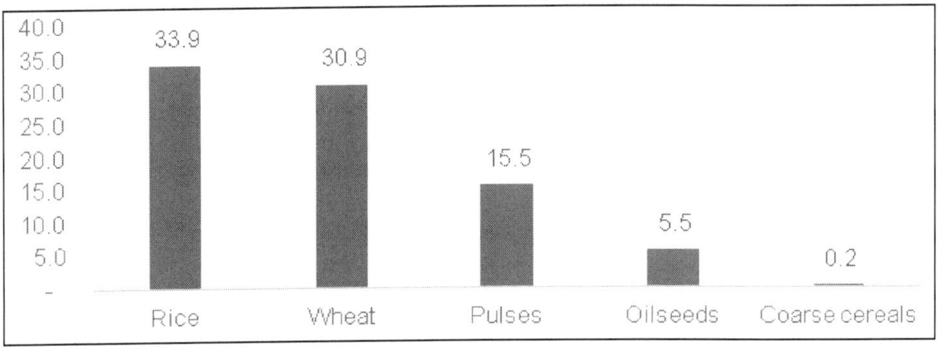

Figure 8.4: Average Procurement (as per cent of production) in 5 Years: 2015 to 2019.
***Source*: CACP, Price Policy Report.**

Recommendations

Agriculture in India for several decades has focussed only on few staple crops. This has been largely due to the discriminatory policy of the government. High procurement of paddy and wheat from farmers has led the shift towards these crops while hugely disincentivised nutri-cereal crops. The shift towards paddy, in particular, have resulted in huge disbalance in the agri ecosystem. A study conducted by ICAR, 2019, identified 145 districts unsuitable for rice production. Clearly, the shift away from rice in these unsuitable rice growing areas is crucial for a balanced agri eco-system. However, diversification in agriculture is needed not just to correct the imbalance in the agri eco-system but also to include crops of high nutritious value in the production system.

Several initiatives by both central and state governments have been taken on having a diversified agriculture system. Crop Diversification Programme (CDP) was launched in original Green Revolution States in 2013-14 to promote technological innovations and encourage farmers to choose crop alternatives. However, the allocation under the programme has declined from, Rs. 40 crores during 2017-18, to Rs.13.3 crores in 2019-20. Government of Haryana launched, "Jal Hi Jeevan Scheme" in 2019 to encourage farmers to shift from paddy to maize and tur with a target of 50,000 hectares, under which farmers were assured of Rs. 2000 per acre and hybrid seeds would also be provided free of cost. The Uttar Pradesh government's 'One District, One Product' programme to encourage indigenous specialized products and crafts, and Central Government's "One Product, One District" for horticulture crops with focus on better marketing and export would also help in crop diversification. However, not much progress has been made so far on crop diversification in the region because of low economic returns from alternative crops, lack of assured marketing and remunerative prices, absence of appropriate technology for alternative crops.

The paper suggests two possible ways of crop management which can lead to higher food security.

- ☆ Firstly, it is being increasingly recognised that future growth in agriculture will primarily be knowledge and technology driven. This demands more investment in agricultural R and D, innovations and institutions to bring about significant change in technologies for resource conservations and productivity enhancement. With the optimum level of mechanisation at each stage of crop production, profitability of crops with respect to paddy is expected to increase. For certain crops, like maize, profitability at an optimum level of mechanisation is expected to be significantly higher than that of paddy.
- ☆ Secondly, diversification is possible if the farmers are assured of higher returns and stable demand. This is possible if the government expand its food distribution system to include food items that are of high nutritious value. This has been recognised by the Government in the early 2000s. However, low procurement of nutri-cereals, pulses as compared to paddy and wheat indicates that the inclusion of nutritious food in the public distribution system has a long way to go.

REFERENCES

Anjum, Shabana and Md Tarique (2017), "Agriculture and Poverty Reduction in India: An Empirical Study", Asian Journal of Research in Social Sciences and Humanities, Vol 7, No. 9, Sept 2017, pp – 35-48.

Bouis, H.E., Welch, R.M., 2010. Biofortification—A Sustainable Agricultural Strategy for Reducing Micronutrient Malnutrition in the Global South. Crop Sci. 50, S20-S32.

Commission for Agriculture Cost and Prices (2020), "Kharif Price Policy Report", Ministry of Agriculture and Farmers' Welfare, Govt. of India.

Engr. Ndubuisi, Christian Ohakwe (2019), "The Role of Mechanized Agriculture in a Developing Economy", paper presented to 'School of Engineering Technology, Federal Polytechnic Nekede Owerri, Imo State

Global Food Security Index (2019), https: / /foodsecurityindex.eiu.com/.

Graham, R.D., Knez, M., Welch, R.M., 2012. How much nutritional iron deficiency in humans globally is due to an underlying zinc deficiency. Adv. Agron. 115, 1-40.

Graham, R.D., Welch, R.M., Saunders, D.A., Hyphen, T.G., Bouis, H.E., Bonierbale, M., de Haan, S., Burgos, G., Thiele, G., Liria, R., Meisner, C.A., Beebe, S.E., Potts, M.J., Kadian, M., Hobbs, P.R., Gupta, R.K., and Twomlow, S. (2007), "Nutritious Subsistence Food Systems", in: L.S.Donald (Ed.), Advances in Agronomy. Academic Press, pp. 1-74.

H Pathak, R Tripathi, NN Jambhulkar, JP Bisen and BB Panda (2020). "Eco-regional Rice Farming for Enhancing Productivity, Profitability and Sustainability". NRRI Research Bulletin No. 22, ICAR-National Rice Research Institute, Cuttack 753006, Odisha, India. pp28.

Krebs, N.F., Mazariegos, M., Tshefu, A., Bose, C., Sami, N., Chomba, E., Carlo, W., Goco, N., Kindem, M., Wright, L. L., Hambidge, K. M., Complementary Feeding Study Group, 2011. Meat consumption is associated with less stunting among toddlers in four diverse low-income settings. Food Nutr. Bull. 32, 185-191.

Loayza N.V. and Raddatz C. (2010), "The composition of growth matters for poverty alleviation", Journal of Development Economics 93, Vol-1, pp-137–151.

Masset, E., Haddad, L., Cornelius, A., Isaza-Castro, J. (2012), "Effectiveness of agricultural interventions that aim to improve nutritional status of children": systematic review. BMJ. 344, d8222.

Miller, Dennis D and Ross. M. Welch (2013), "Food System Strategies for Preventing Micronutrient Malnutrition", Food and Agriculture Organisation of the United Nation, ESA Working Paper No. 13-06

NABARD (2018), "Farm Mechanisation", https: //www.nabard.org/auth/writereaddata/file/NSP per cent 20Farm per cent 20Mechanisation.pdf

Piesse, Mervyn (2019), "Agricultural Mechanisation, Farm Productivity and Food Security in Sub-Saharan Africa", Future Directions International

Prabhakar C, Devi KS, Selvam S (2011), "Labour scarcity: its immensity and impact on agriculture", Agr Econ Res Rev 24: 373 -380.

Prasad, Sarda (2017), "Shortages in Agriculture Labour Market and Changes in Cropping Pattern" in: Bathla, Seema and Amaresh Dubey (Ed.), Changing Contours of Indian Agriculture. Springer Press, pp. 181-204.

Chapter 9

Strategies for Doubling Farmer's Income in Hilly Terrains by Adapting Horticulture based Integrated Farming Agri-Entrepreneurship Model

Udit Joshi[1], *T.S. Bisht*[1], *Rajendra Bhatt*[2] *and Vatsala Tewari*[2]

[1]*Department of Horticulture, H.N.B. Garhwal University (A Central University) Srinagar (Garhwal) – 246174, Uttarakhand,*
[2]*Department of Vegetable Science G.B.P.U.A&T, Pantnagar – 263145, Uttarakhand*

Uttarakhand being a predominantly hilly state consists of the maximum area under wastelands and forests that leave only 14 per cent under cultivation. In hilly regions of Uttarakhand, people are not able to follow large scale farming as 89 per cent out of total farmers of the state come under the small and marginal category. The objective of the present study is to provide a Horti-based Agri-entrepreneurship model for the villages of the Garhwal region for doubling farmer's income in these regions through agricultural development. The present model was formulated based on the studies conducted at 3 villages of Garhwal region i.e., two villages from district Pauri Garhwal and one village from district Tehri Garhwal. Almost every family in every village has considerable cultivable land hence, all are dependent upon agricultural and other allied activities for livelihood sustainment. Major problems observed were damage to the crops by wild animals, prevalence of migration among local youth, lack of technical know-how, very low idea about several government schemes and incentives, lesser extension activities involvement by government agencies like KVK's etc. Most of the horticultural crops are seen mostly confined to only kitchen gardens. Hence, a Horticulture based Agri-entrepreneurship model was formulated for employing more youth manpower in villages itself, as the post COVID era is going to provide a great opportunity in grasping the effect of reverse migration. The active participation of the youth of the villages, Co-operative societies, Government organizations like KVK's and SAU's, NGO's can lead to some successful achievements. An actual ground basis work by these organizations in the context of technology transfer through extension and other activities can ensure overall agricultural transformations, rural development and promotion of these types of Agri-entrepreneurship models.

Keywords: *Agri-entrepreneurship, Horticulture, Migration, Post- COVID era and Reverse migration.*

Introduction

Horticulture sector has a tremendous potential for improving the farmer's socio-economic conditions hence considered as a priority sector in agriculture. Several agro-based industries are supplied by the horticulture sector which generally carry high avenues for skilful employment generation and creation of self-employment opportunities in rural areas Shehrawat (2006). There is commendable contribution of agriculture in our country's livelihood and economy as 56.4 per cent of the total population is involved in agriculture and allied sectors, contributing 15.4 per cent to the nation's GDP. According to FAO (2019), around 70 per cent of rural households in India depend primarily on major agricultural activities for their sustainment of livelihood. Small and marginal farmers having a land value less than two hectares generally account for around 82 per cent of all farmers of India, but own just 47.3 per cent of cropped area while semi-medium and medium land holding farmers (2-10 hectares of land) accounts for 13.2 per cent of all farmers owning 43.6 per cent of cropped area (10th agriculture census 2015-16). Uttarakhand state comprises most of hilly areas hence, most of its area are forests and wastelands. About only 14 per cent of area (0.741 million hectares) out of the total reported area of 5.672 million hectares is cultivated. About 33 per cent of the total area in Uttarakhand is either rocky or snow-covered/glaciated or otherwise unproductive. Only 33 per cent of the cultivated land has irrigation facilities, of which 90 per cent of the land is used for growing cereals, fodder, and vegetables. The prevalence of farmer fields in patches, hilly terrains, very low farm mechanization, lesser irrigation water availability, no integrated farming approach, damage caused due to wild animals, low productivity, low processing, low knowledge about government schemes, *etc.* has caused a condition of large-scale migration of local youth to large cities for seeking jobs.

During the first decade of the 21st century, two contrasting trends have been noticed;

1. There is a great potential of horticulture development in Uttarakhand being recognized as the horticulture state and
2. Its slow growth observed in the horticulture sector is causing concerns for future food and nutritional security demand.

It is a paradox that even after seven decades of independence hill regions are still not well developed in respect of technologies. It is a well-known fact that horticulture in hilly areas lacks appropriate technologies, infrastructure support, marketing, extension network, organized planning, and execution of research/ education and diversity in the farming system. In Uttarakhand, latitude, longitude, altitude are the aspects that play a pivotal role in determination of agro-climate ranges which generally add to the habitat's multiplicity. Horticulture has been emerging and has emerged over the years as one of the potential enterprises in accelerating the growth of the economy in the state. Role of horticulture is becoming increasingly important in poverty alleviation through employment generation programs and nutritional security. It is continuously providing great scope and opportunity for sustaining a large number of Agro based industries that generate various employment opportunities and along with this, also gives wide range of crop diversification options to the farmers. A focused attention has been given to research and development in the field of horticulture during the previous two Plan periods. The encouraging result has been obtained through these attentions and efforts. A Golden Revolution is in the offing because of significant production increases in horticultural crops across several states, various state has emerged and are emerging as a leading player in the horticulture-based sector and sub-sectors. Today, horticulture in India, has become a sustainable and viable venture for the small and marginal farmers as a result of several thoughtful research, technological and policy initiatives and inputs. The matter of satisfaction is that the farmer's food consumption levels and household income is increasing because of horticulture. Besides, the horticulture sector is continuously attracting entrepreneurs from various sectors for taking up horticulture as a sustainable and continuously growing commercial venture. Hence, horticulture industry has a great scope in its grow and flourishment.

Major Constraints and Productivity Gap

Various abiotic (frost, hail storms, drought, floods, and insufficient chilling) and biotic (monkeys, diseases, insect pests, nematodes, stray/wild animal and bird) constrain the productivity of horticultural crops. Monkey menace along with damage by the wild/stray animal is leading to giving up of crop cultivation by many farmers in various parts of the state. The availability of food grains in the public distribution system and the tendency among the youth to look for white-collar jobs are the major reasons for lack of interest among the youth in farming. Lack of irrigation facilities (as <20 per cent of the total cultivable area is irrigated) is one of the major reasons for slow growth in the adaptation of precision farming in horticultural crops. Lack of transport, storage, and marketing are the major bottlenecks in the adoption of large cultivation of horticultural crops.

In the mid and high hills of the state temperate fruit has made a huge impact but, it has not been taken up on a commercial scale in the lower hills, where the majority of the state population resides. In order to increase the production and productivity of horticultural crops, it is important to adopt modern technologies such as high-density orcharding, protected cultivation, of vegetable and flower crops, balanced use of organic and synthetic fertilizers, and need-based application of micronutrients for increased productivity. There is a need to integrate crop production with sericulture, bee-keeping, and mushroom cultivation and adopt an integrated horticultural-livestock-medicinal plant production system. This requires an in-depth understanding of the nature of interactions of horticultural crops with the organic inputs, fertilizers, and the climate for sustainable crop production.

In the present scenario, in a COVID-19 and post COVID-19 era, we all are witnessing a large-scale reverse migration due to the loss of jobs in almost every sector. Now, this situation can be turned into a great opportunity in the revival of hilly rural agriculture by generating employment via the creation of a Horticulture based entrepreneurship model that helps in-situ employment of reverse migrated youth. This will certainly boost the revival of rural agriculture in hilly terrains which is a need of the hour. Thus, to stop turning villages into ghost villages and even revive the villages which have already turned into ghost villages.

Hence the present literature, based on surveys conducted in various villages of Uttarakhand, selecting some villages of Garhwal region named Village Dhulet, Saiji, and Bhaiswara State Uttarakhand, a rural Agri entrepreneurship model was developed and various ideas for the smooth running of the model are given to boost the rural agriculture and generating employment in hilly terrains in the field of agriculture to stop the migration and turn the reverse migration into the opportunities to revive the ghost villages and turning ghost villages into the prosperous villages.

Objectives of the Model

The key objectives of this model are to double the farmer's income in the hills of Uttarakhand through incorporating a horticulture based integrated farming model to improve production, productivity, and quality of the products to enhance its consumption, availability, and exports.

- ☆ To turn reverse migration into the opportunity for in-situ employment generation of local youths for boosting hilly rural agriculture.
- ☆ To create a representative and a motivating model for all the farmers of hilly regions of Uttarakhand so that they can learn and practice the agricultural techniques more extensively.
- ☆ To meet the domestic requirement of protective foods in the diet of rural hill population.
- ☆ To provide raw materials for the rising agro-industries.
- ☆ To reduce post-harvest losses and improve marketability by developing certain systems of processing and value-addition.
- ☆ To make optimum use of the basic natural resources.
- ☆ Diversification of agriculture with horticultural crops.
- ☆ To enhance export potential and opportunities for enriching our foreign exchange.

☆ To Promote better utilization and consumption of the product and even waste utilization for ensuring a higher return to the grower and better nutritional health to the people.

Methods and Components

The present study was conducted at 3 villages of Garhwal region *i.e.*, Village Saiji and Dhulet under Pabo Block District Pauri Garhwal located 14 km from KVK Bharsar and College of Horticulture Bharsar and 5 km from NH-121 at an altitude of 1937 m, and Village Bhaiswara under Kirtinagar Block District Tehri around 10 Km from Srinagar. The study of village Bhaiswara was conducted by collecting various Primary data, Secondary data, and individual questionnaires. The study of village Dhulet and Saiji was conducted as an experiential learning program for UG students conducted in the final year of every agricultural university affiliated by ICAR, named RHWEP (Rural Horticulture Work Experience Programme). It is an awareness of horticulture graduates about the actual situation and status of the farmers and farming practices. The program was conducted in the year 2017 and 2018 from 29th September to 4th December (70 days) at village Dhulet and Saiji respectively under the ICAR program READY (Rural Entrepreneurship Awareness Development Yojana).

The whole program consisted of 5 parts mentioned below

1. **Rural economics:** (Survey of the village and socio-economic condition of families)
2. **Horticultural crop production:** (Giving knowledge to villagers about different improved techniques of crop production, improved varieties, intercultural activities, *etc.*)
3. **Horticultural crop protection:** (Giving knowledge to villagers about various crop protection techniques use of biological agents *etc.*)
4. **Post-Harvest technology:** (About best utilization of their resources and harvested produce)
5. **Extension education:** (Creating a medium by which they can gain knowledge about new and improved techniques)

General Information of the Village Dhulet

To increase the socio-economic condition of farmers it is essential to know about their economic conditions particularly Landholding, cropping patterns and occupation, and problems threatening to their production sector.

Table 9.1: General Information of the Village Dhulet

Village:	Dhulet
Tehsil	Pauri
District	Pauri Garhwal
Block	Pabau
Patti	Baalikannarsyun

Head post office	Pauri
Sub post office	Chipalghat
Branch post office	Bharsar
Altitude	1937m
Krishi upaj Mandi	Kotdwara
Name of MLA	Mr. Dhan Singh Rawat
Name of Pradhan	Mrs. Kalpa Devi
Name of Up-Pradhan	Mrs. Heera Devi
Ward Members	Gabbar Singh, Heera Devi, Shakambari Devi, Ramli Devi, Suman Devi, Bhaga Devi.
Aasha worker	Mrs. Vimla Devi
Anganwadi workers	Mrs.Sangeeta Devi, Mrs. Rajni Devi.

Table 9.2: Transport, Education, and Health Facilities of Village Dhulet

Bus stand	Nautha (5 Km from the village)
Nearest Railway Station	Kotdwara (105 Km from the village)
Sub Post- Office	Chipalghat (25 Km from the village)
Condition of road	Metalled road (under the scheme of Pradhan Mantri Gram SadakYojna)
School	One Primary and one Junior High School
Senior Secondary School	Jagteshwar (3 Km from the village)
Private and government Dispensary	Nautha
Government Hospital	Pabau
Anganwadi	Yes

Table 9.3: Marketing Facilities at Village Dhulet

Sl.No.	*Type*	*Availability within Village*	*Nearest Place*	*Distance (km)*
1.	Weekly Haat	No	Not in trend	No
2.	Regulated Market	No	Nautha	5
3.	Daily Need/Grocery Shop	Yes (5 shops)	-	-
4.	Vegetables and Fruits stall	No	Nautha	5
5.	Restaurant/Dhaba	Yes	-	-
6.	Flour Mill	Yes	-	-

Table 9.4: Transport Means Available

Sl.No.	*Type*	*No*
1.	Cycle	No
2.	Motor cycle	8
3.	Cars/Jeep	1(commercial)

Table 9.5: Other Village Organizations

Sl.No.	Organizations	Yes/No
1.	Gram Panchayat	Yes
2.	Mandi	No
3.	Co-operative Society	No
4.	Kisan Club	No
5.	Youth Club	No
6.	Mahila Club	No

Table 9.6: Availability of Agro-inputs

Sl.No.	Agro- inputs	In the Village	At the nearest place	Distance (km)
1.	Implements	Yes (self-made)	Nautha	5
2.	Seeds	No	COH Bharsar, Sainji, Chipalghat, and Pabau (Agri. Dept.)	14,7,25 and 30
3.	Fertilizers	No	Sainji, Chipalghat and Pabau (Agri. Dept.)	same
4.	Insecticides	No	Sainji, Chipalghat and Pabau (Agri. Dept.)	same
5.	Herbicides	No	Sainji, Chipalghat and Pabau (Agri. Dept.)	same

Table 9.7: Level of Literacy (Age wise)

Sex	Category	Literate	Illiterate	Total
Males	Age> 35 years	88	24	112
	Age 18-35 years	113	8	121
	Age< 18 years	144	23	167
Females	Age> 35 years	9	81	90
	Age 18-35 years	90	34	124
	Age< 18 years	99	25	124
Total	543	195	738	
Total Literacy Rate: - 73.57 per cent				
Male Literacy Rate: - 86.25 per cent				
Female Literacy Rate: - 58.57 per cent				
Sex Ratio: - 845 females/1000 males				

☆ From the literacy point of view, the highest literacy rate was observed in males of age group 18-35years and that was lowest in females of age group > 35 years. The overall female literacy rate was 58.56 per cent compared to that of males 86.25 per cent and overall, 73.57 per cent.

Classification of Families (Occupation wise)

By taking several data and collection of information from the village it was found that there was a total of 128 families, among them, 19 families have total dependency over agriculture for their livelihood, 7 were self-employed by certain small businesses, 2 were government teachers, 34 were private sector employed, 51 were government employee (most of them were on deputation basis) and 15 were army persons. Although many of the families have a different source of income certain members of every family were somehow involved in agricultural practices as they all have a good piece of land value.

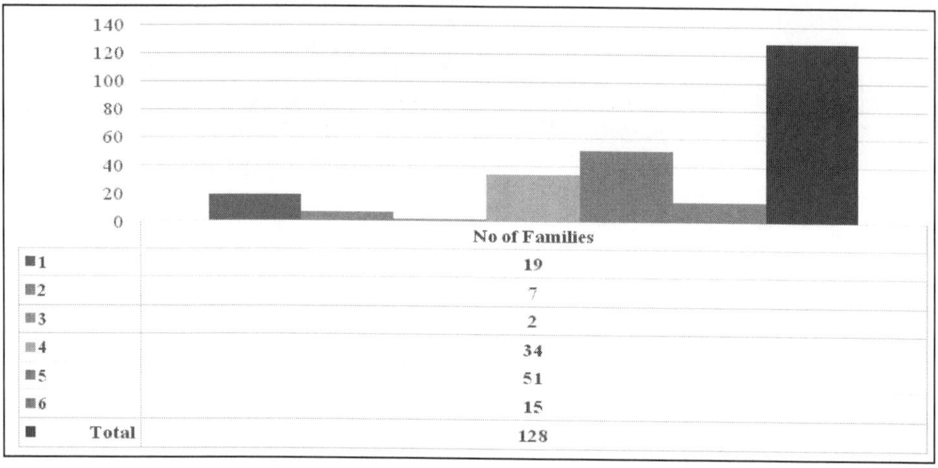

Figure 9.1: Classification of Families (Occupation wise).

General Information of the Village Saiji and Bhaiswara

Area and Land Utilization Pattern of Village Saiji and Bhaiswara

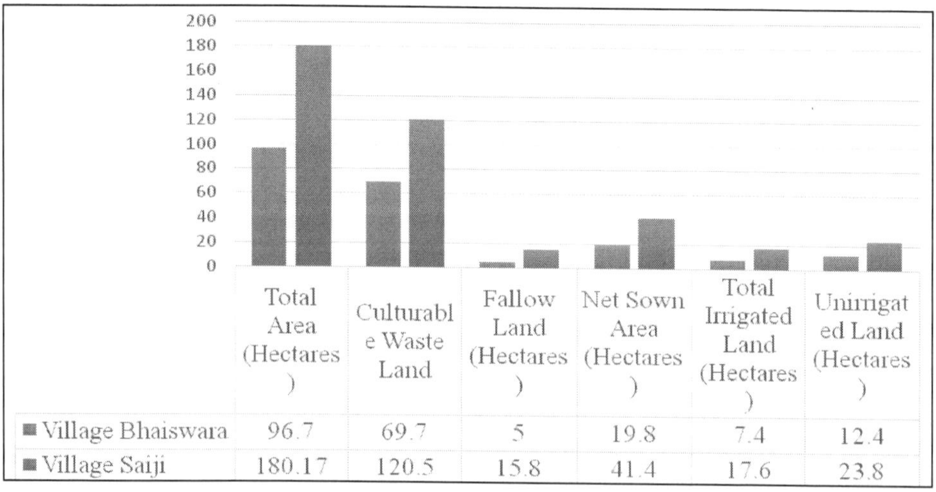

Figure 9.2: Area-wise Statisics of Village Saiji and Bhaiswara.

Table 9.8: Status of availability of market nearby village Saiji and Bhaiswara

Sl. No.	Type	Availability within Village Saiji	Availability within Village Bhaiswara	Nearest Place to Village Saiji	Nearest Place to Village Bhaiswara	Distance in km from Village Saiji	Distance in km from Village Bhaiswara
1	Weekly Haat	**No**	**No**	**No**	**Srinagar**	-	**10**
2	Regulated Market	**Yes**	**No**	-	**Srinagar**	-	**10**
3	Daily need groceryshop	**Yes**	**No**	-	**Srinagar**	-	**10**
4	Vegetables and FruitStall	**Yes**	**No**	-	**Srinagar**	-	**10**
5	Restaurant and Dhaba	**Yes**	**No**	-	**Srinagar**	-	**10**
6	Flour Mill	**Yes**	**No**	-	**Srinagar**	-	**10**

Table 9.9: Availability of Agricultural Inputs in Village Saiji and Nearby

Sl.No.	Agro- inputs	In the village	At the nearest place	Distance in Km
1	Implements	Self-Made	Saiji	0.5
2	Seeds	No	COH Bharsar, Sainji, Chipalghat, and Pabau (Agri. Dept.)	12,0.5,10 and 25
3	Fertilizers	No	Sainji, Chipalghat, and Pabau (Agri. Dept.)	0.5,10 and 25
4	Insecticides	No	Sainji, Chipalghat, and Pabau (Agri. Dept.)	0.5,10 and 25
5	Herbicide	No	Sainji, Chipalghat, and Pabau	0.5,10 and 25

Table 9.10: Availability of Agricultural Inputs in Village Bhaiswara and Nearby

Sl.No.	Agro- inputs	In the village	At the Nearest Place	Distance in Km
1	Implements	Self-Made	Srinagar	10
2	Seeds	No	Srinagar	10
3	Fertilizers	No	Srinagar	10
4	Insecticides	No	Srinagar	10
5	Herbicide	No	Srinagar	10

Table 9.11: Cropping Pattern Followed in the Villages

Seasons	Agronomic Crops	Horticultural Crops
Kharif	Rice, Maize, Pigeon Pea, Sesame, Kala Bhatt, Horse Gram, Jhangora (Barnyard Millet), Mandua (finger Millet), Grain Amaranth, Black Gram	French Bean, Sword Bean, Bottle Gourd, Bitter Gourd, Sponge Gourd, Cucumber, Pumpkin, Summer Squash, Tomato, Brinjal, Chilli, Capsicum, Colocasia
Rabi	Wheat, Barley, Rye/Mustard, Lentil, Broad Bean	Pea, Onion, Garlic, Cauliflower, Cabbage, Coriander, Palak, Radish, Carrot, Fenugreek

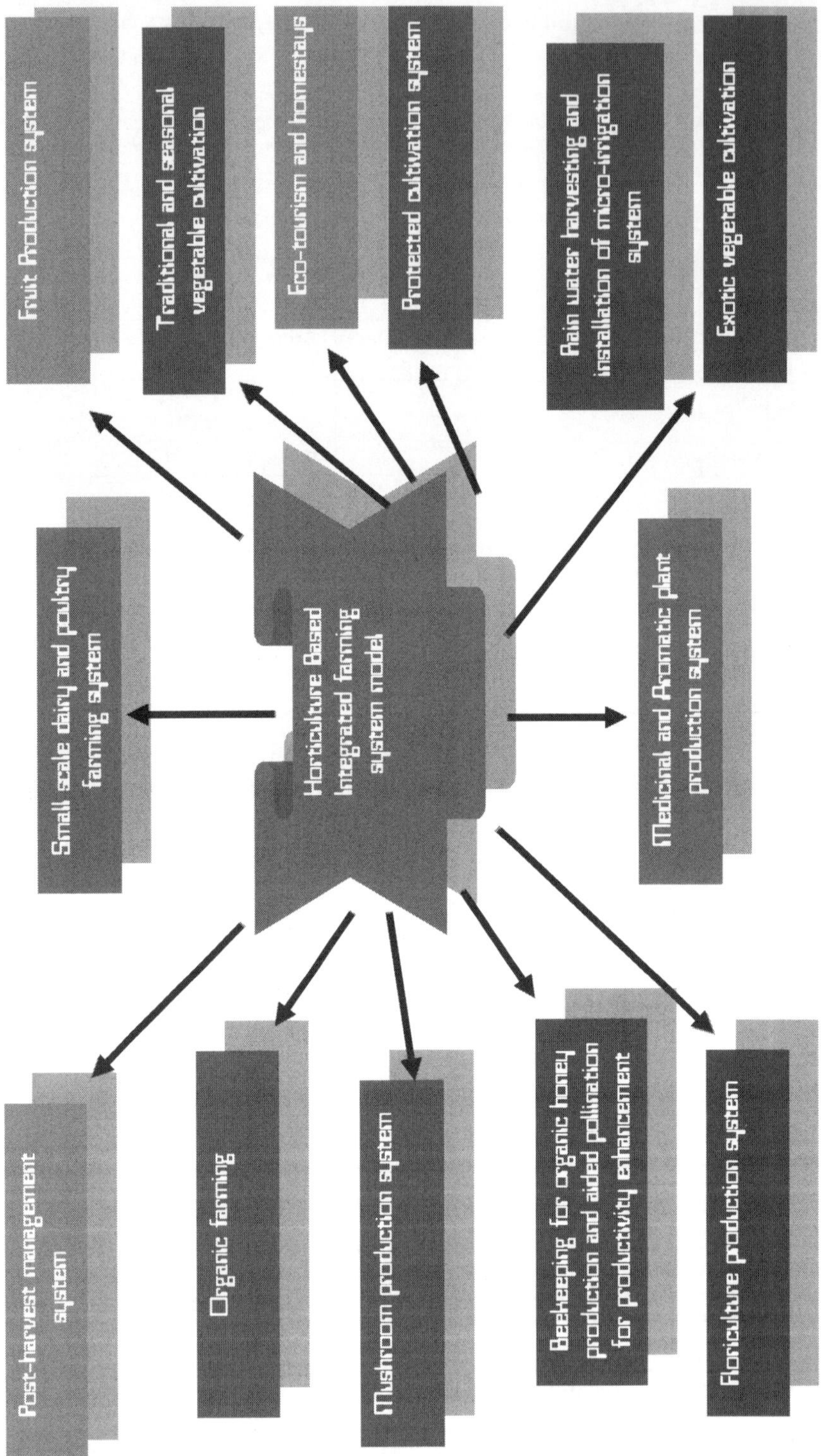

Figyre 9.2: Components of the Present Agri-entrepreneurship Model.

The Expected Outcome of the Horti-based Agri-entrepreneurship Model

- ☆ Promotion of horticulture sector and opportunities for unemployed youths.
- ☆ Old farmer seed replacement at a large scale resulting in an increase in production and productivity per unit area.
- ☆ Pre-harvest loss reduction.
- ☆ Increase in the cropped area and cropping intensity.
- ☆ An industry-supported and industry-driven farming activity.
- ☆ Increased adoption of new and advanced farming system and technologies also reduced yield gaps.
- ☆ Increase in achieving the research and developmental program goals.

SWOT Analysis of the Model

Strengths

- ☆ Local Himalayan conditions are best endowed with various natural resources and climatic conditions yet they remain unexploited; hence, they need to be exploited for in-situ employment generation of youth.
- ☆ Rural hilly terrain villages have a good hard-working workforce that can work in extremely harsh conditions.
- ☆ Climatic conditions specific for various temperate fruit trees, exotic vegetable, and high quality medicinal and aromatic herbs is a great strength and needs to be exploited.
- ☆ Villagers in rural terrains by-default practice farming which is organic; if land certified it can act as a boost to their income.
- ☆ There are some excellent opportunities in the field of eco-tourism and home-stays which attracts a lot of foreign tourists.

Weaknesses

- ☆ Prevalence of migration among youth is the biggest problem in the case of villages of hilly terrains
- ☆ Lack of basic transportation, health and education facilities in the villages aids to the migration of youth
- ☆ Crop failure due to harsh climatic conditions (like heavy erratic rainfall, hailstorms) and combats with wild animals like monkeys, wild deer, and wild pigs
- ☆ Lesser farm mechanization in hilly terrains
- ☆ Lesser adaptation of Horticultural Crops and their confinement to just kitchen gardens

- ☆ Prevalence of rainfed agriculture and low irrigation facilities and almost nil micro-irrigation techniques
- ☆ Low availability of elite cultivars and their lesser adaptation
- ☆ Lack of technical know-how
- ☆ Lesser adaptation of plant protection measures leading to crop failures and less productivity
- ☆ Lesser awareness of various government schemes
- ☆ Lack of extension education
- ☆ Due to very low processing, there is a lot of post-harvest losses
- ☆ Low proximity of the market leads to difficulty in the marketing of horticultural produce.

Opportunities

- ☆ At the present COVID-19 era, reverse migration has created an excellent opportunity for starting such Agri-entrepreneurship models
- ☆ Various government incentives and schemes are there for support of rural agriculture and allied activities
- ☆ More youth are nowadays getting interested in Agri-based start-ups which is a great sign for Agri-based rural development in the hilly terrains
- ☆ The government is aiding and easing the process of organic certification which can help in export of the produce fetching a good market price
- ☆ The high demand for organic products and other hilly products like coarse millets and their processed products, squashes, jam, chutney, juices from indigenous fruits and flowers like citrus, aonla, bael, rhododendron
- ☆ Excellent opportunities in the eco-tourism sector and homestays

Threats

- ☆ Lesser availability of irrigation water is a major problem which can lead to crop failure
- ☆ Monkey menace and other wild animal problem is prevalent in every village of hilly terrains
- ☆ Horticultural crops being a highly perishable commodity aided with lesser proximity to marketing facilities leads to various post-harvest losses
- ☆ A lethargic and inefficient government system has also created and going to create a lot of problems in the implementation of various policies.
- ☆ The interest of youth towards big and metro cities is going to aid the migration process.

Conclusion

To achieve the objectives of increasing the farmer's income through intensive cultivation of horticultural crops, it is necessary to overcome major hurdles for increasing crop production such as improved efficiency in use of inputs, shortage of roads in the farms, lack of irrigation facilities, lesser quality planting material availability, and development of facilities for farm produce marketing. For gaining a higher boost in income of farmers, it is important to develop new technologies and innovations and invest in them as well as eliviate the collaboration between different institutions, and promote public-private partnerships for large-scale production of horticultural crops for the national and international markets.

Table 9.12: Major Activities to be Followed in Starting Three Years of the Model

Year/ Type of Activity
First-year
1. Conducting Front Line Demonstrations of various technologies and practices to convince farmers by showing the difference between the traditional methods and improved practices and technologies along with the benefit-cost analysis and motivating them by giving various success stories of the farmers of hilly terrains.
2. First-year should focus mainly on the establishment and management of the units and components mentioned above.
3. Formation of a village-level committee or SHG (Self-help groups) for starting the Agri-entrepreneurship model in this employing more youth and reverse migrants in the village and assigning them their duties according to their skills and qualifications.
4. Preparation of a base-line survey as given in head 3.1-3.4.
5. As there are 12 components in this model there should be around 6 SHG's each allocated with 2 components. Each SHG can have around 10 members in it.
6. Reporting and analyzing the economic aspects of the model including benefit-cost ratio and future viability and progress of the model.
7. Further, the component should be divided into the modules. For example, if we must allocate Fruit production components it should be divided into
a) Nursery raising for both marketing and orchard plantation
b) Orchard establishment (During initial stages intercropping)
c) Crop production practices
d) Crop protection practices (Emphasis should be given to eco-friendly approaches)
e) Natural resource management
8. Undertaking the land for starting-up the enterprise.
9. Land leveling land cleaning for making land suitable for cultivation.
10. Forming basic facilities like
a) The first focus should be on practicing vegetable production both of traditional and exotic (for export to bigger cities like Dehradun and Delhi) for fetching immediate benefits from the model.
b) Rainwater harvesting system and tanks
c) Micro-irrigation systems
d) Protected cultivation structures
e) Mushroom cultivation units
f) Small scale processing units
g) Establishing small scale poultry and dairy farming units

Year/ Type of Activity
h) Establishing orchards and practicing inter-cultural practices until the ending of the gestation period of orchards i) Designation of a little area for exploitation of local conditions for the cultivation of Medicinal and Aromatic crops which are confined upper hilly terrains j) Designation of a small area for the production of flower crops that have high storage life and export potential k) Forming some beekeeping structures for harvesting some honey and productivity enhancement of fruit and vegetable crops l) Designating some area under organic farming and applying for certification for fetching good price by the export of the produces m) Waste management to minimize the external output in consecutive years.
Second-year 1. In this year more focus should be given on maximum government departments linkage to the farmers to get maximum benefits of government policies and schemes. 2. Replacement of old varieties and cultivars with new ones and hybrids. 3. The focus should be on making the village a seed village so that making village independent of seed purchase instead of making village self-sufficient to meet out the seed requirement and even also producing seeds for sale. 4. Making value-added composts (added with bioagents) to enhance their efficacy also focus should be on their bulk production for use in organic farming production practices and supply to nearby villages. 5. Designation of a little area for exploitation of local conditions for the cultivation of Medicinal and Aromatic crops which are confined upper hilly terrains. 6. Application for State organic certification. 7. Conducting diagnosing visits by various KVK's and state agricultural departments. 8. Conducting exposure visits of the farmers to various Farmer fairs, KVK's, and agricultural departments for creating awareness among farmers. 9. Reporting and analyzing the economic aspects of the model including benefit-cost ratio and future viability and progress of the model.
Third Year 1. The focus should be on technological rectification and refinement for getting ready for future problems and demand-based solutions. 2. Application for central organic certification for the export of organic produce for fetching higher prices of organically produced crops. 3. Conducting various pieces of training, exhibitions, motivational support regularly from 2nd year onwards with government departments, KVK's and farmers to attract more farmers and also farmers from other nearby villages. 4. There should be a complete linkage between government departments and villages for maximizing the benefits of government policies and subsidies. 5. For the analysis of the Agri-entrepreneurship model, there should be a farmer's feedback impact analysis based on various questionnaires prepared. 6. Reporting and analyzing the economic aspects of the model including benefit-cost ratio and future viability and progress of the model.

Realizing the overall scenario of horticulture production; if the above-mentioned modern innovative technologies are taken care of no doubt, the horticulture industry, especially the fruit, vegetable, flowers, and medicinal and aromatic plant-based industry in Uttarakhand would flourish very well. However, there is an urgent need to take up to adopt these valuable modern innovative technologies in horticultural crop production supplemented with using others. Networking of institutions already engaged in technology demonstration, but there is a need to develop strategies and quick for quick implementation of aforesaid technologies for the farmers of the state. These techniques are now successfully applied to a range of various horticultural crops. Effective implementation of mentioned measure and model in a sincere manner at the state-level can lead to a doubled farmers' income in the coming years in Uttarakhand State.

REFERENCES

Anonymous[1], (2020). www.fao.org.in

Anonymous[2], (2020). www.agriculture.uk.gov.in

Chattopadhyay, M., and Bhattacharyay, G. (1986). Growth of Indian Agriculture: A reappraisal. *Indian Journal of Agricultural Economics,* 42(1): 67-75.

Ghosh, S., and Munsi, S.P. (2015). Report on Rural Agricultural Work Experience Program.

Jairath, M. S. (1996). Agro-Processing and Infrastructural Development in Hilly Area: A case of Fruits and vegetable processing in Himachal Pradesh. *Indian Journal of Agricultural Marketing,* 10(2): 28-47.

Kashyap, O. P., and Tomar, G. S. (2005). Rural Agricultural/Horticultural Work Experience Program Practical Manual for B.Sc. (Agri./Horti.).

Purohit, V. K., Nautiyal, M. C., Nautiyal, A. R. and Gairola, K. C. (2016) Successful cultivation of highly valued and endangered medicinal herb Kutki (*Picrorrhiza Kurrooa*): A case study from Ghesh village of Uttarakhand, Himalaya. *Indian Journals: Medicinal Plant,* 8(3): 207-212.

Rao, N. C., and Dasgupta, S. (2009). Nature and employment in the food processing sector. *Economic and political weekly,* Pp: 109-114.

Roy, B. C. (2008). Fruits and Vegetable Processing in India and its Role in agricultural development. *Indian Journal of Agricultural Economics,* 63(3): 14-20.

Shehrawat, P. S. (2006). Agro-Processing Industries-A Challenging Entrepreneurship for Rural Development. *Journal of Asia Entrepreneurship and Sustainability.* 2(3): 10-22.

Srivastava, U. K. (1989). Agro-processing Industries: Potential, Constraints, and Task Ahead. *Indian Journal of Agricultural Economics,* 44(3): 242-256.

Chapter 10

Impact of Organic Farming on Biodiversity

Rajni Chaudhary[1] *and J.I.S. Khattar*[2]

[1]*Department of Agriculture, Baba Farid College, Deon, Bathinda, Punjab*
[2]*Department of Botany, Punjabi University, Patiala, Punjab*

Biodiversity is the one of most important resource on earth. A rich biological diversity is essential for preserving natural processes contributing to man's ability to live, such as natural pest regulation, pollination of fruit blossoms by insects, and the decomposition of organic matter into humus. Various human activities endanger the total number of species such as intensive farming, introduction of foreign species, habitat change and fragmentation. Farming is the big reasons to lose of biodiversity. Agriculture helps in enhance and protect the biodiversity but it is estimated that intensive farming is one of the main causes of species decline. So organic farming is an alternative method to better food production as well as in protection of biodiversity. Organic farming helps in ecosystem management by excludes of agrochemicals such as inorganic fertilisers and most pesticides. These factors enhance not only biodiversity, but strengthen natural cycles and improve environmental performance that in turn increase the sustainability of organic farms. At present times it is necessary to develops more such sources which help in enhancement of richness of biodiversity. So present study concerned with the impact of organic farming on biodiversity.

Introduction

Biodiversity is defined as the variability among the living organisms including terrestrial, marine and other ecosystems. Biodiversity comprises at many levels: species diversity, genetic diversity as well as habitat and ecosystem diversity. Various natural processes by human activities such as natural pest regulation, pollination of fruit blossoms by insects, and the decomposition of organic matter into humus essential for rich biological diversity. Biodiversity is the one of most important resource on earth and human activities endangers the total number of species. According to Rio conference 1992, it is estimated that biodiversity losses to zero upto 2020 (Gabriel *et al.*, 2010) in united nations. Farming is the big reasons to losses. Agriculture can also help in enhance and protect the biodiversity. An enormous decline in biodiversity has been the result of intensive farming, introduction of foreign species, habitat change and fragmentation (Hardman *et al.*, 2016) According to report of threatened species by Red Lists intensive farming is one of the major causes of species decline in cultivated landscapes. Drastic loss of biodiversity mainly contributed by pesticide use, synthetic nitrogen fertilizer, land consolidation drainage and the use of heavy machinery. In historical times, a more diverse landscape unfolded through farming from what was once an undifferentiated landscape dominated by forests (Crowder *et al.*, 2010). Several mechanical and chemical practices in agricultural in conventional farming create negative impact on biodiversity (Tscharntke *et al.*, 2012). These include:

The Impact of Tillage

Tillage affects the soil's physical environment like: soil water, aeration, compaction, porosity and temperature which results in organic matter and nitrogen in the top layer of soil deplete. Soil losses can be quite substantial, intensive agriculture can lead to devegetation, erosion and desertification as occurred in the experience of the Dust Bowl in 1935, when the U.S. Department of Agriculture alone estimated that 40 million ha of arable land on the Great Plains were ruined for agriculture and another 40 million ha were severely damaged (Burger, 1985). The use of surface tillage implements kills or injures nesting birds and small mammals (Rodent house), densities of waterfowl nests on untilled land may be 12 times those on croplands.

The Impact of Drainage

Active drainage for the creation of farmland decreases the wetlands, which has led to direct damage and decline of many species of plant, mammal and insects. It is estimated that half of the original wetlands of southern Canada have been lost through agricultural land drainage. Most of marsh inhabitant species like certain reptiles, amphibians, birds, mammals, fish declined due to habitat loss whenever marsh drainage to increase or improve agricultural land. It also results in increases of pollutant and sediment inputs accumulated in drainage water. Wetlands are also the habitat of about one third of wildlife species currently identified as endangered, threatened, or rare by the Committee on the Status of Endangered Wildlife in Canada (FON, 1987). Which may be diminishes by continuously conversion of the wetlands into farmlands.

The Impact of Intercropping

Although the Intercropping provides many benefits as compared to monoculture like pest control, weed control, reduced wind erosion and improved water infiltration but Some combinations such as combined growing of corn and soybean can affects the biodiversity like reduced some particular crop pests *etc.* Even during the intercropping a particular crop used a particular nutrients from the soil which can affects the soil fertility as well as growth of the beneficial microbes.

Impact of Pesticides Use

Pesticides, which comprise insecticides, herbicides, fungicides and others, are designed to kill something, therefore affects species diversity at least in the area where they are applied. With the uses of pesticides in agriculture population of non target species decline due to habitat loss. For example, loss of rare plant Purple twayblade from few site of Canada due to a combination of shade, slug predation and spraying of a herbicide. Large use of insecticide carbofuran are very efficient for killing a large proportion of the songbird population breeding on the edge of fields where they are applied also liquid form shown to have an impact on at least one endangered Prairie species, the Burrowing Owl (*Speotyto cunicularia*). Also use of the insecticide diazinon on turf also considered as important source of mortality for the population of Brant geese *Branta beriitcfa*) wintering in the mid Atlantic states of the U.S. Insecticides such as carbofuran, phorate and terbufos are considered to be more toxic than herbicides to soil fauna, earthworms and soil arthropods (Chopra, Sharma and Chamoli 2011). The intensive use of conventional insecticides generally reduces diversity through direct toxic effects. Many of the widely used classes of conventional insecticides, including organophosphates and pyrethroids, have been shown to adversely affect a broad range of non target species, including species of economic importance. Local extinctions are common where these insecticides are frequently used. Such insecticides have been shown to eliminate important predator and parasitoid species from agricultural systems. (Geiger, *et al.*, 2010). Thus increased use of all these chemicals not only affects the soil fauna but also disrupt the soil fertility.

Impact of Fertilizers

Fertilizers consists of substances and chemicals like methane, carbon dioxide, ammonia, and nitrogen, the emission of which has contributed to a great extent in the quantity of greenhouse gases present in the environment. This results in global warming and weather changes. Nitrous oxide, which is a by product of nitrogen, is the third most significant greenhouse gas, after carbon dioxide and methane. It will results in acid rain which not only affects the soil fertility as well the diversity of microbes, earthworms *etc.* Nitrogen from agricultural fertilizers leaches through soils to groundwater and runs off into rivers and lakes, increasing aquatic dissolved nitrate. Excessive quantity of nitrate in the water can results in excessive growth of aquatic algae i.e algal bloom, which can use up most of a water body's oxygen (Erisman, *et al.*, 2013). With the deficiency of the oxygen in water bodies the life of the other fauna and flora also comes in danger. Even aquatic algae die and are decomposed it creates "dead zones" on the surface of water that cannot support life of other organisms in water. Due to these increases in fertilisers, stocking densities and nutrients there has been an increase in pollution of rivers leading to loss of biodiversity, especially in sensitive rivers such as Salmonid rivers. Even the phosphorus (P) pollution from fertilizers cause eutrophication of water bodies. It is estimated that nutrients run off from fertilizers has contributed to more than 500 hypoxic zone due to eutrophication around the world (Rabotyagov, *et al.*, 2014).

Impact of Livestock Production

Intensive, or so called landless, large scale production systems are on the rise leading to an increase in both demand cultivated systems, with consequently an increased demand for water and for animal feeds and of site concentrated livestock wastes. The increased demands for feed puts increased pressures on nitrogen, other fertilizers and other chemical inputs. Emphasis in modern systems on quantity of yield, has led to selection and breeding for high production and the loss of traditional breeds that held other traits, qualities and adaptations, now lost. To meet the increased demand of the land human beings cleared the forests areas which results not only the habitat loss of the animals but also affects the valuable fauna also. With the continuously use of the fertilizers, pesticides as well as other chemical pollution of an environment increases which disturb the life of living organisms.

Impact of the Land Reclamation

Land reclamation including the removal of small scale farmland habitats. The removal of habitats such as trees, hedges, dry stone walls, woodland and scrub, together with land reclamation, create a decline in the native biodiversity of areas.

Impact of Silage Making

Silage making is generally more intensive than haymaking leading to a decline in grassland habitats creating problems for many species, most publicly the corncrake (Crex crex).

Impact of Crop Rotation

The abandonment of small scale crop rotation and the intensification of large scale farming increased the amount of nutrients being added to the land resulting in eutrophication and pollution of water courses. Eutrophication decreases the amount of the dissolve oxygen in water bodies which affects the life of both aquatic flora and fauna

Impact of Overgrazing

Overgrazing of cattle and sheep's caused the destruction of many habitats especially peatlands. Overgrazing results in the exposure of bare rock and the widening of rivers, and increased braiding of channels, making their course highly unstable from one flood event to another. Also with overgrazing compactness of the soil decreases which not only losses in the habitat of earthworms and arthropods but also affects the fertility of soil.

Organic Farming

Organic farming is considered as ecofriendly form for food production as well as helps in protection of biodiversity. Organic farming, which aims at ecosystem management and excludes agrochemicals such as inorganic fertilisers and most pesticides (Crowder *et al.*, 2010), can be an alternative to conventional farming.

Organic farming can influence biodiversity directly and indirectly. Direct effects of organic farming; include reduced exposure to pesticides and inorganic fertilisers. According to a European study increased use of pesticides reduced the diversity of plants, carabids and birds (Geiger *et al.*, 2010). Indirect effects include effects of changed farming practices that results from the restrictions on agrochemical use, such as the use of organic manure and changed crop choice which can increase the local habitat diversity on organic farms. Organic farming increases the local biodiversity with on average about 30 per cent as compared to convention farming. This variation may be due to different responses of organisms to changes in farming practice. Comparing biodiversity between organically and Intensive farming it is estimated that Intensive farming is one of the main causes of species decline.

Biodiversity play an important role in ecosystem's function. Species richness in habitats is shown to better adapt of environmental changes. For example, during dry season mountain meadows species erode less and show more stable yields. Critical ecological processes are influenced by the higher biodiversity and larger population densities of various species seen on organic farms. Various significance of Organic farming are:

Effect on Diversity of Organisms

Organic farming has positive effect on flora and fauna on field and also farm level. A comprehensive analysis of 66 scientific studies shows that organically farmed areas have on average 30 per cent more species and 50 per cent more individuals than non-organic areas (Schneider *et al.*, 2014). This is due to more environment adapted organisms respond to the to organic farming in better way, such as the reduction in pesticide use and crop choice. Effect of organic farming in Cleared landscapes seem to be most significant as compared to structurally rich regions. Also organic farming on affect some other organism such as mammals, reptiles, amphibians and protozoa,

Enhanced the Biodiversity of Birds

Birds are generally positively affected by organic farming because organic farms were associated with habitats favouring corvids, which are important nest predators. Landscape structure are more important determinant of bird species richness than farm management (Wilcox *et al.*, 2014). Due to organic farming bird species richness in both arable fields and in meadows increased. Also organic farming has been found to have larger effects on bird species richness in landscape with low amounts of semi natural habitat. Organic farming benefits the birds in winter, at least in simplified agricultural landscapes.

Enhance the Pollinating Organisms

Populations of wild bees, honeybees, and bumble bees increased in organic farming areas than in the surrounding farmland and semi-natural areas. Biodiversity increased three times and the number of bees seven times higher in organic farm area than in conventional areas. Thus Organic agriculture also improves the pollination of flowering plants in an environment. As insect depends on nectar and or pollen of flowering plants, thus increased in and around organically managed fields. High pollinator diversity on organically managed farmland contributes the support insect pollinated plants, both wild and cultivated. The use of insecticides to control pest, insects in conventional farming can have negative effects also on non target organisms so the limited use of such agrochemicals in organic farming may contribute high pollinator diversity.

Enhance Plant Diversity

The strong effect of organic farming on plant diversity is primarily due to the direct negative effects of herbicides in conventional farming. Organic cereal fields generally have higher plant diversity compared to conventional fields, both on cereal farms and on farms with both cereal and animal production. Plant diversity remains higher in uncultivated field borders and hedges adjacent to organically managed fields compared to adjacent to conventional fields (Winqvist *et al.*, 2012). Six times more species are found within the crop on organic farms compared to conventional farms.

Enhance Weed Population

Several valuable ecological functions in agro ecosystems are controlled by the biodiversity of weeds populations (Jastrzebska, *et al.*, 2013). Both the conventional and integrated production systems tend to be similar in intensity management and field biodiversity, but organic production tends to support greater density, species number and biological diversity (Edesi, *et al.*, 2012). At the field level, species richness and greater abundance of weeds is found in organic agricultural for e.gs in root crops, red clover/grass mixtures and in winter triticale. But species composition of weeds was affected more by crop species and cultivation regime, compared to different agriculture practices. The maintenance diversity of weed community is one of the most important step for sustainability of an agroecosystem. Weeds diversity can be improved by various steps like nutrient cycling, pest control, improved soil chemical and physical properties, and reduction of soil erosion. An important aspect in the evaluation of the environmental impact of production systems is the biodiversity index for weed species. Organic farming had the lowest impact on the environment than integrated farming. Input of agrochemicals affects more weed flora as compared to seed bank, which is buffered by the persistence of weed seeds in the soil. Different farming practices such as organic, integrated and conventional exert selection pressure on the species composition of weeds.

Enhanced Earthworms Population

Organic farming systems are not only increase biological activity but also increase biodiversity below ground. The main impacts of organic farming is to maintained the biological fertility by increasing amount and quality of the soil organic matter which usually disrupt by integrating farming well as the disruptions of soil habitat using different tillage tools. Organic farming is based on the principle of the maintenance and enhancement of soil life and natural soil fertility, soil stability and soil biodiversity for preventing and combating soil compaction and soil erosion, and for the nourishing of plants primarily through the soil ecosystem. Even within the constraints of organic farming practices, farmers can also improved soil biological quality by using less tillage (Verbruggen,et al., 2010). An important part of soil biodiversity is arbuscular mycorrhizal fungi, which can provide several benefits to plants and ecosystems which can be enhanced by Organic farming. Their richness increased significantly over time since conversion to organic agriculture (Verbruggen,et al., 2010). Many soil microorganisms and other biota including earthworms are also important factors for improve the soil fertility. The number and mass of earthworms increased in organic farms as compared to conventional, integrated farms. The abundance of earthworms, as well as their total body mass, was affected by plant species occurring in crop rotation. Oil pumpkins were revealed to have a beneficial effect on earthworms. There was also a significant production system and plant species interaction concerning the population of small earthworms (Bavec *et al.*, 2011). In addition to a production system, tillage is also a major driver for altering communities of earthworms and microorganisms in arable soils. The use of reduced tillage provides an approach for eco-intensification by enhancing inherent soil biota functions in organic arable farming (Kuntz, *et al.*, 2013).

Enhance Herbivores Diversity

The effect of organic farming on the diversity of herbivores is not well known, partly because many applied studies rather focus on effects of organic farming on abundances of single pest species instead of community level analyses. True bug communities (mainly consisting of herbivorous species) shows higher species richness on organic farmland than conventional farm. taxonomic distinctness of species under conventional farming reduced due to intense loss of phylogenetics.

Enhanced Predaceous Arthropods

The main groups of arthropods like predaceous beetles and spiders seem to more benefit from organic farming. Effects on the diversity of these taxonomic groups are less predictable, as predaceous species groups within these functional groups can be negatively affected by organic farming (Ponisio *et al.*, 2010). Organic farming also alters the composition of ecological traits in generalist predator communities compared to that in conventionally managed fields (Diekotter *et al.*, 2010, Rusch *et al.*, 2014).

Enhanced Soil Life Biodiversity

Conventional farming more affects the biomass of soil bacteria and fungi as compared to integrated or organic farming. Biomass of Soil microbes and their activity increased in organic systems. Root length of the plants colonised by mycorrhizae 40 per cent higher in organic than in conventional systems. Arbuscular mycorrhizal fungi spore and species diversity significantly higher in organic system. Important soil animal groups like annelids (earthworms and pot worms) and nematodes increased in organic farming also biomass and abundance of earthworms become higher by a factor of 1.3 to 3.2 in organic plots as compared to conventional farming.

Increased Mammal Biodiversity

Activity levels of small mammals (wood mouse, bank voleand common shrew) greater in organic than conventional fields. on organic farms total bat and foraging activity significantly higher by 61 per cent and 84 per cent respectively.

Enhanced Landscape

Organic farms are located in more diverse landscape types, have smaller field sizes, less frequent hedgerow management, use rotations including grass and are more likely to be mixed. Even within diverse landscapes, organic systems have greater field and farm complexity than non-organic systems Greater total areas of semi-natural habitat (woodland, field margins and hedgerows) on organic farms (Smith *et al.*, 2010).

Reduction of Water Pollution

As with increased use of synthetic in agriculture pollution of groundwater becomes a major problem. In organic farming use of these synthetic fertilizers and pesticides replaced by organic fertilizers (*e.g.* compost, animal manure, green manure) which enhance soil structure and water infiltration. Well managed organic systems with better nutrient retentive abilities, greatly reduce the risk of groundwater pollution. In some areas where pollution is a real problem, conversion to organic agriculture is highly encouraged as a restorative measure (*e.g.* by the Governments of France and Germany).

Enhance the Climate Change

Organic agriculture reduces non renewable energy use by decreasing agrochemical needs (these require high quantities of fossil fuel to be produced). Organic agriculture contributes to mitigating the greenhouse effect and global warming through its ability to sequester carbon in the soil. Many management practices used by organic agriculture (*e.g.* minimum tillage, returning crop residues to the soil, the use of cover crops and rotations, and the greater integration of nitrogen-fixing legumes), increase the return of carbon to the soil, raising productivity and favouring carbon storage. A number of studies revealed that soil organic carbon contents under organic farming are considerably higher. The more organic carbon is retained in the soil, the more the mitigation potential of agriculture against climate change is higher. However, there is much research needed in this field, yet.

Genetically Modified Organisms

The use of GMOs within organic systems is not permitted during any stage of organic food production, processing or handling. As the potential impact of GMOs to both the environment and health is not entirely understood, organic agriculture is taking the precautionary approach and choosing to encourage natural biodiversity. The organic label therefore provides an assurance that GMOs have not been used intentionally in the production and processing of the organic products. This is something which cannot be guaranteed in conventional products as labelling the presence of GMOs in food products has not yet come into force in most countries.

Main Causes of Higher Biodiversity

Organic farming has positive influences on biodiversity. The following practices carried on organic farms promote biodiversity are:

- Forgo use of herbicides
- Forgo use of chemically synthesized pesticides
- Less and purer organic fertilizer
- Fewer cattle per square meter
- Higher percentage of semi-natural areas
- Higher percentage of arable and ecological areas

These factors enhance not only biodiversity, but strengthen natural cycles and increase the sustainability of organic farms.

Conclusion

There are various benefits of organic farming. Organic farms has many benefit in providing antibiotic, and pesticide free food which are good for our health as well as environmental is beneficial. Organic farming is a more natural and efficient way to not only preserve wildlife biodiversity but also in reducing the amount of waste and energy. Eating organically will help promote the organic farms, and decrease in prices because of support and popularity that these farms would be gaining over time.

REFERENCES

Bavec M., Prašnièki M., Grobelnik Mlakar S., Turinek M., Robaèer M., Bavec F. (2011). Influence of different production systems on body mass and number of earthworms. V: 46th Croatian and 6th International Symposium on Agriculture, Opatija, Croatia. Pospišil, Milan (ed.). Proceedings. Zagreb: University of Zagreb, Faculty of Agriculture.**23**; 61-65.

Boyce, J.K. A Future for Small Farms? Biodiversity and Sustainable Agriculture. Working paper series. 2004: No 86. http: //scholarworks.umass.edu/cgi/ viewcon tent.cgi?article=1071 and context=peri_workingpapers (accessed 24 April 2014).

Chopra, A. K., Mukesh Kumar Sharma, and Shikha Chamoli. (2011). "Bioaccumulation of organochlorine pesticides in aquatic system—an overview." *Environmental Monitoring and Assessment* 173.

Crowder, D.W., Northfield, T.D., Strand. M. and Snyder, W.E. (2010) Organic agriculture promotes evenness and natural pest control. Nature 46: 109-112.

Diekötter, T., Wamser, S., Wolters, V and Birkhofer, K (2010). Landscape and management effects on structure and function of soil arthropod communities in winter wheat. *Agriculture, Ecosystems and Environment* **137**: 108–112.

Edesi, L., Jarvan, M., Adamson, A., Lauringson, E., Kuht, J. (2012). Weed species diversity and community composition in conventional and organic farming: a five-year experiment. Zemdirbyste-Agriculture. **99** 339-346.

Erisman, Jan Willem, James N. Galloway, Sybil Seitzinger, Albert Bleeker, Nancy B. Dise, A. M. Roxana Petrescu, Allison M. Leach, and Wim de Vries. (2013). "Consequences of human modification of the global nitrogen cycle." *Philosophical Transactions of the Royal Society of London B: Biological Scie*

Gabriel, D., S. M. Sait, Jenny, A., Hodgson, Schmutz, U., William E. K and Schmutz, W. E. (2010). Scale matters: the impact of organic farming on biodiversity at different spatial scales. Ecology Letters **13**(7): 858-869.

Geiger, Flavia, Jan Bengtsson, Frank Berendse, Wolfgang W. Weisser, Mark Emmerson, Manuel B. Morales, Piotr Ceryngier, *et al.*, 2010. "Persistent negative effects of pesticides on biodiversity and biological control potential on European farmland." *Basic and Applied Ecology* 11 (2): 97- 105. doi: 10.1016/j. baae.2009.12.001.

Hardman C. J, Harrison D.P.G., Shaw P. Nevard, T. M Hughes, B. Potts, S.G and Norris, K. (2016). Supporting local diversity of habitats and species on farmland: a comparison of three wildlife-friendly schemes. *Journal of Applied Ecology* **53**: 171–180.

Jastrzebska, M., Jastrzebski, W.P. Holdynski, C. Kostrzewska, M.K. (2013). Weed species di versity in organic and integrated farming system. Acta Agrobotanica. **66**: 113-124.

Ponisio L. C. M., Gonigle L. K., Mace K. C, Palomino, J., Valpine P and Kreme, C. (2015) Diversification practices reduce organic to conventional yield gap. Proceedings of the Royal Society B **282**: 20141396.

Rabotyagov, S S, C L Kling, P W Gassman, N N Rabalais, and R E Turner (2014). "The Economics of Dead Zones: Causes, Impacts, Policy Challenges, and a Model of the Gulf of Mexico Hypoxic Zone." doi: 10.1093/reep/ret024.

Rundlof M, Andersson G. K., Bommarco R (2015). Seed coating with a neonicotinoid insecticide negatively affects wild bees. Nature **521**: 77–80.

Rusch A, Birkhofer K, Bommarco R, Smith, H.G., Ekbom, B. (2014). Management intensity at field and landscape levels affects the taxonomic and functional structure of generalist predator communities. *Oecologia* **175**: 971–983.

Schneider M. K., Luscher, G., Jeanneret, P (2014). Gains to species diversity in organically farmed fields are not propagated at the farm level. *Nature Communications* **5**: 4151.

Smith H. G., Ockinger, E and Rundlof, M (2010). Biodiversity and the landscape ecology of agri environment schemes. *Aspects of Applied Biology* **100**: 225–232.

Tscharntke, T., Clough, Y., Wanger, T, C. Vandermeer, V., Whitbread, A. (2012). Global food security, biodiversity conservation and the future of agricultural intensification. *Biological Conservation* **151**: 53–59.

Wilcox J. C., Barbottin, A., Durant D, *et al.* (2014) Farmland birds and arable farming, a meta-analysis. In: Lichtfouse E (ed) *Sustainable Agriculture Reviews*, vol. **13**, pp. 35–63. Cham: Springer.

Winqvist C, Ahnström J and Bengtsson J (2012) Effects of organic farming on biodiversity and ecosystem services: taking landscape complexity into account. *Annals of the New York Academy of Sciences* **1249**: 191–203.

Chapter 11

Prospects of Organic Dairy Farming in India: An Overview

N.D. Chethan Patil[1], Bindu K. Ramalinannanavar[2] and Rahul Dundesh Bellagi[1]*

[1]Dairy Extension Division, ICAR- NDRI, Karnal – 132001, Haryana
[2]Microbi Agrotech Pvt Ltd. Kattigenahalli, Yalahanka, Bangalore – 560053, Karnataka

Dairy farming sector India has enormous potential to generate additional income for farmers. Globally, it is the country producing the greatest amount of milk. In addition, nowadays consumers are becoming more aware of adulteration, the presence of residual chemicals, etc. in milk and milk products. As a result of that, the demand for organic milk and milk products has increased. While the roots of the organic movement have long been organic fruits, vegetables, grains and some poultry, organic dairy farming is a relative newcomer. Indian conditions promote organic farming in different ways because of agro-climatic, cultural and economic influences, but the involvement of small dairy farmers presents problems in this sector's rapid growth. Natural dairy entered the agricultural market in the nineties and became an important player. Over the past few decades, the demand for organic food and milk has risen rampantly as customer awareness and awareness of wellbeing has increased. Since most organic consumers and customers can pay a premium for organic goods in developed countries, organic farming is a specialty sector with excellent export prospects. The section of this chapter provides an overview of the various aspects of organic dairy including the advantages and challenges facing us today. It also addressed the choices, construints and specifications for organic dairy production.

Introduction

Despite the duty to feed the earth, farming is an intrinsic part of man, industrial and technical advancements together with unintentional and inefficient usage for higher efficiency of chemical fertilisers, pesticides, and irrigation also created several big health and environmental problems. The negative and needless effect of agriculture on the environment and the conservation of natural resources created public awareness, particularly in developing countries. Sustainable agriculture came to attention to ensure that agriculture is economically and ecologically productive and organic farming became increasingly significant. Good impacts of organic farming on food quality, balanced diet, human health and environmental conservation have raised demand both within and outside the world for these products (MFAL, 2013). Indian agriculture was based on traditional cultivation methods (*e.g.* organic) until the middle of the last century. The Green revolution movement, which was the backbone of India's farming progress, was introduced in India during the 1960s, turning India from food deficits into self-sufficiency. However, over time production grew fourfold, indiscriminate and wasteful use of chemical compounds raised a question mark on the sustainability of agriculture, covering health care, human wellbeing, environmental protection and sustainable agriculture. The method of organic farming, invented by British agronomist Sir Albert Howard in 1990, was instrumental in the UK's extension of the programme for coping with the present problematic situation.

Brief History of Organic Dairy

It is well known that in the early 1940s in England the organic farming revolution began. The first experienced the idea of organic farming was Sir Albert Howard, who learned about the organic activities in India during the 1920s. J.I. Rodale usually gets credited with the birth of the organic movement in the U.S. The justification for the cultivation and purchasing of organic food is focused on consumer decisions, which requires complex professional decision making. So, most of them fell into three categories: well-being, society and the environment. Since Rachel Carson published Silent Spring in 1962, there has been growing concern about the new agricultural model that relies on synthetic inputs to increase yields; poses threats to the environment; and disconnects producers, property, and their populations. The organic movement is based on one central principle: good soils lead to healthy food, healthy animals, healthy people and a healthier environment. To create a safe, diverse ecosystem for nutritious food and feed processing, organic crops and livestock production rely on soil organic matter and biology building. Organic farming is often used as a way of sustaining and preserving family farmers relative to super- and industrial agricultural models that are rapidly growing. While organic fruits, vegetables, grains and some livestock have long been the cornerstones of the organic movement, organic milk is a relative newcomer. Natural dairy burst into the mainstream industry in the 1990s and became a major force. The popularity of organic dairy can be traced in large part to many crucial events including a successful reaction to Monsanto's 1994 launch of genetically modified or recombinant Bovine Growth Hormone (rBGH). The expansion of rBGH use — along with a rise in market awareness of genetically modified soybean, maize and other crops treated with a variety of synthetic pesticides fed to livestock; feeding ruminants with slaughter by-products and mad cow disease concerns; and expanded use of synthetic pharmaceutical drugs including hormones, antibiotics and steroids — has caused multiple congenital reactions. Many consumers depend on the reliable supplier of unadulterated dairy products, focused on certified organic milk guarantees. Around the same time, organic food growers and manufacturers' coordinating and marketing activities have built an infrastructure and retail reach that makes high-quality organic dairy products readily available, inexpensive, and attractive in both niches natural and mass-market grocery stores. Organic dairy products are also used as "gateway items," whereby consumers make their first attempts to buy organic dairy products through the purchasing of organic dairy products, gradually growing their commitment to organic products as they become more food-savvy.

Organic Dairy Farming

Organic dairy farming refers to raising dairy animals on organic feed, *i.e.* pastures grown without any chemical inputs, in farmlands or grazing lands, along with restricted use of any antibiotics or growth hormones. One of the main aims of organic dairy farming is to provide opportunities to the animals to express their natural behaviour. It involves high food quality and welfare of animals, along with emphasising upon animal health promotion, prevention of diseases and natural lives of the dairy animals.

A system of production that prevents or precludes the use of synthetic substances, pesticides, regulators and feed additives are known as organic farming. Organic farming systems rely as much as possible on crop rotation, crop residues, animal manure, vegetables and green manure, farm organic waste and organic pest control aspects to maintain the full potential of organic farmers (Lampkin,1990). The Codex Alimentarius Commission reported that organic farming is a holistic method of controlling development, cultivating and enhancing biodiversity, ecological processes and soil conservation of the agro-ecosystems. This promotes sustainable management practices rather than the use of off-farm tools, taking due account of local adaptation strategies suited to different geographic circumstances. Organic processing must be knowledge-intensive, highly scalable, and farmers must understand organic quality requirements, principles and processes that demand high expertise and experience and do so using agronomic, biological and mechanical methods whenever possible to fulfill a particular device purpose, as opposed to the use of synthetic materials (Codex Alimentarious, 2007). The approved certification bodies or organisations must review and approve the organic processing not just of the finished product but also of the whole phase of production. Organic milk production postulates the rearing of cattle on organically raised feed, *i.e.* pastureland without fertilisers or pesticides, with limited use of pharmaceuticals, antibiotics with hormones. Hence, organic farming is beginning to grow, and further effort is required to make it sustainable.

Scope of Organic Dairy Farming

Farms with poor, small- and semi-medium-scale agricultural properties have a greater chance of being organic agriculture, as organic farming is practiced in most Indian farms, not by choice but as a conventional method. India, with an abundance of livestock capital, comprises approximately 20.45 per cent of the buffalo population, 35.94 per cent of the cattle population, 13.87 per cent of sheep, 27.80 per cent of goats and approximately 1.69 per cent of pigs (20th Livestock Census) with export earnings from beef, poultry and related goods, approximately 30.632 crores during 2019-20. Although organics are established under the Indian agricultural framework, certification criteria are mandatory, recommended by several recognised agencies. Indian agriculture is produced organically by customs but only to some degree through organic processing methods. Other specific requirements such as organically certified animal nutrition, record-keeping, design animal treatment, bedding supplies, animal nutrient supply, and trace elements are also required. Farmers did not actively follow organic farming; in fact, most Indian farmers did not understand the organic concept "in their own way." Since India is the world's largest producer of 187 million tonnes of milk each year (2019-20), it is expected to comply with international requirements set by the Food and Agricultural Organisation (FAO), the World Health Organization (WHO), the Codex Alimentarius Group, *etc.* The General Agreement on Trade and Tariff (GATT) requires the Government to bargain with commercial agreements in order to protect human, animal and plant life. India must therefore tackle the problems that occur concerning exports of dairy products, as advanced countries with quality requirements are a big obstacle for developing countries like India. Several countries around the world have increasingly advocated for the implementation and compliance of agricultural traceability systems to track the concentrations of chemical contamination, prescription products, aflatoxin, heavy metals and dioxin toxins in foods, such as food safety issues such as Foot and Mouth Disease (FMD), *etc.* India has ample opportunities to prosper in the global organic market, as items such as skimmed milk powder (44.07 per cent), milk fat and butter (35.00 per cent) whole milk powder (8.14 per cent), cheese and fermented products (2.93 per cent), processed cheese (2.81 per cent) and whey and other products 2.04 per cent are already exported to countries including UAE, Nepal, Bangladesh, *etc.* The demand for organic food is escalating by 20-22 per cent with a compound annual growth rate, with 2000 organic farms and NGOs selling organic food goods (Organic Trade Association, 2012).

The Indian government introduced a National Programme for Organic Production (NPOP), which was formally authorised in October 2001 under the Foreign Trade and Research Act (FFDR Act), to allow for focused and well-directed organic farming research. The Indian Agricultural Research Council (ICAR), Agriculture and Manufactured Goods Export Development Authority (APEDA), State Agricultural Universities (SAUs), State Department of Animal Husbandry, State Departments of Agriculture in collaboration with NGOs are now working to encourage organic farming. Madhya Pradesh has the largest bio-registered region in the world, followed by Rajasthan and Uttar Pradesh. The word "Organic State" was given to the state of Sikkim in India to support organic farming and cultivation practices by strict policies to minimise the use of inorganic or chemical inputs in agriculture.

Overview of Organic Dairy Farming

Prospects of Organic Farming

Organic dairy farming has promising prospects in the world, as it essentially consists of a low-input and crop-residue fodder-based production mechanism dominated by small and marginal farmers contributing 70 per cent of the country's milk output (Kumar *et al.*, 2005). Numerous tests have found that organic milk is healthier than inorganic milk because it has protective fatty acids, including complete polyunsaturated fatty acids. Organic milk products are nutritious and environmentally friendly and hence create massive consumer demand and prospects for the production of organic milk products (Bodapti, 2013). It is also observed that because of the biogas system built in the organic milk production system, it saved 46 per cent more power and energy with low GHG emissions (Kimming *et al.*, 2014).

Major Limitations of Organic Dairy Farming

Organic dairy farming has various limitations in it. Berentsen *et al.* (2012) reported that owing to higher costs and low milk production per cow, the risk contained in the gross margin per cow for organic farming is considerably greater. Pathak *et al.* (2002) noted a low level of literacy among dairy farmers in small farms. Lack of knowledge, a huge number of herd supplies, poor quality of feed and forage, high cost of certification, lack of marketing infrastructure have been major obstacles in turning traditional dairy farming into organic agriculture. Sangkumchalianga and Huang (2012) also observed that lack of customer knowledge was the biggest obstacle for the successful marketing of organic foods due to the limited availability of market information. Berentsen *et al.* (2012) recorded that higher organic food yields of organic farming defined higher veterinary costs per cow and low average milk production as major adverse relative to traditional dairy farming. Mongeon and Summerhayes (2010) recorded that dairy cow production and productivity in organic farming was smaller, but had better fertility than traditional herds. And hence milk yield, protein and fat content in the case of organic farming is comparatively lower. It is also stated that in the case of the organic dairy production scheme, the overall number of inseminations for the cow was lower with greater size at initial insemination. Organic dairy farming is highly labour intensive, requiring about 10 to 20 per cent more labour than conventional dairy farming (Wolde and Tamir, 2016).

Perception of Dairy Farmers towards Organic Dairy Farming

The perception of dairy farmers towards organic dairy farming is crucial to know about the farmer's mindset about organic dairy production. Garmo *et al.* (2010) found that more milk was provided by traditional cows than organic cows. For organic cows, the milk yield and ratio are smaller than for typical cows. Also, the study found that there were no major differences between conventional and organic farming in reproductive and udder safety. Similarly, in their research on organic farming practices, Subraahmanyeswari and Chander (2008) suggested that since most farmers were rearing indigenous cattle species, they could well adapt to local systems with low to average production rates and disease resistance, and now most dairy farmers have neglected veterinary measures. Cranfield *et al.* (2010) indicated that the main factors for conversion to organic dairy farming were health and safety problems, such as reduced contaminant contamination, and improved food quality and environmental issues. In their research, Chander *et al.* (2011) indicated that dairy farmers in tropical countries that grow native animal species were found to be less vulnerable to stress and disease. The study also found that all policy-makers in those countries were considering financing organic certification to encourage environmentally sustainable production of organic milk. In another study, Herath and Wijekoon (2013) stated that strong knowledge of organic farming; a positive understanding of environmental conservation and the desire to devote more time to agriculture contributed to a favourable attitude towards organic farming.

Major Constraints

In spite of the sustainable nature of organic dairy farming, it has many constraints also. Pathak *et al.* (2003) found that the transition from traditional to organic farming would be hampered by small-scale land ownership, low literacy levels, shortage of expertise, high stock size, inadequate feed and forage production, high certification costs and lack of labeling facilities. Odlong *et al.* (2014) reported that inefficiency of organic inputs to combat external and internal pests, lack of curative steps for diseases such as mastitis, availability of organic Feed and inaccessibility to customers were primary hurdles to the shift to organic dairy farming. Also, as many farmers sold their milk locally; there was little hope of organic milk production receiving better prices. Not all farmers kept regular records, lack of knowledge of organic farming, lack of funds to fulfill their unique certification standards, lack of visibility, low income and other local certification facilities were major constraints that impeded the move. Nemes (2009) argued that organic farming faces unequal competition from the existing subsidy-based funding structure for conventional farming, insufficient study and extension efforts in the field of organic farming and the failure to recognise clear environmental, social and health externalities at the expense of traditional foodstuffs.

Basic Norms of Organic Dairy Farming

Since the inception of the National Organic Program (NOP) of the US Department of Agriculture (USDA) in October 2002, organic farming and labeling have become increasingly common in the USA. Before that, numerous private and state certifiers have played the role of overseeing and selling inspections of organic goods. Now all goods marketed as "normal" in the United States must be made, processed and shipped in accordance with one standard — the "Ultimate Standard" as recommended by NOP (USDA, 2000). It's crucial to remember that organic farming is a development method, a set of goal-based regulations that allow farmers to cope individually with their own unique circumstances while preserving organic credibility. It is also important to consider the recent new uncertainties in the organic farming system and recommendations as laid out in the National Organic Programme. However, and though the scheme matures, insurances are included in all organic dairy products approved by the USDA.

Those include the following:

- Both the cows and the calves are fed nutritious pure organic feed. Restricting the use of chemical fertilizers and pesticides for field crops, hay and pasture which have not been properly tested and certified for organic farming.
- Allowing non-natural feed additives and nutrients such as vitamins and minerals to be used in organic products is also important.
- Genetically modified species (GMOs, referred to as 'Excluded Processes' in the regulation) are explicitly forbidden. Land used for organic crops to be at least three years before the first organic harvest free of all inorganic and contaminating materials.
- The usage of synthetic milk replacers is banned. The young calves must be fed daily with organic milk. Both the dairy cattle have to be introduced to the natural climate, *i.e.* under the barn lighting.
- Animals older than six months must have access to grass grazing during the growing season. With approval from the veterinarians, only approved items for animal health care can be used.
- In organic dairy farming antibiotics can be generally avoided. No organic animals can be fed by-products of slaughter, urea or manure. Taking into account animal health, certain practices are forbidden; such as tail-docking. Other measures need to be done to decrease the burden on the horse, such as dehorning.
- An organic farmer must maintain proper records to ensure he or she fulfills the requirements needed. Any farm is inspected and audited annually. Any farm can be audited for inspection at any time by unanimity.

Standards for Organic Milk Production

It should be registered by an organic certification authority to preserve the organic dairy farm, and the approved production method must follow the organic requirements. Currently, five organic principles are essential and agreed all over the world, *viz.* The United States Organic Food Products Act (OFPA), the European Union Legislation (1804/1999), the draught guidance Codex/WHO/FAO, the United Kingdom Organic Food Quality Registry (UKROFS) and the International Federation of Organic Agricultural Movements (IFOAM). It has been recorded that 468 organisations providing organic certification services are present globally (Yadav, 2008). Most credential bodies work in Europe (37 per cent), followed by Asia (31 per cent), and North America (18 per cent). The nations with the most approval bodies are the US, Japan, China and Germany. The European Union approves 40 per cent of the certification bodies, 32 per cent have ISO 65 accreditation and 28 per cent are accredited under the United States National Organic Programme. Certification steps include the registration of producers and processing industries, the provision of basic crop and farm information and the inspection and verification of the farm, processing unit, production methods and production practices by the inspector appointed by the certifying agency, such as APEDA (Agricultural Products Export

Development Agency); NSOP (National Standards for Organic Products), USOCA (Uttarakhand State Organic Certification Authority) appointed by the Uttarakhand Government, ECOCERT appointed by Agriculture Ministry, Govt. of India, *etc.* For organic milk production the following recommendations should be considered (Alexander, 2010):

1. **Conversion to organic from conventional farming**: Strategy for conversion from traditional to organic production is very important. Either the whole farm will be turned into one unit, or over several years the conversion can be phased out. It takes a minimum of two years for the property to be transferred to organic status. Organic dairy farming can be done from the day that the soil reaches its maximum organic status. To gain organic status, the herd must have begun nine months and feed six months before the date of production of the expected organic milk.
2. **Feeding**: All feedstuffs used at the farm must be produced and certified to organic standards from the beginning of the conversion. The farm should be provided with all the appropriate feed, and the maximum use should be made of grazing land. A minimum of 60 per cent feed should be collected from the farm or related organic farms and up to 30 per cent from in-conversion sources. The structure of the ration must meet complete organic criteria. Mixtures and compound rations received must be 100 per cent organic. Mineral supplementation is allowed only when trace element criteria cannot be fulfilled by the organic husbandry activities. Many synthetic vitamins can be included but would be subject to approval from the control board. Centered on clover fodders are essential to the sustainability of organic dairy farms as they are the main source of nitrogen. If used, molasses too must be sustainable.
3. **Soil fertility**: Soil fertility can be improved by careful irrigation, alternating silage and pasture soil wherever possible, and selectively using recycled manure and slurry. Synthetic fertilisers are not approved for use in organic farming, but the use of lime or some natural sources of nutrients is applied as needed.
4. **Livestock manures**: While it is possible to carry in manure from other organic farms, the amount does not exceed 250 kg N/ha/yr extended to either sector. Even poultry litter can be purchased from certified organic farms. Consent to the use of manure produced on conventional farms can be sought from the authorities concerned.
5. **Housing**: Space parameters can vary from the control body to the control body. There has to be a comfortable, dry bedded lying area for cows. Well, loose bedding accommodation is favoured. Dairy cows should be kept to a minimum of 6 m per horse. Youth stock room needs will vary from 1 to 1.75 m per 100 kg live weight. Slats used by each type of stock should not be more than half the usable floor space.
6. **Animal health**: Both cohorts and progeny of Bovine Spongiform Encephalopathy cases must be eliminated from the herd before the start of organic conversion. Preventive surgery, and homeopathic treatments,

are still welcomed. Veterinary medicinal products and antibiotics shall not be used as a preventive medicinal device, but must be used at least twice the prescribed withdrawal time in case of sickness or accident, in order to avoid discomfort. Good control methods may be used for controlling mastitis, including dipping teat and high-cell count cow culling. Parasitic protection can be accomplished by proper monitoring of the grazing to mitigate parasite risk. Any anthelmintic brand negotiated with the control body and could be used to treat animals with health problems as part of a treatment plan. In situations where there is a suspected risk of infection, vaccination is allowed under derogation. Ethical status is revoked if more than three treatment programmes, with the exception of vaccine, parasite care and other obligatory eradication programmes, are provided to animals within one year.

7. **Sources of stock**: Purchased cattle do not emerge from herds that have had a Bovine Spongiform Encephalopathy event over the past six years. When a farm is converted to organic farming, the current animals can be kept but can never be sold as organic, although the milk and milk products produced from these cows and their progeny can be branded as organic dairy products during the necessary conversion times. Per year, up to 10 per cent of the breeding herd can be supplemented from traditional herds. Whole organic milk can constitute at least 51 per cent of the overall ration for calves up to 12 weeks of age. Either organic or conventional farmers will market the surplus calves. Stock bulls can be bought from conventional farmers, or hired bulls can be used when they arrive on the organic farm as per the organic requirements.
8. **Selling of organic milk**: To get the high rates for organic milk, it is important to sell the milk to an organically certified manufacturing outlet. Marketing should often be considered before starting organic development. Authorized and certified sterilisers can be used in milking and dairy parlours.
9. **Dairy bred beef calves**: If the dairy farm does not have a beef company, caution should be taken to build relations with organic beef rearers and finishers who may be involved in purchasing weaned calves. It should also be considered when considering a Bull breed.

Benefits of Organic Milk

Fresh milk is made of more protective Omega-3 (Lairon and Huber, 2014), and less toxic Omega-6 (Benbrook *et al.*, 2013). Omega-3 is an essential fatty acid required for healthy development, and its deficiency has given rise to numerous health problems that have begun to intensify in recent years. Regular consumption of Omega-3 fatty acids help to decrease the occurrence of heart disease, acne (in skin disorders such as eczema), cancer and arthritis (Annon, 2014). Larger quantities of Conjugated Linoleic Acid (CLA) also exist in organic milk (Mercola, 2014). Healthy cows are grazed on organically farmed pastures. Therefore, their milk is not polluted with toxic substances such as traces of pesticides, fertilisers and hormones (Singh *et al.*, 2011). Furthermore, this organic nutrient-rich milk does not contain traces of pesticides, GM feed, urea, or growth hormones, since they are not fed to the cows to increase their milk yield. Lutein is particularly important for the protection of the eyes and is effective in reducing multiple ocular diseases including macular degeneration and cataracts. Zeaxanthin is essential to healthy eyes, too. It protects the eye from UV damage and the free radical effects. Thus, cataract treatment, diabetic retinopathy, glaucoma, and macular degeneration have been considered dangerous. Organic milk also has a larger concentration of vitamins than regular milk, such as vitamin A and vitamin E.

Production Costs for Organic Dairy Farming

Feed costs are the key source of investment in organic dairy production, as dairy farmers need to generate their own energy and protein feed supplies (*e.g.* grain and protein supplements). The recent price increase for organic products such as milk would make this a profitable option for many organic dairy farmers. It would be more commercially viable in regions where these feeds are easily available and don't require high transportation costs. Animal health care is another aspect that affects the cost of production in an organic dairy. Job expenses also play a crucial role in the table on investment.

Market Dynamics of Organic Dairy Farm Products

Like other organic foodstuffs, the demand for organic milk products was connected to the increasing customer desire for natural and organic products. Demand for organics is attributed to many reasons, such as the view of anything organic as clean and natural, improved respect for the environment and concern for farm animals. Both of these factors are key to increasing demand for organic milk products, with labels like 'grass-fed' and 'pasture-raised' being common. Price is the basic restriction for the organic milk industry, as organic milk is almost three times more expensive than conventional milk. This large difference limits the market's scope only to urban and high-aware consumers.

Potential of Organic Dairy Farming in India

India has enormous promise in the manufacture of organic milk. In Indian dairy production operations are not highly intense relative to other industrialised dairy countries. Some of the country's agro-climatic regions are ideally suited for organic milk production. Such regions include Rajasthan, Gujarat, Madhya Pradesh, Himachal Pradesh, Uttaranchal, Jammu and Kashmir, Tamil Nadu and the Northeast as a whole. The methods of farming practiced in many of these areas are still quite natural, and the milk produced today is almost organic. There are many places in the world (particularly mountainous places) and communities (some tribes) where green revolution technologies have not yet penetrated and have not accepted the use of agrochemicals. These regions have been known as "flood zones" (Singh, 2007). The North-East Indian area provides substantial potential and possibilities for organic farming owing to the least use of chemical inputs where it is estimated that 18 million hectares of such land are available which can be used for systematic organic production (Ghosh, 2006). Small farmers in these areas who produce a few liters of milk a day are unable to sell it as organic milk because of confusion and unavailability on the sale of local organic products. The Trans-Gangetic Punjab area, Haryana, Western U.P., And parts of Rajasthan improved crop husbandry through intensive crop rotations and heavy use of inorganic and agrochemical fertilizers. Even in this area, however, as has been the case with advanced countries, milk farming has not received much intensification and is therefore likely to be transformed with little effort into organic. Organic dairy farming has a wide reach in the world as it is the low-input, crop-residue-fodder-based production method for smallholder farmers contributing 70 per cent of total milk production in the country (Kumar *et al.*, 2005). It is recommended that the cooperative association should come forward to authorise, purchase, store, and distribute organic milk to access the organic milk supplied in indoor rural areas.

Constraints in Organic Dairy Farming

Price is a major challenge in the organic milk market as organic milk costs nearly thrice as much as conventional milk. The majority of Indian dairy farmers are small and marginal. But due to increasing demands of organic products and India's success in the co-operative movement, co-operative dairy farming and contract farming can be taken up to overcome these barriers (Maji, 2017). Kamboj and Prasad (2013) have identified some of the constraints on the growth of organic dairy farming, including lack of expertise and understanding, restrictions on soil-free organic dairy farming which are not allowed under the National Organic Production Standards (NSOP), restricted availability of organic feed ingredients for formulating organic compound feed, issue of proper record-keeping, limited range of certification services and lack of adequate procurement, processing and marketing infrastructure and networks in organic dairy farming.

Future of Organic Dairy Farming

Demand for organic livestock products in the USA, the EU, Japan, Argentina and Brazil is on the rise. Belgium, Luxembourg, the Netherlands and the United Kingdom, as these countries manufacture large amounts of organic produce. Consumers pay a significant premium for organic produce in Austria, Belgium, Germany and the UK. Livestock native breeds are less vulnerable to trauma and disease and thus the need for allopathic medical drugs and antibiotics is much smaller. There is enormous potential for conversion into organic livestock husbandry by grass-based, robust production systems and forest-based animal production systems that are prevalent in many of these countries. Literacy is on the increase, and public awareness and interest in global media for animal rights issues and safe eating is rising. This will contribute to the increased household intake of organic foods.

Conclusion

Organic dairy farming plays a crucial role in any grown, refined and branded chemical or residue-free milk and milk products. It's a smart idea to improve a safe lifestyle for people. In addition to environmental advantages, the regulations on organic livestock production include some prerequisites for ensuring good living conditions for farm animals and today more farmers are converting to organic agriculture. In dairy farming, organic means in all areas of dairy animal breeding, feeding and welfare need to be carefully handled and reported. Owing to recognition of the benefits of organic food products, demand for organic milk and milk products has risen significantly in the last few years. While organic dairy farming is a newer idea, it has gained considerable significance in growing health problems and adulteration levels of dairy products. Different organisations have since made concerted attempts around the world to include a fillip for organic dairy products and to resolve these issues; various requirements are also pursued proper execution. Organic dairy production thus acts as a support to the creation of safe and balanced dairy products.

REFERENCES

Alexander, D. (2010). Organic Milk Production. Department of Agriculture and Rural Development.UK, http: //www.dardni.gov.uk/index/farming/livestock/organic/organicdair ying/organic-milk production.htm

Annon (2014). Health benefits of organic milk. Organic Facts, Organic Information Services Pvt. Ltd. Karnataka, India. http: //www.organicfacts.net/organic-animal-products/organic- milk/health-benefitsof-organic-milk.html

Berentsen, P. B. M., Kovacs, K. and Van Asseldonk, M. A. P. M. (2012). Comparing risk in conventional and organic dairy farming in the Netherlands: An empirical analysis. Journal of dairy science, 95(7), 3803-3811.

Bodapti, S. and Chander, M. (2013). Integrating indigenous knowledge of farmers for sustainable organic farming: An assessment in Uttarakhand state of India. Indian Journal of Traditional Knowledge, 12(2), 259 264.

Chander, M., Bodapati, S., Mukherjee, R. and Kumar, S. (2011). Organic livestock production: an emerging opportunity with new challenges for producers in tropical countries. Rev. sci. tech. Off. int. Epiz., 30(3), 569-583.

Codex Alimentarius (2007). Organically produced foods, 3rd Ed. Codex Alimentarius Commission, World Health Organization (WHO)/Food and Agriculture Organization of the United Nations (FAO), Rome, pp. 51

Cranfield, J., Henson, S. and Holliday, J. (2010). The motives, benefits, and problems of conversion to organic production. Agriculture and Human Values, 27(3), 291-306.

Department of Animal Husbandry, Dairying and Fisheries, Ministry of Agriculture. Retrieved from http: //www.dahd.nic.in/documents/statistics/livestock-census on 13/02/20.

Garmo, R. T., Waage, S., Sviland, S., Henriksen, B. I., Osterås, O. and Reksen, O. (2010). Reproductive performance, udder health, and antibiotic resistance in mastitis bacteria isolated from Norwegian Red cows in conventional and organic farming. Acta Veterinaria Scandinavica, 52(1), 11.

Ghosh, S. K. (2006). Scope of integrated organic farming in India with special reference to North Eastern States, In: Organic Animal Husbandry, Concept,

Standards and Practices, Division of Extension Education, IVRI, Izatnagar, UP; pp. 54-56.

Herath, C. S. and Wijekoon, R. (2013). Study on attitudes and perceptions of organic and non- organic coconut growers towards organic coconut farming. Idesia, 31(2), 5-14.

Kamboj M.L. and Prasad, S. (2013). Organic Dairy Farming in India: Prospects, Practices and Constraints. In: New Paradigms in Livestock Production from Traditional to Commercial Farming and Beyond. (Eds: S. Prasad, A. Kumaresan, S.S. Lathwal, M. Bhakat and A. Manimaran). Agrotech Publishing Academy, Udaipur, India. pp. 634- 647.

Kimming, M., Sundberga, C., Nordberga, Å., Bakyb, A., Bernessona, S., Hanssona, P. (2014). Replacing fossil energy for organic milk production – potential biomass sources and greenhouse gas emission reductions. Journal of Cleaner Production doi: 10.1016/j.jclepro.2014.03.044

Kumar, N., Sawant, S., Malik, R.K., Patil, G. (2005). Development of analytical process for detection of antibiotic residues in milk using bacterial spores as biosensors (Patent Reg # IPR/4.9.1.41 05074114791 del/2006).

Lairon, D. and Huber, M. (2014). Food quality and possible positive health effects of organic products. Food Quality and Possible Positive Health Effects of Organic Products. In: Organic Farming, Prototype for Sustainable Agricultures (Eds. S. Bellon and S. Penvern). Springer, Netherlands. pp. 295-312.

Lampkin, N. (1990). Organic Farming, Farming Press Books, Ipswich, UK.

Maji, S., Meena B. S., Paul P and Rudroju, V (2017). Prospect of organic dairy farming in India: A review. Asian J. Dairy and Food Res, 36 (1): 1-8.

Mercola, J. (2014). Health Benefits of Organic vs. Conventional Milk. Mercola.com. http: //openlibrary.org/

MFAL (2013). The National Action Plan for Organic Farming 2013- 2016. Ministry of Food Agriculture and Livestock, Ankara, Turkey. Available from: http: //www. tarim.gov.tr/BUGEM/Belgeler/Duyurular/UlusalOrganikTarimEylemPlani.

Mongeon, M., and Summerhayes, B. (2010). *Organic dairy production*. Guelph: Ontario Ministry of Agriculture, Food and Rural Affairs.

Nemes, N. (2009). Comparative analysis of organic and non-organic farming systems: A critical assessment of farm profitability. Food and Agriculture Organization of the United Nations, Rome.

Odhong, C., Wahome, R. G., Vaarst, M., Kiggundu, M., Nalubwama, S., Halberg, N. and Githigia, S. (2014). Challenges of conversion to organic dairy production and prospects of future development in integrated smallholder farms in Kenya. Livestock Research for Rural Development, 26(7), 14-21.

Pathak, P. K., Chander, M. and Biswas, A. K. (2003). Organic meat: an overview. Asian- Australasian journal of animal sciences, 16(8), 1230- 1237.

Sangkumchalianga, P. and W. C. Huang (2012). Consumers' perceptions and attitudes of organic food products in Northern Thailand. Int. Food. Agribus. Manage. Rev. 15: 87-102.

Singh, A. K. (2007). Conversion to Organic Agriculture, published by, Intemational Book Distributing Co., Lucknow, U.P.

Singh, M., Sharma, D. K., Mishra, U. K. (2011). Need of Organic Farming. In: Organic Dairy Farming. Satish Serial Publishing House, New Delhi, India.

Subrahmanyeswari, B. and Mahesh, C. (2008). Livestock production practices of registered organic farmers in Uttarakhand State of India. In Cultivating the Future Based on Science. Proceedings of the Second Scientific Conference of ISOFAR.

Wolde, D. T and Tamir, B (2016). Organic Livestock Farming and the Scenario in the Developing Countries: Opportunities and Challenges. Global Veterinaria, 16: 399-412.

Yadav, A. K. (2008). The strategic significance of organic farming to India, In Global Organic Agribusiness India Arrives! Papers presented at India Organic 2008 Seminar, Westville Publishing House, New Delhi.

Chapter 12

Status of Production, Availability and Strategies for the Promotion of Organic Inputs in India

Arshdeep Singh[1] and Amanpreet Singh[2]*

[1]Department of Economics and Sociology, Punjab Agricultural University, Ludhiana, Punjab
[2]Department of Agronomy, Khalsa College, Amritsar, Punjab

In India, after green revolution usage of synthetic chemicals is increased day by day. The higher usage of chemicals, effects the crop yield and soil health negatively. Now a days, the different type of organic manures are popular in India. The scenario regarding usage of organic inputs helps to understand the dynamics of organic farming over the conventional farming. It also helps to understanding the increasing pattern of organic inputs and biofertilizers.

Keywords: Organic inputs, Biofertilizer, Organic manure, Availability, Production.

Background

The government of Indian started the "National Project on Development and Use of Biofertilizer" during 1983 to promote sustainable agriculture, cost-effective organic inoculant like Azospirillum, Azotobacter, Rhizobium, and Phosphorous-solubilizers. Biofertilizers are the preparation of living beneficial microorganisms which is the component of integrated nutrient management. Various promotional programs and activities are organized under the mission, targeting the promotion of biofertilizers and organic farming on large scale since a lot of efforts done under the National Project on Development and Use of Biofertilizers from 1983-84 to September 2004, biofertilizers are now broadly used in the cultivation of cereals, cotton, oilseeds, sugarcane, legumes, millets, fruits, vegetable, horticultural and plantation crops. Understanding the importance of the scheme, In October 2004 GOI started a new Central Scheme entitled "National Project on Organic Farming". Due to severe attempts are made under the schemes and overall Biofertilizer production in the country, which remained at less than 500 tons/annum in 1984-85, was increased to approximately up to 10,000 tones/annum in 2003-04.

In a vision of the increasing demand for biofertilizers in the country several Centre Govt. Institute, State Department of Agriculture, Public sector, State Agricultural Universities (SAUs), NGOs, and private organizations are involved in the production of biofertilizers on a large scale. All these official doms are directly or indirectly associated with the National Centre of Organic Farming in terms of strain supply and other technical difficulties relating to production and promotion of biofertilizers. The present level of biofertilizer production in the country is estimated to be about 1,48,499 tons per annum while the production capacity is around 3,35,303 tons per annum in the year 2017-18 in the country.

Introduction

India has cultivated a large number of crops and has different agro-climatic regions so specific crops are cultivated according to their scenario (Butterworth *et al.*, 2003; Reddy, 2010). In many regions, the application of synthetic chemicals is quite low and has a huge area that enjoys organic by default. These farmers use traditional practices for cultivation of crops and due to huge investment and research in the field of organic farmers formulate standardize, and collection of these traditional and indigenous techniques or practices from all over the country for the cultivation of crops. These techniques contain farm inputs and some items from the local area due to the production of these organic inputs are quite cheaper than synthetic chemicals. Moreover, these organic inputs are eco-friendly, pollution-free, improve soil fertility and health, cost-effective, easy to produce, easy to use, and having a low maximum residual limit. With the increase in time, both area and production started increasing in organic farming due to the higher demand for organic products in both national and international market and farmers earns remunerative price of organic products. With the help of the NCOF scheme, govt. aware of the farmers regarding benefits of organic inputs, how to prepare them, and when to apply it. Thus, many public and private players come in the field production of biofertilizer and organic manures due to control and regulate wrong practices in trade, price, and distribution of fertilizers and to ensure quality production of organic inputs in the country thus government introduce a regulatory mechanism for biofertilizers, organic inputs and manures under Fertilizer Control Order (FCO), 1985 and to the continuous supply of fertilizers to the farmer's Govt. put fertilizers as an essential commodity under Essential Community Act 1955.

Definition by Fertilizer Control Order, 1985

Biofertilizers

"Biofertilizer means the product containing carrier-based (solid or liquid) living microorganisms which are agriculturally useful in terms of nitrogen fixation, phosphorus solubilization, or nutrient mobilization, to increase the productivity of the soil and/or crop".

Organic Fertilizers

"Organic fertilizers mean substances made up of one or more unprocessed material(s) of a biological nature (plant/animals) and may include unprocessed mineral materials that have been altered through microbiological decomposition process".

Need for Organic Inputs

In organic agriculture, the initial stage is to improve the fertility and health of the soil. This can be accomplished by a complete prohibition on the application of inorganic inputs and amplified the application of organic inputs and fertilizers. For soil fertility, health and integrated nutrient management are improved by crop residue, forest leaf litter, slaughterhouse waste, bone meal, blood meal, animal dung, and green manures are significant organic materials. All organic sources are required to be composted accurately for maximizing the effect. The nutrient value of the raw material and composting process, procedure, and methodology decided the quality of produce. Biofertilizers and other microbiological inputs paying a lot of attention and production are being encouraged on a large scale. Under the National Centre on Organic Farming (NCOF) funding are available for establishing production of vermiculture, agro-waste compost, Vegetable market waste, and biofertilizers. Under different schemes of Central and State Governments, aids are available for the establishment of vermicompost production units.

Status of Biofertilizer Production in India

Different categories of biofertilizers are being produced by numerous firms in the nation. Different categories are Azospirillum, Rhizobium, Acetobacter, Azotobacter, Phosphate Solubilizing Bacteria (PSB), Potassium Mobilizing Biofertilizers (KMB), Zinc Solubilizing Bacteria (ZSB), Vesicular ArbuscularMycorrhizal (VAM) and NPK consortium are most predominant and determined as biofertilizers whereas other inoculants include all other microbial formulations such as compost enrichers (Trichoderma, Paceliomyces, *etc.*), Blue-Green Algae (BGA), Azolla, Plant Growth Promoting Rhizobacteria (PGPR), and biopesticides. Overall production details of these Biofertilizers and biofertilizers + other inoculants from the year 2002-03 to 2017-18 are given in Table 12.1.

Total production of biofertilizer and other inoculants was observed huge growth in production in the year 2002-03 it was 9065 MT to 157198 MT in the year 2017-18. It is the result of the untiring efforts of extension staff, scientists, and other govt. official's efforts to achieve the goals and huge achievement in the field of biofertilizers. Now further emphasis will be given on the production of biopesticides to protect the crops from insect pest attacks.

Table 12.1: Total Production of Biofertilizers and other Inoculants in India from 2002-03 to 2017-18

Year	*Biofertilizer Production*	*Other Inoculants Production*	*Total Production of Biofertilizer and other Inoculants*
2002-03	9064.79	NA	9064.79
2003-04	9798.89	NA	9798.89
2004-05	10600.69	12065.24	22665.93
2005-06	10764.42	9027.45	19791.87
2006-07	16145.26	11211.66	27356.92
2007-08	20111.05	18517.80	38628.85
2008-09	25065.03	38497.48	63562.51
2009-10	20040.35	45475.79	65516.14
2010-11	37997.61	69137.39	107135
2011-12	40324.21	28854.13	69178.34
2012-13	46836.82	13003.44	59840.26
2013-14	65527.87	162.01	65689.88
2014-15	68910.77	28165.95	97076.72
2015-16	101521.3	11222.20	112743.5
2017-18	148499.2	8699.10	157198.3

Note Production in MT;

Source: Compiled by author from the various annual reports of NCOF.

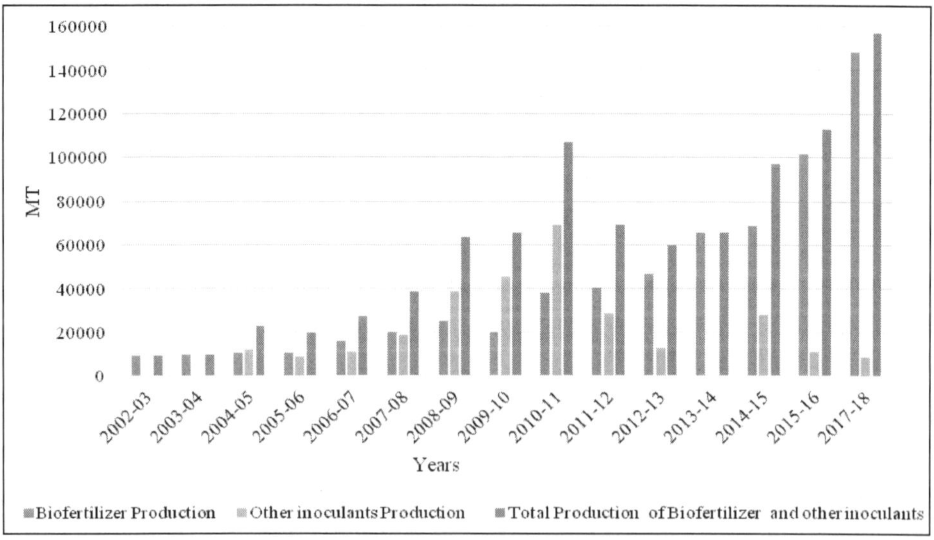

Figure 12.1: Year-wise Status of Biofertilizer Production.

Many organic inputs are prepared and formulated on farmer farms these are *beejamrit, Matkakhad, ghanjeeva, jivamrit,* Fermented buttermilk, *vermiwash, etc.* These organic inputs are prepared by farmers. They are properly trained under several schemes by SAUs and the Department of Agriculture to formulate these organic inputs to ensure less expenditure on organic nutrients and plant protection measures in ultimate for higher return and wellbeing of farmers.

Bio-fertilizers production has enhanced from 9064.79 Metric Tons (MT) to 157198.3 MT from 2002-03 to 2017-18 in the country. The production of other inoculants in the year 2004-05 was 12065 MT with continuous up and down to 8699 MT annual production based on demand. It is cleared that data provided by NCOF for production of biofertilizers is divided into two parts liquid base production and carrier base production after the year 2013-14 and later on but total production is calculated by the formula given in annual report 2013-14.

Status of Organic Manures and Fertilizers Production in India

Different types of Organic Manures and Fertilizers are being produced by many companies in the country. Between Rural compost, City compost, Farmyard manure (FYM), Vermicompost, and Phosphate rich organic manure (PROM) are most main organic manures and fertilizers. Overall production details of these organic manures and fertilizers from the year 2007-08 to 2017-18 are presented in Table 12.2. Rural compost production was 1693 lakh tones in 2007-08 and declined tremendously in 2017-18 with annum production of 377.36 lakh tones due to increasing production of vermicompost which was only 31 lakh tonnes in the year 2007-08 and highest production in the year 2017-18 with 775 lakh tones annum production. Urban compost production was at a peak in the year 2009-10 with an annual production of 619 lakh tones which is all-time high approximately 3 to 6 times than normal production of urban compost.

Table 12.2: Total Production of Organic Manures and Fertilizers in India under Various Sub Categories from 2007-08 to 2017-18

Year	*Rural Compost*	*Urban Compost*	*Farm Yard Manure*	*Vermi-compost*	*Other Manures*	*Total Manure Production*
2007-08	1693.24	152.64	1862	30.95	92.08	3830.91
2008-09	486.39	255.88	1302.38	55.79	35.2	2135.64
2009-10	804.49	618.96	1220.65	73.72	40.62	2758.44
2010-11	797.54	111.24	2234.44	114.16	131.82	3389.2
2011-12	939.85	140.86	1860.64	268.5	35.43	3245.28
2012-13	1039.35	143.04	2850.5	53.7	29.16	4115.75
2013-14	224.97	63.33	9647.8	582.27	24.27	2294.14
2014-15	225.88	64.22	1400.39	583.2	24.93	2298.62
2015-16	231.4	79.45	1409.59	591.76	126.74	2438.94
2017-18	377.36	110.46	1885.92	775.01	238.45	3387.2

Note: Production in lakh tones;

Source: Compiled by author from the various annual reports of NCOF.

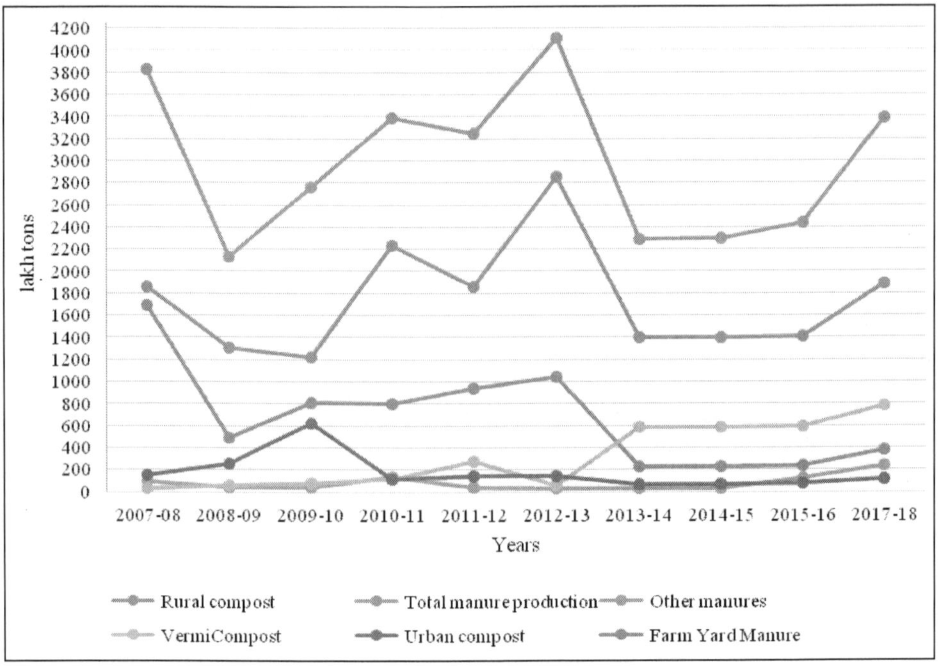

Figure 12.2: Total Production of Organic Manures and Fertilizer.

Farmyard manure is an important organic manure in rural areas and is widely used by farmers before sowing of crops and easily available and produced as compared to other compost, manures, and fertilizers. It is also used by an inorganic farming system but farmers are unable to apply the recommended application of farmyard manure in both organic and inorganic farming. FYM production was lowest in the year 2009-10 and highest in the year 2010-11 with annual production of Farmyard manures 1221 and 1886 lakh tones, respectively.

Vermicompost production was very less in initial years but due to increasing awareness, the importance of vermicompost production start inclining and reach its highest level in a year in 2017-18 with an annual production of 775 lakh tones which was 25 times higher production than the year 2007-08. This is a great achievement and follows this trend to fulfill demands and continuously adds more area under organic farming. Other manures include sheep and goat manure, poultry manure, and others which is not added above. The production is the highest level in the year 2017-18 and lowest in the year 2013-14 with production of 238 and 24 lakh tones per annum, respectively.

Total manure production shows less fluctuation in production, it was highest in the year 2012-13 and lowest in the year 2008-09. Total manure production shows steady growth, there are many reasons for steady growth such as demand is quite low only organic growers used organic fertilizers and most of the organic farmers prepare most of organic manures and fertilizers on their farm to reduce their production cost moreover they also sell these organic inputs in the local area to other farmers. Government of India encourage farmers to produce organic inputs on their farms due to this several training camps and other activities organized by state university and Department of Agriculture from time to time to aware farmers from new formulations and their production methods. These on-farm productions of organic inputs are no added into data due to they are having any license and use their inputs on their farm not commercially. So the production of this organic manure and compost having many times higher than reported in NCOF annual reports moreover some institute doesn't provide data on time and NCOF reports it as zero.

Promotion of Organic Input Use

India has also initiated a vermicomposting program of Municipal Solid Waste (MSW). In recent years it is emergent as a part of 'sustainable non-chemical agriculture', 'village and farm waste management', jointed with 'poverty eradication' program (Sinha *et al.*, 2010). It has improved the lives of poor peoples of India and created self-employment opportunities for the unemployed population. In many villages of India, NGO's are freely distributing cement tanks and 1000 earthworms and urging men and women to collect waste from villages and farmers, do both composts them and sell earthworms and compost to the farmers (Suhane, 2007; NIIR, 2009). Department of Agriculture is giving a 50 per cent subsidy for farmers towards the construction of vermisheds and beds to encourage organic farming through composting. India, especially in U.P., Punjab, and Haryana, residue management techniques were developed and associated benefits were spread, thus many farmers are adopting in-situ incorporation of crop residues. With the advent of mechanization and availability of rotavators, it has become a regular practice of in-situ incorporation in the maize-groundnut system in South India among growers. These practices ultimately result in enhanced C-Sequestration.

Area Covered by Organic Manures

Different organic manures have a different composition and have a different portion of nutrient content. Thus keeping this point in mind, all manure have the different recommended quantity to provide a certain amount of nutrients to plant in direct and indirect ways. It is cleared from Table 12.3 area covered by all manures declines continuously the main reason behind this is not due to a decline in total production of all manures but due to the use of the higher application as compared earlier. Before the start of NCOP and NCOF program there is not any recommended limit to apply various manures but with the time researcher and scientist prepared package of practices for different crops and by doing various experiments give clear cut recommendation for the application of different manures in different crops and also a time of application based on the duration of crops. Different crops have different nutritional requirements moreover scientists also standardized these manures for better outcomes.

Table 12.3: Area Covered by different Manures from 2007-08 to 2015-16

Year	*Area under Green Manure*	*Area Covered Organic Manures**	*The Total Area Covered by all Manures*
2007-08	13.03	1694.82	1707.85
2009-10	26.57	1186.10	1212.67
2010-11	20.62	526.07	546.69
2013-14	12.34	458.83	471.17
2014-15	12.81	229.86	242.67
2015-16	12.82	328.73	341.55

Note: Area is in Lakh hectare, Manures* means Rural compost, Urban compost, Farmyard manure (FYM), Vermicompost, Phosphate rich organic manure, and other manures.

Source: Compiled by author from annual reports of NCOF

Green manures are generally grown in the period of two main crops to nourish the soil health, soil fertility, maintain soil microorganism, optimize organic content of the soil, reduce bulk density, improves soil carbon content, increase micronutrient content and mobility. Generally, green manure crops are grown and incorporate in the soil called a "green manure" and process called "green manuring".

It is of two types

i) Green manure crop

ii) Green leaf manures

Crops which generally grown in India as Green manure crops are dhaincha, sunhemp, cowpea, wild indigo, and Green leaf manures are glyricidia, karanji, neem, calotropis, avise, subabul, and other shrubs, respectively.

In India, green manure was grown on 13 lakh hectare in the year 2007-08 which is more or less similar in the year 2015-16 with an area of 12.82 lakh hectare. In 2010-11 green manure was cultivated in an area of 2.1 million hectares. (Chanda *et al.*, 2011). The utilization of organic manures on area basis declines from 1695 to 329 lakh hectare from 2007-08 to 2015-16, respectively. The total area covered by all manure including green manure is also declined with a similar trend as a decline in green manure and organic manure*. The total area covered by all manures was 1708 lakh hectare in the year 2007-08 which falls continuously and in the year 2015-16 it was 342 lakh hectare.

Conclusions

Soil and climatic conditions are the two suitable factors favoring organic farming in Indian drylands. Marginal and small farmers don't respond well to intensive farming system. Instead they prefer low input and cost effective farming system which make abundant use of biodiversity. This helps to improve the economic health of that farm. Along with economic health it helps to improve the ecological health of the farm by improving the soil health, water holding capacity of the soil and by enhancing their ability to supply nutrients. All these factors account in removing constraints in cultivation crops in certain zones and helps in improving the ecological and economic health of the farms. Semi-arid and arid dryland soils generally poor in water-holding ability as well as organic fertilizers and sources. In certain zones, the depth of the soil is another constraint for the cultivation of crops. Accumulation of organic inputs will not only enhance the physical condition of these dryland soils but also extraordinarily improve their capability to supply balanced plant nutrients. The selection of organic farming practices can assist with improving soil fertility, health, and productivity improper land use, leading to desertification.

REFERENCES

Anonymous. Annual reports of the National Centre of Organic Farming from 2004-05 to 2015-16 and 2017-18 (https: / /ncof.dacnet.nic.in/ reports.html)

Butterworth, J., Adolph, B., and Suresh Reddy, B. (2003) How Farmers Manage Soil Fertility. A Guide to Support Innovation and Livelihoods. Andhra Pradesh Rural Livelihoods Programme, Hyderabad/Chattam: Natural Resources Institute, UK.

Chanda G.K., Goutam B., and Chakraborty S.K. 2011. The effect of vermicompost and other fertilizers on the cultivation of Tomato plants. *Journal of Horticulture and Forestry*, 3(2): 42–45.

NIIR. 2009. The complete technology book on vermiculture and vermicompost, Publication of National Institute of Industrial Research, New Delhi, India, ISBN 8186623817.

Reddy Suresh, B. (2010) Assessment of Economic and Ecological Returns from Millet- based Bio-diverse Organic Farms vis-à-vis Conventional Farms, CESS Monograph Series No.8, Centre for Economic and Social Studies, Hyderabad.

Sinha, R.K., Herat, S., Valani, D., and Chauhan, K. (2010) Earthworms - the environmental engineers: a review of vermiculture technologies for environmental management and resource development, *Int. J. Global Environmental Issues*, Vol. 10, Nos. 3/4, pp.265–292.

Suhane R.K. (2007) Vermicompost (In Hindi). Publication of Rajendra Agriculture University, Pusa, Bihar, 88 pp. (www.kvksmp.org)

Chapter 13

Organic Farming: An Ecofriendly Approach Towards the Sustainable Agricultural Development

Kishor Kumar Sahu

Department of Soil Science and Agricultural Chemistry, College of Agriculture, CSK Himachal Pradesh Agricultural University, Palampur – 176062, Himachal Pradesh

Agriculture plays a vital role in a developing country like India. Apart from fulfilling the food requirement of the growing Indian population, it also plays a role in improving economy of the country. The development of the organic farming worldwide had gone through three stages, emergence, expansion, and growth. Organic farming through sustainable agriculture meets not only the food requirements of present generation in an environment friendly way but also the requirements of future generations and maintains our environment. Modern agriculture involving high-yielding varieties are being used with infusion of irrigation water, fertilizers, or pesticides. It causes negative impact on environment by affecting soil fertility, water hardness, development of insect resistance, genetic variation in plants, increase in toxic residue through food chain and animal feed thus increasing health problems and many more serious health concerns and degradation of environment. Organic farming provides macronutrients and micronutrients to the plants and also improves soil physical, chemical and biological characteristics of soil. Organic farming is a holistic production management system which promotes and enhances agro-ecosystem health, including biodiversity, biological cycles and soil biological activity. The production of these organic crops and products is viewed with regard to sustainable agricultural development. There are existing problems for the further development of the world's organic farming, as well as the development status, problems and strategies of the organic farming.

Keywords: *Organic farming, Economy, Sustainable agriculture, Ecosystem and Biodiversity etc.*

Introduction

Organic farming is an option in contrast to ordinary or conventional farming. It utilizes manure, compost, green excrement, bone feast instead of utilizing fertilizers and pesticides. This framework utilizes organic squanders and yields are brought up in such a way, that it keeps the soil an alive entity. Microorganisms are utilized as bio-composts to build creation without dirtying nature. Organic farming advances eco-accommodating rural practices without utilizing manufactured sources of info and significantly depends upon the utilization of organic squanders to raise crops. Organic farming professes to can possibly give benefits as far as ecological security, preservation of non-inexhaustible assets, improved food quality, decrease in yield of surplus items and the reorientation of agriculture towards zones of market request (Auobamiri and Ahmad 2014). Some European governments have perceived these likely advantages and reacted to them by urging farmers to embrace organic farming practices, either straightforwardly through money related motivating forces or in a roundabout way through help of examination, expansion and advertising activities. Be that as it may, farmer's choices on whether to change from conventional to organic farming have not been concentrated broadly hitherto.

The organic development in India has its birthplace in crafted by Howard who figured and conceptualized the vast majority of the perspectives which were later acknowledged by those individuals who got dynamic in this development. Organic farming is a creation framework which dodges, or to a great extent bars, the utilization of engineered manures, pesticides, development controllers, and domesticated animals feed added substances. The targets of ecological, social, and financial maintainability are the fundamentals of organic farming. The key attributes incorporate ensuring the drawn out fruitfulness of soils by keeping up organic issue levels, cultivating soil natural action, cautious mechanical mediation, nitrogen independence using vegetables and natural nitrogen obsession, successful reusing of organic materials including crop buildups and domesticated animals squanders and weed, and ailments and nuisance control depending principally on crop revolutions, characteristic predators, assorted variety, organic manuring and safe assortments. An extraordinary accentuation is set to keep up the soil fertility by restoring all the losses to it mostly through manure to limit the hole between NPK expansion and Removal from the dirt.Today, the prospering populace pressure has constrained numerous nations to utilize synthetic compounds and fertilizers to expand the homestead efficiency for meeting their ever-expanding food prerequisites. The delayed and over use of synthetic substances has, in any case, brought about human and soil wellbeing dangers alongside ecological contamination. Farmers in the created nations are, in this way, being urged to change over their current homesteads in to organic homestead.

Agriculture assumes a crucial job in a creating nation like India. Aside from satisfying the food necessity of the developing Indian populace, it additionally assumes a job in improving economy of the nation. The Green Revolution innovation appropriation between 1960 to 2000 has expanded wide assortments of rural harvest yield per hectare which expanded 12-13 per cent food gracefully in creating nations. Sources of info like fertilizers, pesticides helped a great deal in such manner.

However, regardless of this reality, food frailty destitution despite everything wins unmistakably in our nation. Employments of substance biopesticides and manures have caused negative effect on condition and expanding the medical issues and some more. India has been customarily rehearsing organic agriculture yet present day agriculture rehearses have pushed it to dividers. This impact sly affects crop development and wellbeing and treats organic waste in a domain agreeable manner.

History of Organic Farming

Since the start of the 21st century, the improvement of organic farming worldwide has indicated a solid development. Very nearly 23 million hectares are overseen organically worldwide in 2002. Contrasted with 2000 this is an expansion of 3 1.4 per cent. The all out world retail deals of organic items came to US$25 billion of every 2002, which expanded by 42.9 per cent contrasted with 2000. Organic agriculture is rehearsed in practically all nations of the world, and a lot of agricultural land and farms are developing. The market for organic items is developing, not just in the significant markets like Europe, North America, and Japan, yet in addition in numerous different nations, including developing nations.

Emergence and Development of Organic Farming

The development of the organic farming worldwide had gone through three stages

a. Emergence
b. Expansion
c. Growth in chronological sequence.

Stage of Emergence (1924-1970)

The start of organic farming could follow back to 1924 in Germany with Rudolf Steiner's seminar on Social Scientific Basis of Agricultural Development, where his hypothesis considered the individual as an integral part of an astronomical balance that he/she should comprehend so as to live in congruity with the earth. In this manner, equalization must be struck between the profound and material side of life.

H. Pfeiffer applied these hypotheses to agriculture and brought forth biodynamic agriculture. It was created toward the finish of the 1920s in Germany, Switzerland, England, Denmark and the Netherlands. In Switzerland in 1930, legislator Hans Mueler offered impulse to organic-natural agriculture. His objectives were on the double monetary, social and political as they imagined autarchy of the farmer and a considerably more immediate and less jumbled association between the creation and utilization stages. Maria Mueler applied these speculations to plantation creation. Austrian specialist Hans Peter Rush adjusted these thoughts and fused them in a technique established on maximum use of sustainable assets. Hans Mueler and Hans Peter Rush established the hypothetical framework for the organic-natural agriculture and its advancement in the Germanic talking nations and districts.

Sir Albert Howard was the organizer of the organic farming development. His book "An Agricultural Testament" summed up his exploration works of 25 years at Indore in India, where he built up the celebrated Indore treating the soil procedure, which put the old craft of fertilizing the soil on a firm logical premise and clarified the connection between the soundness of the soil, the wellbeing of plants and the strength of creatures.

J.I. Rodale started his exploration and practice on organic farming in the United States of America. His essential objective was to create and show useful techniques for modifying regular soil fruitfulness. By 1942, he distributed the magazine "Organic Gardening". Woman Eve Balfour began the experiment - the primary investigation contrasting customary and regular farming techniques. Her thoughts motivated the development of the Soil Association that was established in 1946 in England. The Soil Association endeavored to return humus and soil fertility to their fundamental spot in the organic equalization and biological balance. It was established on the speculations spread by Sir Albert Howard in his agrarian confirmation of 1940.

During 1950-1960s, because of specialists and customers whose mindfulness continually developed with respect to food and its impact on wellbeing, organic fanning started to grab hold in France. Nature and Progress Association was established. Mokichi Okada began characteristic agriculture in 1935 in Japan. His fundamental contemplations were to regard and underscore the capacity of nature and soil in the farming creation, and to facilitate the relationship between person and nature through expanding soil humus to get the yields without fertilizers and agrarian synthetic concoctions. The ecological and medical problems exacerbated during the 1950s-1960s of the only remaining a very long time in Japan encouraged the advancement of normal agriculture. The basics of characteristic farming turned into the significant substance of Japanese agrarian, standard of organic rural items.

Stage of Expansion (1970-1990)

The exploration and practice of organic agriculture extended worldwide after the 1960s. Specifically, the extension and double extremity of organic horticulture began with the oil emergency of 1973 and the developing affectability to agro environmental issues. This was a period of new thoughts, critical sociological changes, fight developments and the multiplication of elective ways of life. The new contemplations as far as utilizing common assets sanely, ensuring the earth, acknowledging low in-put and high proficiency, guaranteeing food security, coming back to the earth and keeping up a feasible improvement of agriculture, for example, organic, organic-biological, bio-dynamic, ecological and natural agriculture, were strikingly evolved in their ideas, research and useful exercises. William Albrecht gave a meaning of ecological agriculture in 1970, in which the natural standard was acquainted with the creation arrangement of organic agriculture. In England the Soil Association made a logo and in equal presented the thought of legitimately defined particulars and quality controls that gave a lawfully restricting assurance for the purchasers (Behera *et al.*, 2011).

The biggest non-administrative association of organic agriculture on the world - IFOAM (International Federation of Organic Agriculture Movements) was established in 1972. The significant organic agriculture affiliations and examination organizations on the world, for example, FNAB (Federation Nationale d' Agriculteurs Biologiques), FiBL (Forschungs institute for Biologischen Landbau), presently the biggest organic exploration foundation around the world, were established during 1970s-1980s. These associations assumed a significant job in normalizing the creation and market of organic items and advancing examination and customer's mindfulness. The administrative activity on organic farming began continuously in the various nations and areas as the rules for organic farming.

In the United States the guideline on organic farming was executed in the province of Oregon in 1974 and in the territory of California in 1979, separately. The United States Department of Agriculture (USDA) made an examination for an enormous scope on organic farming in the 69 organic farmes of 23 states and distributed the Report and Recommendations on Organic Farming, wherein the improvement status and potential stayed as issues and the exploration headings were broke down. In this report the definition and rule for the organic farming were given, and an activity plan for the improvement of organic farming was called for. The distribution of this report was an achievement in enactment and improvement of organic farming in the United States. In France, the organic farming guideline was executed in 1985.

Stage of Growth (since 1990)

The organic farming overall entered another phase of development during the 1990s. The exchange associations for organic items were established, organic farming guidelines were executed, and organic farming development was advanced by legislative and non-administrative associations. In 1990, the first BioFach Fair - presently the greatest reasonable for organic items around the world, rose in Germany. The national legislature of the United States distributed the guideline for organic food items in 1990 (Behera *et al.*, 2011).

The European Commission received EU guideline 2092/91 on organic farming in 1991. This guideline became a law in 1993 and was conceded in practically all European Union nations since 1994 (IFOAM 2010). In the North America, Australia and Japan the significant markets for organic items, distributed and actualized organic guidelines in progression (Behera *et al.*, 2011). The International Federation of Organic Agriculture Movements (IFOAM) and the Food and Agriculture Organization of the United Nations (FAO) set out Guidelines for the Production, Processing, Labeling and Marketing of Organically Produced Foods in 1999. This rule is of significance to universal harmonization of the organic farming principles.

Organic farming had quickly evolved overall during this stage. The fundamental drivers of consistent market and creation development were the responsibility of numerous retail chains just as positive arrangement conditions. Together these had made conditions preferring an amicable increment in flexibly and request. The state support for organic farming examination and lawful structure was progressively picking up significance since the finish of the 1990s.

Organic agriculture is all encompassing creation the executives frameworks which advances and improves agro-eco-framework wellbeing, including biodiversity, natural cycles, and soil natural movement. It underlines the utilization of the executives rehearses in inclination to the utilization of off-farm inputs, considering that local conditions require privately adjusted frameworks. This is cultivated by utilizing, where conceivable, social, organic and mechanical strategies, rather than utilizing manufactured materials, to satisfy a particular capacity inside the system. Terms, for example, Organic, Biological, Biodynamic, and Ecological are perceived as organic farming in the EU guidelines (Behera *et al.*, 2011). Albeit organic farming is one among the expansive range of techniques which depend on the particular and exact norms with various names, for example, organic, organic, organic-natural, bio-dynamic, common and biological agriculture, there are some common followed standards in the organic agriculture.

These standards are summed up as follows (IFOAM 2010):

- ✰ Maintain long-term soil fertility though biological mechanism.
- ✰ Recycle wastes of plant and animal origin in or- der to return nutrients to the land, thus minimizing the use of external inputs outside systems, and keep the nutrients cycle within the system.
- ✰ Prohibit the use of synthetic materials, such as pesticides, mineral fertilizers, chemical ingredients and additives.
- ✰ Using natural mechanism and rely on renewable resources to protect the natural resources.
- ✰ Raise animal restricted in areas and guarantee the welfare of the animals.
- ✰ Adapt local environment and diversified organization.
- ✰ Produce high quality products

Organic Farming

As per the definition of the USDA study team on organic farming "organic farming is a system which avoids or largely excludes the use of synthetic inputs (such as fertilizers, pesticides, hormones, feed additives *etc.*) and to the maximum extent feasible rely upon crop rotations, crop residues, animal manures, off-farm organic waste, mineral grade rock additives and biological system of nutrient mobilization and plant protection". In another definition FAO suggested that "Organic agriculture is a unique production management system which promotes and enhances agroecosystem health, including biodiversity, biological cycles and soil biological activity and this is accomplished by using on-farm agronomic, biological and mechanical methods in exclusion of all synthetic off-farm inputs".

Method of Organic Farming

Organic farming includes different procedures which are ecofriendly and by rehearsing it the richness of soil is preserved for long time. There different strategies in organic farming some of them are crop rotation, utilization of green manures, natural irritation control and treating the soil, these likewise give work to agriculture workers (Madhusudhan, 2016).

Employment to agriculture labors -In present day machinery are replacing man power and making them unemployed but with organic farming it provides employment because many techniques are used, from preparation of manure to cop harvesting.

Crop rotation - It is a technique of growing different crops in same area according to the seasons and it is practiced to avoid agriculture pests, and to maintain soil fertility.

Green manures - Green manures are the plant leaves and waste material of plant which cover the soil and stuffed in to soil and become as nutrient to the soil and increase the soil fertility.

Vermicomposting- It is a process of composting using different worms like white worms, earth worms and red wrigglers for preparation of compost with mix of kitchen waste and other vegetable waste. This is rich in nutrients and used as fertilizers in the agriculture fields.

Biological pest control - Living organisms are used to protect plants from peats without synthetic chemicals.

Soil management- Soil is the foundation of terrestrial life. Specific soil management practices are needed to protect and conserve the soil resources. Natural microorganisms such as mycorrhiza form a symbiotic relationship with plant roots and take nutrients from the soil which the plant roots are not able to access to. The act of growing different type of crops in a particular area season after season is known as crop rotation. If same crop is grown every season in a particular area, it will deprive the soil of one particular nutrient but if we will grow different crops every season, it will help in maintaining the nutrient balance of soil. Thus, crop rotation helps in maintaining soil quality (Kumar and Singh, 2015). Leguminous plants are also used to increase soil fertility as the fix atmospheric nitrogen with the help of rhizobia bacteria. Addition of manure also enhances soil quality as it contains nutrients such as nitrogen.

Weed management- Weeds are the undesirable plants that compete with the crop plants. Organic farming promotes weed management in a number of ways. Weed growth is blocked using plastic films and the process is known as mulching. Mowing and cutting removes the top growth of weeds. Grazing is another method which helps in reducing weed growth. Organic crop rotation also promotes weed suppression.

Crop diversity- In ancient times, the practice of growing only one type of crop was followed. But now-a-days, poly-culture is coming in trend *i.e.* the practice of growing multiple crops within the same place. It improves the quality of soil by supporting beneficial soil microorganisms.

Other organisms' control - Following methods can be adopted for control of other organisms and encourage beneficial predatory insects that feed on pests and helps in controlling them

- Make use of organic pesticides and herbicides causing less pollution to the environment

- ☆ Field sanitization to keep it pest free
- ☆ Growing crops in different location also helps in controlling pests as it disturbs the reproduction cycle of pests.

Components of Organic Farming

Significant parts of organic farming are biological nitrogen fixation, crop rotation, residues of crops, biopesticides, biogas slurry *etc.* Vermicomposting has risen as a significant part in organic farming which is extremely successful in upgrading soil fruitfulness and development of harvests in an economical manner.

The different components of organic farming are (Kumar *et al.*, 2017):

1. **Crop rotation** - For rehearsing supportable agriculture there ought to be revolution of yields on a similar land over a time of two years or more for keeping up soil fruitfulness and control of bugs, weed and sicknesses. For instance utilization of vegetables in turn improves soil fertility.
2. **Crop Residue** - India has incredible capability of utilizing buildups of yields and straw of oats and heartbeats in reusing of supplements during organic farming. Harvest deposits when immunized with parasitic species improve physico-chemical properties of soil and crop yields.
3. **Organic manure** - The organic manure is obtained from biological sources (plant, animal and human residues). Organic manure helps in increasing crop growth directly by improving the uptake of humic substances and indirectly promoting soil productivity by increasing availability of major and minor plant nutrients through soil microorganisms.

   A. **Bulky organic manure -** Bulky organic manure includes compost, FYM and green manure having less nutrients in comparison to concentrated organic manure.

   - ☆ **FYM** - Farm Yard Manure (FYM) refers to the well decomposed combination of dung, urine, farm litter and leftover materials (roughages or fodder).
   - ☆ **Compost** - Large quantities of waste material (vegetable refuse, weeds, stubble, bhusa, sugarcane trash, Sewage sludge, animal waste, human and industrial refuse) can be converted into compost manure by anaerobic decomposition. Compost is used in the same way as FYM and is good for application to different type of soils and crops.
   - ☆ **Green Manuring** - Green manuring is practice of adding organic matter to the soil by ploughing and adding into the soil undecomposed green plant tissues for improving physical structure and fertility of the soil. The green manure crop (legume crop) supplies organic matter and additional nitrogen. Commonly used green manure crops are such as Sun hemp (*Crotalaria juncea*), Dhaincha (*Sesbania aculeata*), Cowpea, Cluster Bean, Senji (*Melilotus parviflor, Vigna sinensis*), Berseem (*Trifolium alexandrium*) *etc.*

B. **Concentrated Organic Manure** - Oilcakes, blood meal, fishmeal, meat meal and horn and hoof meal (Concentrated organic manures) that are organic in nature made from raw materials of animal or plant origin and contain higher percentage of vital plant nutrients such as nitrogen, phosphorous and potash, as compared to bulky organic manures.

4. **Waste**
   - ☆ **Industrial waste** - Industrial by products such as spent wash and coir waste can be used as manure.
   - ☆ **Municipal and Sewage waste** - It is an important component of organic waste.

5. **Biofertilizer:** Biofertilizers are defined as preparations containing living cells or latent cells of efficient strains of microorganisms that help crop plants' uptake of nutrients by their interactions in the rhizosphere when applied through seed or soil. They accelerate certain microbial processes in the soil which augment the extent of availability of nutrients in a form easily assimilated by plants. Very often microorganisms are not as efficient in natural surroundings as one would expect them to be and therefore artificially multiplied cultures of efficient selected microorganisms play a vital role in accelerating the microbial processes in soil. Use of biofertilizers is one of the important components of integrated nutrient management, as they are cost effective and renewable source of plant nutrients to supplement the chemical fertilizers for sustainable agriculture. Several

Sl.No.	Groups	Examples
	N_2 fixing Biofertilizers	
1.	Free-living	*Azotobacter, Beijerinkia, Clostridium, Klebsiella, Anabaena, Nostoc*
2.	Symbiotic	*Rhizobium, Frankia, Anabaena azollae*
3.	Associative Symbiotic	*Azospirillum*
	P Solubilizing Biofertilizers	
1.	Bacteria	*Bacillus megaterium* var. *phosphaticum, Bacillus subtilisBacillus circulans, Pseudomonas striata*
2.	Fungi	*Penicillium* sp., *Aspergillus awamori*
	P Mobilizing Biofertilizers	
1.	Arbuscular mycorrhiza	*Glomus sp.,Gigaspora sp.,Acaulospora sp.,Scutellospora sp.* and *Sclerocystis sp.*
2.	Ectomycorrhiza	*Laccaria sp., Pisolithus sp., Boletus sp., Amanita sp.*
3.	Ericoid mycorrhizae	*Pezizella ericae*
4.	Orchid mycorrhiza	*Rhizoctonia solani*
	Biofertilizers for Micro nutrients	
1.	Silicate and Zinc solubilizers	*Bacillus* sp.
	Plant Growth Promoting Rhizobacteria	
1.	Pseudomonas	*Pseudomonas fluorescens*

microorganisms and their association with crop plants are being exploited in the production of biofertilizers. They can be grouped in different ways based on their nature and function.

6. **Biopesticide** - Biopesticides are of plant origin and include plant products like alkaloids, phenolics, terpenoids and some secondary chemicals. They are biologically active against insects, fungi, nematodes affecting their behaviour and physiology. Commonly known insecticides are Pyrethrum, Nicotine, Neem, Margosa, Rotenone *etc.*
7. **Vermicompost** - Vermicompost is organic manure or compost produced by the use of earthworms that generally live in soil, eat organic matter and excrete it in digested form. These are rich in macro and micronutrients, vitamins, growth hormones and immobilized microflora essential for plant growth.

Benefits of Organic Farming

The various benefits provided by organic farming are (Moore, 2013) given below:

- It keeps up wellbeing of condition by diminishing contamination.
- It helps in expanding farming creation in a supportable way.
- It helps in improving soil wellbeing.
- Agriculture items acquired from organic farming are better in quality. (Greater in size, flavor, size and smell)
- Water holding limit of the soil is expanded through organic farming.
- It improves the accessibility of supplements required and basic for plants. (Macro supplements and Micro supplements)
- Organic farm items are ordinarily of better size, flavor, smell (Quality)
- Underground water of the territory under organic farming is liberated from harmful synthetic compounds.
- Vermicomposting cuts down waste mass thickness.
- Vermicomposting has hormone like substance auxins which expands plant development.
- Maintains C:N proportion in the soil and expands the richness and profitability of the soil.
- **Nutrition** - Organic food is wealthy in supplements and it is liberated from destructive synthetic substances, it additionally expands the supplements in the soil so the developed yield is more advantageous to devour (Sreerag and Jayaprakas, 2015).
- **Free from synthetic compounds** - In organic farming synthetic concoctions are not used to control bugs and other destructive plant ailments, which makes malignant growth and different maladies the buyers. However, organic farming is free from harmful synthetic compounds.

- **Quality food** - The organic food is having quality with supplements and it tastes better than the food developed by utilizing manufactured synthetics and nature of food is dictated by its taste. Brix examination is utilized to quantify the nature of vegetables and organic products. Long Time Store Organic food has the ability of longer time stockpiling because of its metabolic and basic honesty in their cell structure than different harvests developed by utilizing manufactured synthetic substances.
- **Low input cost** - Expenditure on agriculture is low with organic farming since it need animals to work the land, composts which are effectively accessible and they can set up their own, and the bio manures are set up with ease. Unlike regular farming practices, organic farming doesn't utilize costly synthetic compounds and manures. In this way, it decreases the creation costs for farmers. In correlation with ordinary strategies for farming, organic farming utilizes less petroleum product by keeping away from the utilization of manufactured manures and assists with hindering a worldwide temperature alteration (Pandit *et al.*, 2012).
- The utilization of synthetic compounds in farming practices makes hurt the earth as the synthetic compounds dive deep into the dirt and water and debase them. Be that as it may, organic farming maintains a strategic distance from the utilization of synthetic substances and shields the earth from contamination
- Fruits and vegetables developed utilizing the act of organic farming are more nutritious and delicious as they are given the necessary opportunity to develop and no synthetic substances are utilized in their development.

Serious Issues in Organic Product

Apart from advantages organic farming having some demerits. They are as follow:

- Price desires are excessively high corresponding to quality
- Low consistency of value
- Slow shipment, limitations for bringing in Indian organic items
- Time expending and entangled administrative work while managing send out specialists
- The helpless client support from the Indian brokers after deals is the serious issue in trade advertising
- Lack of legitimate promoting system a showcasing execution
- Less exertion to create residential markets
- Organic farming is a period taking procedure in getting the outcome, which makes the farmers to disregard this sort of farming
- It requires more work power and ought to have normal perception contrasted with traditional farming

- Organic farming is an aptitude based work and farmers ought to be prepared opportunity to time as indicated by the seasons and the state of the harvests
- Low profitability is the serious issue in organic farming contrasted with conventional farming, however in customary type of agriculture the fertility of soil is diminishing chance to time with overabundance utilization of synthetic concoctions
- Organic food is more costly than the food developed with customary farming practices. In this way, the buyer needs to pay more for organic food
- Farmer needs to work more earnestly on the off chance that he follows the way of organic farming as this farming practice requires more association of farmer with his harvest
- The practice of organic farming gives the yield sufficient opportunity to develop without the utilization of counterfeit development injectors. Be that as it may, this training won't have the option to satisfy the world's need for food as the organic food sets aside more effort to develop when contrasted and the ordinary acts of developing harvests

Risks and Opportunities Associated with Strategies to Increase Organic Production

There are both risks and opportunities associated with strategies to increase yields in organic production. The final outcome depends largely on management, *i.e.* how strategies to increase yields are implemented and whether tradeoffs are accounted for and managed. Knowledge, skills and system thinking are crucial in this Endeavour, as we demonstrate with numerous examples. The ambition of organic farming to design high yielding farming systems that also care for the environment, people and animals entails a difficult value-based balancing act. In some aspects, current organic agriculture delivers benefits compared with conventional agriculture (Seufert and Ramankutty 2017). Both risks and opportunities associated with strategies to increase yields in organic production for different issues are summarized on the Table 13.1.

Conclusion

Organic farming can give quality food without unfavorably influencing the soil's wellbeing and the environment. There is have to distinguish appropriate harvests/items on territorial reason for natural creation that has worldwide market requests. The entire locale as such can't bear to go for natural at a time as a result of its duties to protect food and wholesome security. This will give sufficient chance to business and acquire thriving and harmony the area. Organic farming is an extraordinary option in contrast to conventional cultivating farmers. It follows eco-accommodating agricultural practices without utilizing destructive synthetic compounds. It helps in keeping up human wellbeing just as shields our condition from unsafe synthetics used to bring manifests in a field. Going for organic is an

Table 13.1

Affected Issue	*Strategies to Increase Yields*					
	Increased N Input in Cropping	*More Intense Weed Control*	*Increased Disease and Pest Control*	*Breeding for Higher Yields (Animals and crops)*	*More Concentrates in Ruminant Diets*	*Reduced Losses (Animals and crops)*
Biodiversity	Risk of reduced floral and faunal diversity	Risk of reduced biological control caused by reduced floral abundance negatively affecting generalist predators	Risk of increased use of substances, with negative effects on non-target organisms. Opportunity for biological control and new control products	Risk of lowered genetic diversity in high-yielding varieties. Opportunity for dual breeds and breeding for high pasture utilization	Risk of reduced utilization of semi-natural pastures and associated species loss	Opportunity for land sparing giving room for elements improving diversity
Emissions of GHG	Risk of increased nitrous oxide emissions	Risk of increased emissions from fossil fuels	Opportunity for lower climate impact due to more efficient production	Opportunity for lower climate impact due to more efficient production	Opportunity for lower methane from ruminant products due to faster growth ratesand higher milk yield. Risk of increased soil C loss or less soil C sequestration due to less ley cultivation	Opportunity for lower climate impact due to more efficient resource use
Nutrient losses	Risk of increased N leaching	Risk of increased N leaching. Opportunity for more vigorous crop growth with reduced leaching	Opportunity for reduced losses due to higher production stability over time	Opportunity for reduced N losses by more efficient crop N uptake, for livestock breeds by more efficient feed N utilisation	Increased N leaching risk due to more annual crops and less ley in rotations, risk of farm P surpluses due to import of concentrates	Opportunity for reduced losses due to more efficient resource use

Affected Issue	*Strategies to Increase Yields*					
	Increased N Input in Cropping	*More Intense Weed Control*	*Increased Disease and Pest Control*	*Breeding for Higher Yields (Animals and crops)*	*More Concentrates in Ruminant Diets*	*Reduced Losses (Animals and crops)*
Soil fertility	Opportunity for increased SOC levels and soil nutrient stocks through more crop residues and organic fertilizers.Risk of contamination by toxic substances in urban wastes	Opportunity for more crop residues which promote SOC levels.Risk of soil compaction	Opportunity for yield increases, giving more crop residues promoting SOC levels	Risk or opportunity depending on amounts of crop residues of new varieties, opportunity with perennial cereals	Risk of lower SOC levels due to lower ley proportion in crop rotations	
Animal health and welfare				Risk of less robust animals with potentially increased health problems	Risk of metabolic disease in cattle, risk of reduced welfare due to less grazing	Opportunity for increased animal welfare, that will make therapeutic medication redundant
Human nutrition and health	Possible risk of slightly lower content of secondary plant metabolites; importance for human health is minor		Risk of pesticide exposure; increased use of compounds with identified human toxicity per se undesirable although exposure is minor	Risk of lower mineral nutrient content (especially in cereals)	Risk of lower content of omega-3 fatty acids in milk, eggs and meat; importance for human health is minor	Opportunity for decreased fungal toxins (DON) in cereals, opportunity for lower pathogen levels in poultry meat
Farm profitability	Risk or opportunity depending on the costs associated with such increases, how it affects the farmer's risk and how the revenues of the farming business are affected, including how consumers respond to such changes.					

incredible method of forestalling synthetic concoctions and securing our wellbeing and condition however there are some of difficulties in this field. Because of the significant expense of organic food, individuals are not yet tolerating the utilization of organic food. The other test in the field of organic cultivating is to fulfill the world's need for food as the development of organic harvests is moderate. Be that as it may, present world a large portion of the expending food contains hurtful synthetic substances which are causing different illnesses unconsciously or dismissed purposely this can be decreased by organic cultivating. The agricultural lands are becoming useless to do agriculture if this continues the coming generations will face a serious problem of food production and they unable to produce quality food. It requires proper practice of organic farming skills with patience. Most of the farmers are doing conventional form of agriculture to get the high yield and quick result, but with conventional agriculture the fertility of the soil is decreasing gradually and if this kind of practice continues the land become useless for agriculture. So, to avoid such a serious problem practice of organic farming helps the soil to maintain the fertility and can get good quality of food products which are also healthier.

REFERENCES

Auobamiri, Ahmad GA. 2014. Effects of conservation tillage in organic brkshavrzy. *IJPAES* 4(4): 263-274.

Behera KK, Alam A, Vats S, Sharma H P and Sharma V. 2011. *Organic Farming History and Techniques. Agroecology and Strategies for Climate Change, 287–328.* doi: 10.1007/978-94-

Essoussi HL and Zahaf M. 2012. The organic food market: Opportunities and challenges 10.5772/30155.

IFOAM. 2010. International federation of organic agriculture movement-The orinciples organic agriculture. http: //www.ifoam.org/about_ifoam/principles/index.html

Kumar S and Singh AB. 2015. Iopesticides: Present Status and the Future Prospects. *J Biofertil Biopestici* 6: 129.

Kumar SM, Reddy GC and Sangwan PS. 2017. A Review on organic farming - sustainable agriculture development. *Int. J. Pure App. Biosci.* 5(4): 1277-1282.

Madhusudhan L. 2016. Organic Farming-Ecofriendly Agriculture. *Journal of Ecosys Ecog*raph. 6(3): 209

Moore J. 2013. Advantages of implementing organic farming methods in agricultural lands. MSc. Thesis, Department of Environmental Sciences, University of Virginia p 40.

Pandit NP, Ahmad N and Maheshwari SK. 2012. Vermicomposting biotechnology: An eco-loving approach for recycling of solid organic wastes into valuable biofertilizers. *J Biofertil Biopestici* 2: 113.

Seufert V and Ramankutty N. 2017. Many shades of gray- the context dependent performance of organic agriculture. *Science Advvances* 3(3): 1-14

Sreerag RS and Jayaprakas CA. 2015. Management of two major sucking pests using neem oil formulation. *J Biofertil Biopestici* 6: 147.

Chapter 14

Essential Mineral Nutrients for Plant Growth: Nutrient Functions and Deficiency Symptoms

Prabhoo Singh and Brook Legese

Department of Natural Resources Management, College of Agricultural Sciences, Bule Hora University, Ethiopia

Plants are absorbing nutrient from air, water and soil. Total 105 elements are discovered but plants require 17 essential elements for normal plant growth and development. The elements required in large quantities are called major/micro nutrients while those required in less quantity called minor/ micro nutrients. All essential elements are constituent of carbohydrates, proteins, fats, nucleic acid etc., and various metabolic processes. The deficiency symptoms are prominent such as Chlorosis, necrosis, stunted growth, impaired cell division etc. Plants absorb mineral nutrient through roots by either passive or active processes. They are carried to all parts of plant through xylem along with water transport. Nutrients are taken up from the soil using plasma-membrane located transporter proteins and excess is stored in the cell vacuole or converted into polymerized storage forms.

Keywords: *Plants, Mineral nutrition, Growth, Nitrogen, Essential elements and Chlorosis.*

Introduction

The study of the chemical elements and compounds is necessary for plant growth, plant metabolism process. The deficiency of the element makes it impossible for the plants to complete the vegetative growth or reproductive stages of its life cycle. All green plants have the ability to production their own food by using energy derived from the sun light to combine chemical elements, taken up in the inorganic ion form, into a multitude of organic compounds. Seventeen elements are considered essential element for plant growth. If any of these 17 elements are lacking, plants cannot complete their vegetative or reproductive life cycles.

Inorganic minerals present in the soil surface are used for nutrition by plants by extracting them from soil or the aquatic environment. These mineral nutrients are formed by the complex interaction involving weathering of rock minerals, decay of organic matter, plant, animals and microbes (Bahadur *et al.*, 2015).

Carbon, hydrogen and oxygen are supplied to the plants through air and water from atmosphere. These three elements comprise about 95 per cent of the total dry matter of the plants. The 17 mineral nutrients required for plant growth are classified as primary, secondary or micronutrients according to the quantity required by plants Primary nutrients and the secondary nutrients are sometimes known as macro or major nutrients. Secondary nutrients are macronutrients but are less frequently deficient in soils (White and Brown 2010).

Mineral nutrients have essential and specific functions in plant metabolism process: they can function as constituents of basic structures, as activators of enzymatic reactions, or as charge carriers and osmo-regulators (Schwambach *et al.*, 2005).

The deficiency of nutrients causes negative effect on plant growth and yield level. The disease inhibiting compounds are produced in plants but their production and transport is dependent on plant nutrients (Meena *et al.*, 2015).

Table 14.1: Essential Plant Nutrients (Arnon and Stout, 1939)

Plant Nutrient	*Plant Usable Form*	*Average Concentration in Plant Tissue*
	Basic elements	
Carbon (C)	CO_2	45 per cent
Hydrogen (H)	H_2O	6 per cent
Oxygen (O)	H_2O, O_2	45 per cent
	Primary elements	
Nitrogen (N)	NO_3^-, NH_4^+	1.5 per cent
Phosphorus (P)	$H_2PO_4^-$, HPO_4^{2-}, PO_3^{3-}	02 per cent
Potassium (K)	K^+	
	Secondary elements	
Sulfur (S)	SO_4^{2-}	0.1 per cent
Calcium (Ca)	Ca^{2+}	0.5 per cent
Magnesium (Mg)	Mg^{2+}	0.2 per cent
	Micro Nutrients	
Iron (Fe)	Fe^{2+}	100 mg/kg
Zinc (Zn)	Zn^{2+}	20 mg/kg
Manganese (Mn)	Mn^{2+}	20 mg/kg
Copper (Cu)	Cu^{2+}	6 mg/kg

Plant Nutrient	Plant Usable Form	Average Concentration in Plant Tissue
Boron (B)	H_3BO_3, $H_2BO_3^-$ BO_3^{3-}	20 mg/kg
Chloride (Cl)	Cl–	100 mg/kg
Molybdenum (Mo)	MoO_4^{2-}	0.1 mg/kg
Nickel (Ni)	Ni^{2+}	mg/kg

Nutrient Uptake

Root structure and membrane transporters are the main factors of available nutrients to plant. Plants can adjust their branching pattern and root hair growth to utilize locally available patches of higher nutrient concentrations. Roots show a hunt growth pattern and when they contact a rich nutrient pocket in the soil their architecture changes to branch locally increasing the area for uptake (Mommer *et al.*, 2012).

The surface layers of soil usually contain the highest amounts of nutrients and so the top surface of soil is frequently densely covered with plant roots. Phosphorus is an example of a nutrient that occurs in the top surface soil and bean roots multiply in this layer to acquire this nutrient (Martinoia *et al.*, 2012).

The plant nutrients are transport across the plasma membrane from the soil by root cells, storage in the vacuole and loading of the long distance transport systems in plants (phloem and xylem), transport into the growing tissues such as leaves or seeds. Each of these transport steps is mediated by specific transporter proteins and they are potential regulatory steps in the pathway for nutrient entry into the plant. This regulation can occur by changes in gene expression or by post-translational regulation of the protein (White *et al.*, 2013). The transporter is results in plants that demonstrate nutrient deficiency symptoms under normal supply conditions that would be adequate for a wild type plant. But sometimes the mutation is hidden because another transport in the family can step up expression to compensate for the missing function. Some of the plasma membrane transporters are also involved in sensing the external availability of nutrients and these have been described as 'transceptors' (Gojon *et al.*, 2011).

Nitrogen

All the essential nutrients Nitrogen is required by plants in large amounts. Nitrogen is absorbed by plant roots as NO_3^- ion and in the case of rice as NH_4^+ ion also. The nitrogen concentration varies 1-5 per cent in plant. Nitrogen is required by all parts of plant, particularly the meristematic tissues and the metabolically active cells. Nitrogen is the major constituents of proteins, nucleic acids, vitamins and hormones.

Nitrogen plays important role in various metabolic process of the plant growth. Nitrogen is an essential constituent of protein and chlorophyll (Meena *et al.*, 2014). In addition, N and P have a stimulating effect on root activity and rooting pattern of the crop. Available nitrogenous compound (also through a starter dose) enables seedlings to make a good start even before nitrogen fixation. Plants fed with organic nitrogen during vegetative periods are much larger by the onset of flowering than those dependent on symbiotic N-fixation.

Factors Affecting Availability of Nitrogen

There are three major forms of nitrogen into soil: a) organic nitrogen, b) ammonium nitrogen fixed by clay minerals and c) soluble inorganic ammonium and nitrate compounds. The process of conversion organic to inorganic form of N is called mineralization and its conversion inorganic to organic form of N is called immobilization. The inherent capacity of soil to produce plant is closely and directly related to their organic matter and nitrogen content. Nitrogen released during mineralization of organic N is converted to NO_3^-. Supply of NH_4^+, population of nitrifying organisms, soil pH, soil aeration, temperature and moisture are the important factors affecting nitrification. The inorganic nitrogen is loss from soils by leaching and volatilization process (Williams and Miller 2001).

Function

Increasing growth and development of all living tissue, Gives the plant their green color, increasing germination of crops. Improve the quality of crop, it increase utilization of P and K, Necessary to develop proteins and chlorophyll.

Deficiency

The **chlorophyll content** of plant leaves is reduced, flowering, fruiting, protein and starch contents are reduced, Yellow-green color, Short stems, small leaves, Slow and dwarfed plant growth.

Impotent Process

Mineralization: The biological conversion of organic forms of nutrient to inorganic or mineral forms is called mineralization.

Immobilization: It is reverse process of mineralization. The conversion of inorganic forms of nutrient by the soil organism into organic forms is called Immobilization.

Plant Metabolism (NO_3^- Assimilation)

$NO_3^- \xrightarrow{\text{Cytoplasm, NR}} NO_2^- \xrightarrow{\text{Chloroplast, NR}} NH_4^+ \rightarrow$ amino acid $\rightarrow$ protein $\rightarrow$ nucleic acid

Aminization

The process of conversion of proteins to amino acids.

Ammonification

The process of conversion of amino acids and amides to ammonia.

Nitrification

It is a biological process where bacteria transform NH_4 into NO_3. Proceed rapidly under warm, moist and well aerated soil conditions.

Denitrification

Bacteria convert nitrate (NO_3) to N_2 or N_2O and lost to the atmosphere under saturated soils (anaerobic conditions). Some anaerobic bacteria in water logged soils get oxygen from NO_3 to produce N_2 which is then released to the atmosphere.

Leaching

Nitrate carries a negative charge and is repelled by the negatively charged soil particles. This allows nitrate to leach with water infiltration. Nitrate loss down the soil structure with water.

Volatilization

Occurs when Ammonia (NH_3) isn't converted to Ammonium (NH_4) lost to the atmosphere as a gas.

Phosphorus

Phosphorus is the second most abundant mineral element in the human body, making up for >20 per cent of the body's minerals (Rattan and Goswami 2009). Phosphorus containing ATP is called the "Energy Currency of the plant". The available form is $H_2PO_4^-$, HPO_4^{2-} and PO_4^{3-}. Phosphorus is a constituent of cell membranes, certain proteins, all nucleic acids and nucleotides, and is required for all phosphorylation reactions.

Phosphorus Uptake and Distribution

In response to persistent Pi deficiency, plants have developed many adaptive morphological, physiological and molecular mechanisms to cope with low Pi. These includes changes in root growth (increased root surface area, fine root hairs, root length) and architecture, induction of high-affinity Pi transporters, increased secretion of acid phosphatase enzyme and low-molecular weight organic acids, symbiotic associations with mycorrhizal fungi and changes in the activity of several key photosynthetic enzymes (reviewed by Lopez-Arredondo *et al.*, 2014).

Factors Affecting Availability of Phosphorus

The availability of inorganic phosphorus is largely determined by soil pH, soluble Fe and Al, presence of Fe and Al minerals, available Ca and Ca-minerals, amount and decomposition of organic matter and activities of soil microorganisms. The pH, organic matter and microorganisms strikingly affect inorganic P availability. The fixation of phosphorus is a major problem at low pH by Al and Fe and at high pH by Ca so it is reduce the availability of phosphorus.

Functions

It promote to root formation and growth, Improve the quality of fruits Increasing nodulation, Seed formation and early maturity, Responsible for the regulation of protein synthesis, Increased N fixing capacity of legumes.

Deficiency Symptoms

Decrease in growth, slow maturity, Leaves is purplish color.

Potassium

Plant roots absorb potassium as potassium ion (K^+). In plants, this is required in more abundant quantities in the meristematic tissues, leaves, buds and root tips. Potassium helps to maintain an anion-cation balance in cells and is involved in opening and closing of stomata, protein synthesis and activation of enzymes and in the maintenance of the turgidity of cells.

Potassium Uptake and Distribution

Plants take up potassium as monovalent cation, K^+. Uptake of K in plant tissues occurs at high rates due to relatively high permeability of plant membranes to K. This high permeability of membranes to K results from ionophores located in membrane that enables facilitated diffusion. Besides the passive uptake, K also enters plant roots via high- and low-affinity transporters (Giertha and Maser 2007; Wang and Wu 2013).

Factors Affecting Availability of Potassium

The various forms of potassium in soils can be classified on the basis of availability in three category a) un-available (feldspars, micas etc), b) available (exchangeable and in soil solution) and c) slowly available (non-exchangeable). The ability of the various soil colloids to fix potassium varies widely. For example, 1: 1 type clay such as kaolinite. The mechanism for potassium fixation is probably the same as that for fixation of the Al ion. Applications of lime materials sometime result in an increase in potassium fixation of soils. Finally, potassium deficiency has been noted in soils with excess amount $CaCO_3$.

Functions

It increasing in resistant in plants against moisture stress, heat, frost and disease, Improve the crop quality, Lodging resistant in cereals, Wilt resistant in cotton and mosaic virus resistant in cotton, Control the opening and closing of the stomata, Important in the biochemical reaction in plant, Activates enzymes responsible for specific function.

Deficiency

Brown **scorching** and curling of **leaf tips** as well as **chlorosis (yellowing)** between leaf veins, Purple spots may also appear on the leaf undersides, Plant growth, root development, and seed and fruit development are usually reduced. High K content in the growing medium does not usually produce any toxic symptoms in plant. This may be because uptake of K in excess represents 'luxury consumption' and it is stored in the vacuole without any increment in growth.

Secondary Nutrients

Secondary nutrients (Calcium, magnesium, and sulfur) are essential plant nutrients and plants require them in small quantities than primary nutrient (nitrogen, phosphorus, and potassium). On the other hand, plants require these nutrients in larger amount than the "micronutrients" such as B, Fe, Mn, and Zn.

Calcium and magnesium both increase with soil pH, but sulfur is reduce with soil pH (Maathuis, 2009). The primary function of calcium in plant is to provide physical structural support or part of cell walls. Magnesium is absorbed as the Mg^{2+} ion and is mobile in plants, moving from the older leaves to the younger leaves. Magnesium is involved in photosynthesis and activates enzymes (McCauley *et al.*, 2009).

Calcium

Plant absorbs calcium by soil in the form of calcium ions (Ca^{2+}). Calcium is essential for meristematic and differentiating tissues. During cell division it is used in the synthesis of cell wall, particularly as calcium pectate in the middle lamella. It is also needed during the formation of mitotic spindle. It is accumulates in older leaves. It is involved in the normal functioning of the cell membranes. It activates certain enzymes and plays an important role in regulating metabolic activities.

Calcium Uptake and Distribution

Plants absorb calcium as a divalent cation, Ca^{2+}. The uptake potential of Ca is lower than other cations such as K^+ even though the Ca concentration of soil solution is ten times higher than K. This is because Ca absorption takes place only through young root tips in which the endodermis cell walls are still unsupervised. Uptake of Ca from soil solution by roots is mainly a passive process, and even within the plant tissues, Ca translocation occurs passively. The upward translocation of Ca in xylem sap occurs with the transpiration stream. Since Ca precipitates as Ca-phosphate in the phloem sap, the downward translocation of Ca is very slow. The identity of the specific protein(s) that mediates Ca^{2+} uptake is still unknown (Demidchik and Maathuis 2007).

Factors Affecting Availability of Calcium

calcium element is largely in the exchangeable form in acid, humid region soils. The activity of Ca^{2+} in solution is decreased by leaching or plant removal. The Ca^{2+} in solution is increased with subsequent adsorption of some of the Ca^{2+} by the exchange complex. The type of clay influences calcium availability; 2: 1clays require higher Ca^{2+} saturation than 1: 1 clays.

Deficiency

Al toxicity and **P deficiency** induces Ca deficiency due to a reduction in root growth. Ca deficiency is more of **tuber crop** and **fibrous grass root crops**. Leaf tip and margin becomes chlorotic, colorless and necrotic; known as **die back or tip burn.** Shoot **terminal buds** and **apical root tips** are **deformed** resulting in the death of growing points. In severe, **root development is reduced**, and **root tip** becomes dark and **die**.

Magnesium

It is absorbed by plants in the form of divalent Mg^{2+}. It activates the enzymes of respiration, photosynthesis and is involved in the synthesis of DNA and RNA, Resistance to unfavorable factors; Magnesium is a constituent of the ring structure of chlorophyll and helps to maintain the ribosome structure.

Magnesium Uptake and Distribution

In plants, the free cytoplasmic Mg^{2+} remains at a concentration of about 0.5 mM but total Mg^{2+} levels may vary from 0.3 to 1.0 per cent . Transport of Mg^{2+} is passive, mediated by ionophores in which Mg^{2+} moves down against the electrochemical gradient. In this transport system, Mg^{2+} faces cation competition; therefore, if an excess of other cations such as K^+ or NH_4^+ is present in the medium, the uptake of Mg^{2+} is adversely affected. Mg^{2+} is very mobile in the phloem and so can be translocated from older to younger leaves or the meristem (Demidchik and Maathuis 2007).

Factors Affecting Availability of Magnesium

Magnesium occurs predominantly as exchangeable and solution Mg^{2+}. The absorption of Mg^{2+} by plants depends on the amount of soil solution, soil pH, the quantity of other exchangeable ions and the type of clay. The availability of magnesium can also be adversely affected by high H^+ ion activity in acid organic soils where exchangeable Al^{3+} is not a major cause of the acidity. Mg deficiencies also can occur in fertile soils with high ratios of exchangeable Ca/Mg (10/1 to 15/1).

Deficiency

Mg is mobile in the plant, so its deficiency symptoms first occur on lower leaves. Interveinal chlorosis which becomes chlorotic to necrotic in a severe case.

Sulphur

Plants absorb sulphur in the form of sulphate SO_4^{2-}. Sulphur is present in two amino acids – cysteine and methionine and is the main constituent of several coenzymes, vitamins (thiamine, biotin, Co-enzyme A) and ferredoxin.

Sulphur Uuptake, Distribution and Assimilation

In addition to S uptake by roots as SO_4^{2-}, plants can also absorb S from the atmosphere in the form of SO2 through stomatal openings. Uptake of SO_4^{2-} is an active process as it is absorbed via plasma membrane sulphate transporters energised by H^+ gradient. The proton-coupled cotransport occurs with $3H^+/SO_4^{2-}$, stoichiometry. Transcription of this type of mechanism is induced when S becomes deficient. Analysis of genome sequences of Arabidopsis and rice has led to the identification of 14 putative sulphate transporter genes in each genome. These genes are subdivided into 4 closely related groups and all having 12 membrane-spanning domains and a STAS domain at their carboxy-terminus (Aravind and Koonin 2000).

Factors Affecting Availability of Sulphur

Sulphur is present in soil in both organic and inorganic forms. The inorganic forms are SO_4^{2-}. Influenced by the following soil properties- 1) Clay content and type of clay mineral, 2) Hydrous oxides, 3) Soil horizon or depth, 4) Effect of pH, 5) SO_4^{2-} concentration, 6) Effect of time, 7) Presence of other anions, 8) Effect of cations, 9) Organic matter.

Function

Promotes root growth, Formation of chlorophyll, Production of proteins, Synthesis of oils in oilseeds, Activation of enzyme.

Deficiency

Unlike N deficient plant, generally S deficient plant show chlorosis of younger leaves because of its less mobility in plants. Its deficiency causes reduced plant growth, stunted plant, thin stem and chlorotic. N and S deficiency symptoms are difficult to distinguish. Cruciferous crops, Leaves cupped inward, reddish color on lower leaf surface which proceeds on both sides of the leaf in severe conditions.

Iron

Plants absorb iron in the form of ferric ions (Fe^{3+}). It is essential larger amounts for plant in comparison to other micronutrients. It is an important constituent of proteins involved in the transfer of electrons like cytochromes and ferredoxin. It is reversibly oxidised from Fe^{2+} to Fe^{3+} during electron transfer. It activates catalase enzyme, and is essential for the formation of chlorophyll.

Iron Uptake and Distribution

The preferred form of Fe by plants is the reduced cation Fe^{2+}. Fe cannot be taken up in ionic form as such; it has to chelate with other compounds that facilitate Fe uptake. Plant species differ in their ability to utilise sparingly soluble inorganic Fe and Fe chelates. Plants grown under Fe stress show physiological and morphological changes that help in Fe uptake (Romheld and Marschner 1986).

Deficiency

Fe deficiency occurs in neutral/alkaline soils but very rare in acid soils (if occurs; due to low total Fe). Deficiency symptoms 1st appear on young leaves because of its immobility in the plant. Deficiency symptoms – Interveinal chlorosis of younger leaves which becomes white and necrotic entirely in severe cases.

Manganese

It is absorbed in the form of manganese ions (Mn^{2+}). It activates many enzymes involved in photosynthesis, nitrogen metabolism and respiration. The best defined function of manganese is in the splitting of water to liberate oxygen during photosynthesis process.

Manganese Uptake and Distribution

Mn uptake is an active process involving metabolic energy. It is relatively immobile in the plants. It is preferentially translocated to meristematic tissues; therefore, young plant organs are rich in Mn. The genes involved in transport of transition metal in plants have been identified which are also responsible for Mn^{2+} transport (Reviewed by Pittman 2005).

Deficiency

Mn is immobile, so deficiency symptoms occur on younger leaves. Common deficiency symptom is interveinalchlorosis(as in Fe) which later converted into chlorotic or necrotic spots, that's why sometimes mistaken by Fe deficiency symptoms.

Cooper

It is absorbed as cupric ions (Cu^{2+}). It is essential for the overall metabolism in plants. Like iron, it is associated with certain enzymes involved in redox reactions and is reversibly oxidised from Cu^{+} to Cu^{2+}.

Cooper Uptake and Distribution

Plants take up Cu in very little quantities, so the Cu content in plant tissues varies from 2 to 20 ppm. Cu uptake is an active process requiring metabolic energy. Under physiological pH, Cu exists in two oxidation states Cu^{1+} and Cu^{2+} and can interchange between these forms (monovalent copper is unstable). This property of Cu is responsible for its redox function in biochemical reactions. Cu has high affinity for N atom of amino groups, sulfhydryl groups, carboxylic groups and phenolic groups. More than 98–99 per cent of Cu is present in these complexed forms in soil solution and in xylem and phloem sap (Kramer and Clemens 2005).

Deficiency

Chlorosis of younger leaves is the common deficiency symptoms. Due to Cu deficiency **die back** occurs. In maize and small grain crops; **chlorosis** of younger leaves and **stunted growth** occurs. At severe deficiency like K leaf margin and tip burns.

Zinc

Plants obtain zinc as Zn^{2+} ions. It activates various enzymes, especially carboxylases. It is also needed for synthesis of auxin and flower setting and proper development of fruits.

Zinc Uptake and Distribution

Uptake of Zn is metabolically controlled, that is, Zn uptake is an active process requiring energy. The form in which Zn is translocated from roots to shoots is not clear. Mobility of Zn inside the plant organs is relatively less, so there is accumulation in root tissues particularly when Zn concentration in the media is high. In older leaves, Zn can become very immobile. Much of the Zn absorbed is localised in the seeds and grains in the protein bodies in the form of globoid crystals. These globoids consist mainly of phytate or salts of phytic acid (Guerinot 2000).

Deficiency

Interveinal yellowish or whitish of middle leaves, starting at the tip and margins of leaves. Short stem internode, **stunted** plant (**bushy**) and **rosette** leaves due to auxin inhibition. Small (**little**) and thickened leaves; sometimes **malformation** of **leaves**, as well as **fruit,** occurs. Zn deficiency reduces RNA synthesis. Khaira of rice due to Zn deficiency was first time reported by **Y.L. Nene, 1966** in tarai region of Uttarakhand.

Boron

It is absorbed as BO_3^{3-} or H_2BO^{3-}. Boron is required for uptake and utilization of Ca2+, membrane functioning, pollen germination, cell elongation, cell differentiation and carbohydrate translocation.

Boron Uptake and Distribution

Uptake of B is a non-metabolic process and its distribution in the plants is also governed by transpiration stream. Boric acid channels, which are major intrinsic proteins, facilitate B transport across membranes (Takano *et al.*, 2006).

Deficiency

B deficiency first cause **cessation of terminal bud growth** and then the death of young leaves. Death of **shoot tips** because of its requirement in DNA synthesis. B affects **reproductive stage** more and causes **sterility** and severely **impaired seed-set**.

Molybdenum

Plants obtain it in the form of molybdate ions MoO_4^{2-} . It is a component of several enzymes, including nitrogenase and nitrate reductase both of which participate in nitrogen metabolism.

Uptake and Distribution

Plants absorb Mo as molybdate ion (MoO_4^{2-}). However, the uptake may be reduced due to the competition by other anions like SO_4^{2-}, whereas ions such as PO_4^{2-} are known to enhance the uptake of Mo. The uptake of Mo in plant cells has been discovered recently (Baxter *et al.*, 2008).

Symptoms of Deficiency

In Mo-deficient plants, particularly legumes, symptoms of N deficiency are common, producing chlorosis in younger leaves and stunted growth. In some dicotyledonous species, *e.g.* cauliflower, there is a drastic reduction in size and irregularities in the formation of leaf blade. Under such severe Mo deficiency, only the midrib of leaf is present giving it a whip appearance called 'whiptail'. Local chlorosis and necrosis along the main veins of mature leaves called 'yellow spots' are common in citrus.

Chlorine

It is absorbed in the form of chloride anion (Cl^-). Along with Na^+ and K^+, it helps in determining the solute concentration and the anion-cation balance in cells. It is essential for the water-splitting reaction in photosynthesis, a reaction that leads to oxygen evolution.

Uptake and Distribution

Chlorine uptake is an active process and uptake occurs very rapidly in considerable amounts. However, its uptake also takes place against an electrochemical gradient by carrier proteins present in plasma membrane. In green tissue, under light Cl^- uptake is enhanced as ATP formation during photosynthetic phosphorylation provides energy source for active uptake (Marmagne, 2007).

Deficiency

The most common deficiency symptoms are **chlorosis of younger leaves** and **wilting of the whole plant**. Deficiency symptoms also observed in young tissue due to variation in mobility as in B and S. **Necrosis**, leaf **bronzing** and root and leaf growth reduction is also found.

Nickel

Ni is taken by plants in the form of Ni^{2+}. Ni concentration in plant ranges from 0.1-1.0 ppm. Ni is essential for plant supplied with urea and those in which ureides are important in N metabolism. Ni is beneficial in N metabolism in legumes.

Deficiency: Ni deficiency causes **chlorosis of younger leaves** that proceeds to become necrotic meristem. In crop plants, Ni toxicity is a major problem rather than deficiency. Under Ni deficiency, accumulation of toxic levels of urea might occur. However, the Ni toxicity symptoms closely related to Fe deficiency symptoms. Acute Ni toxicity gives rise to chlorosis. In cereals, pale yellow stripes along the length of the leaf occur. The whole leaf may turn white and necrosis occurs at leaf margins. Indicots, the Ni toxicity causes chlorotic patches between leaf veins, similar to Mn deficiency.

Table 14.2: Major Physiological Disorders Due to Nutrients Deficiency

Nitrogen	Buttoning in cauliflower
Phosphorus	Sickle leaf disease of cotton
Calcium	Popping of groundnut
	Black heart of groundnut and celery
	Tip hooking in cauliflower
	Blossom end rot (BER) of tomato and peppers
	Cavity spot of carrot
Magnesium	Sand-drawn disease in tobacco
	Grass Tetany (in animals)
	Reddening in cotton
Sulfur	Yellow disease of tea
Zinc	Little leaf of cotton and grape
	Tip disease or white bud of maize and sorghum
	Khaira of rice
	Little leaf of citrus
Manganese	Take all (root rot disease) of wheat and barley
	Grey speaks of oat
	Pahala blight of sugarcane
	Marsh spot of pea
	Speckles hollow of sugarbeet
	Crinkle leaf of cotton (toxicity)

Copper	Reclamation disease of sugarcane and cereals
	Stem melanosis disease of wheat
	Ergot association in wheat and pearl millet
	Die back of citrus
	Exanthema of fruits (citrus)
Iron	Ivory of rice
	Bronzing in rice (toxicity)
Boron	Top sickness of tobacco
	Hollow heart of groundnut
	Browning and hollow stem of cauliflower
	Hollow stem of cauliflower
	Hard fruit in citrus (peal thickness, gummy deposits on fruit)
	Hen and chicken of grapes
	Fruit cracking of tomato and pomegranate
	Brown or black heart of root crops
Molybdenum	Whiptail of cauliflower
	Scald of legumes
	Yellow spot in citrus
Silicon	Freckling of sugarcane

Source: Marschner, H. and Marschner, P. ,2012.

Conclusion

In view of the importance of these macro and micro mineral elements in the nutrition of plants and their metabolic inter-relationships which influences other vital factors needed for the survival of living organisms like enzymes, anti-oxidants, vitamins etc, it is important to regularly obtain up-to-date information on the minerals content of water and the variously commonly consumed plant foods used for human and animal foods and feeds respectively *e.g.* legumes, cereals, fruits and vegetables. Genetic, location and environmental factors could also influence the levels of the mineral elements in plants.

REFERENCES

Aravind, L. and Koonin, E.V. (2000) The STAS domain: a link between anion transporters and antisigma-factor antagonists. Curr Biol, 10: 53–55.

Arnon, D.I. and Stout, P.R. (1939) An essentiality of certain elements in minute quantity for plants with special reference to copper. Plant Physiology 14, 371-375.

Bahadur, B. (2015) Plant Biology and Biotechnology: Plant Diversity, Organization, Function and Improvement, DOI 10.1007/978-81-322-2286-6-20.

Demidchik, V. and Maathuis, F.J.M. (2007) Physiological roles of non-selective cation channels in plants: from salt stress to signalling and development. New Phytol, 175: 387–404.

Giertha, M. and Maser, P. (2007) Potassium transporters in plants – involvement in K+ acquisition, redistribution and homeostasis. FEBS Lett, 581: 2348–2356.

Gojon A, Krouk G, Perrine-Walker F and Laugier E (2011) Nitrate transceptor (s) in plants. Journal of Experimental Botany 62: 2299–2308.

Guerinot, M.L. (2000) The ZIP family of metal transporters. Biochim Biophys Acta Biomembr, 1465: 190–198.

Kramer, U. and Clemens, S. (2005) Function and homeostasis of zinc, copper, and nickel in plants. Top Curr Genet, 14: 215–271.

Lopez-Arredondo DL, Leyva-Gonzalez MA, GonzalezMorales SI, Lopez-Bucio J, Herrera-Estrella L (2014) Phosphate nutrition: improving low-phosphate tolerance in crops. Annu Rev Plant Biol, 65: 95–123.

Maathuis F.J.M. (2009) Physiological functions of mineral macronutrients. *Current Opinion in Plant Biology*, 12: 250-258.

Marmagne, A. (2007) Two members of the Arabidopsis CLC (chloride channel) family, AtCLCe and AtCLCf, are associated with thylakoid and Golgi membranes, respectively. J Exp Bot, 58: 3385–3393.

Marschner, H. and Marschner, P. (2012) Marschner's mineral nutrition of higher plants, 3rd edn. Elsevier/Academic Press, London.

Martinoia E, Meyer S, De Angeli A and Nagy R (2012) Vacuolar transporters in their physiological context. Annual Reviews of Plant Biology 63: 183–213.

McCauley A., Jones C., Jacobsen J. (2009) Plant nutrient functions and deficiency and toxicity symptoms, Montana State University Extension Service, Bozeman, MT. pp. 16.

Meena, J. S., Verma, H. P. and Picholi, P. 2014. Effect of fertility levels and biofertilizers on yield, quality and economic of cowpea. *Agriculture Sustainable Development*, 2: 162–164.

Meena, P.D., Rani, A., Meena, M.C., Sharma, P., Kandpal, B. and Singh, D. (2015) Role of nutrients and lower leaf removal against Alternaria Blight in Indian mustard (B. juncea L.). *Plant Pathology*, 14: 92-96.

Mommer, L., van Ruijven, J., Jansen, C., van de Steeg, H.M. and deKroon, H. (2012) Interactive effects of nutrient heterogeneity and competition: implications for root foraging theory? Functional Ecology 26: 66–73.

Pittman, J.K. (2005) Managing the manganese: molecular mechanisms of manganese transport and homeostasis. New Phytol, 167: 733–742.

Rattan, N.K. and Goswami, N.N. (2009) Mineral Nutrition of Plant. In Fundamentals of Soil Science, Indian Society of Soil Science, New Delhi, pp. 349-386.

Romheld, V. and Marschner, H. (1986) Evidence for a specific uptake system for iron phytosiderophores in roots of grasses. Plant Physiol, 80: 1175–1180.

Schwambach, J., Fadanelli, C. and Arthur G. (2005) Mineral nutrition and adventitious rooting in microcuttings of Eucalyptus globuluS, *Tree Physiology* 25, 487–494.

Takano, J., Wada, M., Ludewig, U., Schaaf, G., Wiren, N., Fujiwara, T. (2006) The Arabidopsis major intrinsic protein NIP5;1 is essential for efficient boron uptake and plant development under boron limitation. Plant Cell, 18: 1498–1509.

Wang, Y. and Wu, W.H. (2013) Potassium transport and signaling in higher plants. Annu Rev Plant Biol, 64: 451–476.

White, P.J, George, T.S., Dupuy, L.X. (2013) Root traits for infertile soils. *Frontiers in Plant Science* 4: 1–7.

White, P.J. and Brown, P.H. (2010) Plant nutrition for sustainable development and global health, *Annals of Botany*, 105, 1073–1080.

Williams, L.E., Miller, A.J. (2001) Transporters responsible for the uptake and partitioning of nitrogenous solutes. Annu Rev Plant Physiol Plant Mol Biol, 52: 659–668.

Pests Management

Chapter 15

Field Diagnosis of Sugarcane Diseases and Suitable Approach for their Management

Md. Minnatullah[1], C.S. Choudhary[2], Shiva Pujan Singh[3], Priya Singh[4], M.S. Ahsan[5] and Lalita Rana[6]

[1]SRI, Pusa, [2]T.C.A., Dholi, [3]SRI, Pusa, [4]SRI, Pusa, [5]Deptt. of PP, [6]SRI, Pusa, Dr. Rajendra Prasad Central Agricultural University, Pusa, Samastipur, Bihar

Sugarcane (Saccharum officinerum L.) is an economically important cash crop grown in the tropical and sub-tropical regions of India.The sugarcane crop is useful in various ways i e. sugar, juggery, khandsari, fuel, cattle feed, trash for mulch and compost while, sugar and other by products plays an important role in the economy of several countries and also molasses is often used as a fertilizer for soils, as a stock feed, ethanol, glycerine and citric acid etc. Sugarcane production is constrained by many biotic and abiotic stresses. Among various stresses sugarcane diseases like red rot, sett rot, wilt, smut, pokkah boeng, ratoon stunting, red stripe, leaf scald, mosaic, leaf spots, Yellow leaf and Grassy shoot diseases are limiting factors in sugarcane production in Bihar. Due to these stresses, 30-100 per cent reduction in yield and 10-15 per cent sugar lost is reported by many workers.

Introduction

Sugarcane (*Saccharum officinerum* L.) is an economically important cash crop grown in the tropical and sub-tropical regions of India. In India, it is cultivated in 5 million ha of land with an annual production of 380 million tonnes and its productivity of 70.0 tonnes/ha. In Bihar, it is cultivated in an area of 2.96 lakh ha of land with a production of 13.04 lakh tonnes and its productivity of 54.42 tonnes/ ha. (ISMA, 2019).

Sugarcane is still only visible cash crop in Bihar and it is a traditional producer in the country and contributes about 70 per cent of world sugarcane production and provides row materials for many other by products (Gawade *et al.*, 2012). The sugarcane crop is useful in various ways *i.e.* sugar, jaggery, khandsari, fuel, cattle feed, trash for mulch and compost while, sugar and other by products plays an important role in the economy of several countries and also molasses is often used as a fertilizer for soils, as a stock feed, ethanol, glycerine and citric acid *etc.* More than 100 diseases in sugarcane reported in India due to which 10-15 per cent sugar is lost (Viswanathan and Rao, 2011). In Bihar, more than 20 diseases are reported affecting the crop, out of them some are of major concern and causes not only severe losses in yield but also generate great concern of their effects on the quality of canes.

Sugarcane production is constrained by many biotic and abiotic stresses. Among various stresses sugarcane diseases like red rot, sett rot, wilt, smut, pokkah boeng, ratoon stunting, red stripe, leaf scald, mosaic, leaf spots, yellow leaf and grassy shoot diseases are limiting factors in sugarcane production in Bihar. Among the diseases, red rot is the most destructive one, which is aptly called as "cancer' of sugarcane (Sharma and Tamta, 2015) and is the chief constriction of sugarcane production, which is responsible for the elimination of several elite clones from the cultivation (Viswanathan, 2010). Due to red rot disease considerable reduction in yield (30-100 per cent) and in sucrose recovery (25-75 per cent) were observed by (Tiwari *et al.*, 2010). Losses due to this disease was also observed by (Minnatullah and Kamat, 2018) and they observed that reduction in Brix 16.60 per cent to 20.80 per cent, sucrose 31.60 to 38.26 per cent and in juice purity 18.00 to 22.10 per cent. It was also reported that pokkah boeng disease play as major constraints in production in China during the recent past (Wang *et al.*, 2016). It was noticed that pokkah boeng ranged between 1-90 per cent in variety S 224/20 and 5-30 per cent in variety CoSe 01434 (Viswakarma *et al.*, 2013). Kailaimani (2000) observed that sett dip with Carbendazim for different duration and foliar spray with Carbandazim and Copper oxychloride inhibit red rot pathogen. Shailbala and Amarendra Kumar (2017) reported that thiophanate methyl suppress red rot infection and improved growth of the sugarcane plant. Bhardwaj and Sahu (2014) also observed that Bavistin fungicide inhibit the mycelial growth of red rot fungi.

Major Diseases of Sugarcane in Bihar

Fungal Diseases

1. Sett Rot/Pineapple Disease (C.O.: *Ceratocystis paradoxa*)
2. Red Rot (C.O.: *Colletotrichum falcatum*)
3. Whip Smut (C.O.: *Ustilago scitaminea*)
4. Wilt (C.O.: *Fusarium sacchari*)
5. Pokkah boeng (C.O.: *Fusarium proliferatum/Fusarium moniliforme*)

Bacterial Diseases

1. Ratoon Stunting (C.O.: *Leifsonia xyli* sub sp. *xyli*)
2. Leaf Scald (C.O.: *Xanthomonas albileneans*)
3. Red Stripe (C.O.: *Acidovorax avenae* sub.sp. *avenae*)

Viral Disease

1. Sugarcane Mosaic (C.O.: *Sugarcane Mosaic Virus*)
2. Yellow Leaf Disease (C.O.: *Sugarcane Yellow Leaf Virus*)

Phytoplasmal Disease

1. Grassy Shoot (C.O.: *Candidatus Phytoplasma*)

Field Diagnosis of Sugarcane Diseases

1. Sett Rot/Pineapple Disease

This disease can be diagnosed as its symptoms first noticed in seed pieces after 3-4 weeks of planting. The pathogen of this disease enters through cut ends of seed piece. Initially, the central portion of affected setts becomes reddish colour which leads to brownish black colour in later stage. Internally the cane pith become red or later black red and emit an odour like pineapple. The setts decay before the bud sprout. Complete death of affected settlings may be noticed. (Viswanathan, 2012). By the infection 80 per cent cane seed does not germinate and decay in the field. Poor germination are due to decay or rotting of seed pieces before sprouting of buds. Some shoots died after attaining the height of 10-15 cm. Occasionally, this disease cause damage in standing crops and it is observed that drought condition in the field are the main reason for the disease development. It is also called pineapple disease because in infected seed cane emits an odor like that of fresh ripe pineapple. This smell is due to the production of ethyl acetate.

2. Red Rot

The disease can be observed in almost all stages of crop in the field and the characteristics diagnosis are noticed after cane formation. Saksena *et al.* (2013) observed that most common symptoms are discoloration and yellowing of the young leaves and it continued from the tip to the leaf base till the whole crown is withered and within 10-15 days plant dies. Initially the affected plants show orange to yellow discoloration in top 2-3 leaves, which can be easily distinguished from the surroundings of the green foliage (Viswanathan, 2010). Important symptoms can also be observed on the midribs as small reddish brown spots appear on the upper and lower surface of the sheath (Viswanathan *et al.*, 2011). During germination phase, the pathogen causes symptoms of pre germination death of the buds and drying of the germinated shoot. In case of post germination infection, death of settling occurs and affected plants show yellow discolouration and drying of entire shoots. Infected cane show drying singly or in clumps. Initially the affected stalks show orange to yellow discoloration in top 2-3 leaves and gradually entire foliage turns yellowish and dry. Brown or reddish brown stripe appear externally at nodal region. On splitting the internal tissue becomes red with white transverse band, tissues emit alcoholic sour smell. In advance stages canes become shrivelled and longitudinally wrinkled, such canes became lighter in weight and easily broken in the process of disease development (Duttamajumadar, 2008).

3. Smut

The initial symptoms are profuse sprouting of the lateral buds, with small and narrow leaves. The diseased canes become thin, slender and dwarf. The characteristics symptom of the disease is the production of black whip like structures from the top portion of the stalk. The whip is covered with a translucent silvery membrane enclosing black powdery mass of spores. The membrane of the whip ruptures releasing the mass of the black spores consisting of millions of smut spores. The most prominent symptom is the appearance of whip like structure, profuse tillering and narrow leaves (Sundar *et al.*, 2014). Whip generally appears at early stage (4 to 5 months) old where diseased setts used for planting. Smutted side shoot can appear on normal stalks as a result of secondary infection through air. The diseased plants produce more tillers than healthy plants and reduction in length and girth of such affected plants may be observed. The buds and root eyes are suppressed of 6-8 nodes below the whip were also observed.

4. Wilt

The symptoms of the wilt disease are mostly noticed after the monsoon. The earliest symptoms of the disease is the stunting, yellowing, unthrifty appearance of the affected clumps and withering of the top leaves. Such types of symptoms mostly observed when the plant attend 5 to 7 months old. According to Viswanathan *et al.* (2012) the symptoms were first observed on 4-5 month old plants gradually yellowing and drying of leaves and shriveling and breaking of affected canes. The midrib of affected leaves become yellow while the leaf lamina remains green. In advance stages, drying of the leaves, shrinking of stalk occur after splitting of such canes the internal tissues get brick red colour and later it becomes brown in colour. In severe case boat shaped cavities were formed and the whole stalk become lighter in weight, cane become hollow making them worthless for milling.

5. Pokkah Boeng/(Top Rot)

The peculiar symptom of this disease is chlorotic patches at the base of young leaves, the diseased leaves become partially white with reddish streaks reduce in size. The wrinkling, twisting and shortening of young leaves are observed at later stages. Malformation along with the twisting, shortening and wrinkling of uppermost young leaves which leads to top rot and death of whole top parts of the plants (Govender *et al.*, 2010). In initial stage the pokkah boeng disease showed whitish colouration, curling, twisting and bending of leaves from the top portion of the canes (Ranjan *et al.*, 2018). In acute infection, distortion of the stalk with internal and external knife cut like lesion and rotting of apical portions are commonly observed. The distortion of stalk and knife-cut stage were also observed by Viswakarma *et al.* (2013). The young affected leaves have several holes and occasionally the new emerging leaves destroyed completely and only pointed spindle core seen on the corns. A ladder like lesion with dark longitudinal edges might also be seen on spindle leaves, which showed dark reddish to brown in colour. The most serious injury is infection of the growing tip of the plant, which resulted in the loss of the entire top of the plant and is referred as top rot.

6. Ratoon Stunting

Germination of diseased setts is very low delayed and root system are poorly developed. Plant show stunted growth, reduced tillering, shortening of internodes and yellowish foliage. The disease is mostly observed in ratoon crops. In diseased fields stunted growth may be observed, after splitting the canes pink/orange-red coloration is observed on the nodes of the affected canes (Rott *et al.*, 2018). The characteristics symptoms of this disease are seen at the nodes of splitted canes and such affected nodes become round in shape. The disease reduces the length, girth and number of canes per clump.

7. Leaf Scald

White stripes appear on the foliage of infected plants and these stripes were seen on the full length of leaf blade. The stripes are narrow and are mostly located over the veins. In later stages, the stripes coalesce and resulting in the death of the leaves. At maturity of cane side buds (lateral buds) sprouts and shortening of internodes is also observed. Sometimes top leaves converted into bunch of leaf and no inflorescence comes out from the diseased canes. On splitting the affected canes showed reddening of vascular strands and oozing from cut ends can be seen. In severe case sudden wilting and drying of whole cane may be noticed. Govindaraju *et al.* (2019) also observed the characteristic symptoms as white pencil line stripes and patches of chlorotic tissues on the leaves, burning of top of leaves and in advance stages the margins of stripes may coalesce and reddish pencil like stripes may be formed. In later stages the internodes of affected canes become short and white streaks may be noticed and in advanced stage necrotic lesions may be seen.

8. Red Stripe

The appearance of the disease commonly observed first on young leaves of the crown near the midrib. Initially the stripes arise as water soaked on the both side of the leaf laminas and within a week turn into reddish brown in colour. The symptoms may be seen as green stripes on both of the leaf side, slowly turn reddish in colour and later become dark red or reddish brown. The bands are observed parallel to the leaf veins as like thin pencil edge (Viswanathan, 2012). In favourable condition the streaks spread very fast and wrapping whole length of leaf lamina. Sometimes two or more thin bands may merge and form straight broad bands close by the mid rib. In advance stage rotting of the top portion of the cane may be observed. The tops dry up and generally break off and drop to the ground and such top rotted canes secrete an undesirable smell which can be distinguished from a long range.

9. Mosaic

Symptoms are more prominently on the basal portion of the young leaves.

The pale yellow patches appear on green tissues of leaves. On leaves, scattered areas of lighter green colorations are seen. The disease account the pulling down in the chlorophyll content of the greenery parts of the crop. In severe case, these pale yellow patches also occur on leaf sheath and stalk exhibit elongated necrotic lesions and internodal parts also get affected and whole crop become stunted and chlorotic. The stunting of crop is depending on the level of varieties, temperature, and virulence of the strain.

10. Yellow Leaf Disease

The main diagnostic symptom is yellowing of the leaf midrib on the lower side in 3 to 6 leaves from the peak. Mostly the symptoms may visible at 6 to 8 months of the plant. The yellowing of mid rib and adjacent areas of laminar lost their primary colour are the sure and usual diagnostic symptoms. In later stages, the yellowing may spread from midrib into leaf blade and yellowing of the leaves can be observed. The spindle axis dries along with the leaves in advance stage of the infection. The characteristics symptom is yellowing of the leaf midrib on the lower side in 3 to 6 leaves from the top. In later stages, the yellowing may expand from midrib into leaf blade and yellowing of whole leaves were noticed by El Sayed *et al.*, 2015.

Origin and Spread of Pathogen

Sl.No.	*Diseases*	*Origin and Spread*
1.	**Sett rot(Pineapple)**	Affected seed setts, crop waste of previous year present in soil, rain water, irrigation water.
2.	**Red rot**	Affected planting seed materials, crop waste of previous year present in soil, wind, rain splashes, irrigation water, insects (borers).
3.	**Smut**	Affected planting setts, crop waste of previous year present in soil, wind, rain splashes, irrigation water, insects, birds.
4.	**Wilt**	Affected planting setts, crop waste of previous year present in soil, rain splashes, irrigation water.
5.	**Pokkah hoeng**	Previous infected seed setts, hot and dry weather followed by wet weather, Wind, rain splashes, irrigation water.
6.	**Ratoon stunting**	Infected seed setts, weed host, infected soil harvesting implements.
7.	**Leaf scald**	Infected setts, stubble, weed host, windblown rains, harvesting and sett cutting implements.
8.	**Red Stripe**	Infected setts, stubble, weed host, windblown rains, harvesting and sett cutting implements.
9.	**Sugarcane mosaic**	Infected setts, aphids, cutting implements.
10.	**Yellow leaf disease**	Infected setts, aphids.
11.	**Grassy shoot**	Infected seed setts, aphids, cutting implements.

11. Grassy Shoot

The diseased symptoms are detected after three to four months of planting. The disease is distinguished by formation of abundant tillers from the bottom of the diseased plants. The leaves turn pale yellow to entirely chlorotic and become small narrow and thin. The length of the internodes is reduced and grassy appearances of the whole clumps were observed. The colour of the entire clumps become whitish and hardly produce one or two canes which do not produce any milleable canes and inflorescence. Tiwari *et al.* (2012) also reported that the affected crops were bushy in appearance and stunted growth, leaves were chlorotic, failure of cane formation and later whole plant dries.

Extensive survey under cane growing areas of Bihar conditions during 2016 to 2019 carried out from Sugarcane Research Institute, Pusa revealed that the following varieties listed below were observed affected with different diseases at different locations during the period mentioned.

Naturally Occurring Diseases on Sugarcane Varieties during 2016-17

Sl.No.	Varieties	Diseases	Locations/Areas
1.	CoLk 94184, CoS 98231, Co 118, Co 0233	Wilt (2-5 per cent)	Sidhwalia, Manjhaulia, Murila (Harinagar)
2.	Co 5011	Pokkah Boeng (T), Mosaic (2 per cent)	Narayan pipri (Hasanpur)
3.	BO 141, BO 136	Smut (T-2 per cent)	Riga, Hatwa Farm
4.	CoS 92423, CoSe 95422	Red rot (2-5 per cent)	Narkatiyaganj and Pusa
5.	Co 0235	Red rot (T) + Grassy Shoot Disease (T)	Manjhaulia
6.	CoSe 95422 and CoS 8436	YLD (T)	Hasanpur

Naturally Occurring Diseases on Sugarcane Varieties during 2017-18

Sl.No.	Varieties	Diseases	Locations/Areas
1.	Co 0238	Yellow leaf disease (Trace)	Hasanpur, Chamua (Harinagar)
		Smut (5 per cent)	Madhopur (Sidhwalia)
		Pokkah boeng disease (5 per cent)	Babhni (Harinagar)
		Wilt (10 per cent)	Hardippur (Narkatiyaganj)
		Wilt (30 per cent), Red rot (20 per cent)	Ahir Tola (Areraj)
2.	CoLK 94184	Ratoon stunting disease (2 per cent)	Bucheya (Sidhwaliya)
		Wilt (5 per cent)	Sahpur (Gopalganj)
3.	Co 0118, BO 110	Wilt (5 per cent), Pokkah boeng disease (2 per cent)	Buchia and Sahpur
		Wilt (10 per cent)	Kharshan (Riga) and Buchia
4.	Co Pant 97222	Wilt (15 per cent)	Madhopur (Gopalganj)
5.	BO 154	Smut (T)	Riga
6.	Co 0235	Red rot (5 per cent), Wilt (5 per cent)	Majhaulia
		Grassy Shoot Disease (2 per cent)	Majhaulia
7.	BO 141	Smut (3 per cent), Wilt (2 per cent)	Hathua (Riga)
8.	Co 0233	Wilt (10 per cent)	Murila (Harinagar)
9.	CoS 8436	Red rot (10 per cent), Pokkah boeng disease (5 per cent)	Pusa
	CoSe 95422	Red rot (10 per cent), Grassy Shoot Disease (2 per cent)	Pusa
	CoV 92102	Yellow Leaf disease (T)	Pusa

Naturally Occurring Diseases on Sugarcane Varieties during 2018-19

Sl.No.	Varieties	Diseases	Locations/areas
1.	CoV 92102,Co0238, Co 0233	Yellow Leaf disease (T), Smut (T), Pokkah boeng disease (2 per cent)	Pusa, Sidhwalia, Majhaulia, Narkatiyaganj
2.	Co 03234	Smut (2 per cent), Wilt (5 per cent), Pokkah boeng disease (2 per cent)	Majhaulia
3.	CoP 2061	Smut (Trace)	Kalyanpur, Pusa, Majhaulia
4.	CoLK 94184	Ratoon stunting disease (Trace)	Sidhwalia
5.	Co 0118	Wilt (10 per cent), Pokkah boeng disease (5 per cent)	Gopalganj
6.	CoPant 97222	Wilt (20 per cent), YLD (2 per cent)	Gopalganj, Hasanpur
7.	Co 0235	Red rot (2 per cent), Wilt (10 per cent),	Majhaulia
8.	BO 141, Co 0233	Wilt (15 per cent), Pokkah boeng disease (2 per cent)	Riga
9.	BO 153, BO 154	Pokkah boeng disease (2 per cent)	Kalyanpur, Pusa
10.	Co 0238	Wilt (10 per cent), Pokkah boeng disease (5 per cent), YLD (T)	Hasanpur

In Sugarcane, preventive plant protection measures are much more needed as comparison to another crop. It is well fact that preventive measures can remove several future diseases. The actual barrier in the complete ending of the diseases have been the absence of proper action for application of control measures in the field. Another complex factor in the emergence of new strains of the pathogen as a result, evolved resistant varieties have failed to contribute against infection for over a large number of periods. Hence, plant protection measures are continuous and necessary process instead of short term pre-requisite for hold up disease level of sugarcane production per unit area, time and input.

The disease management strategies mostly rely upon form of pathogen which is accountable for carrying a disease from one season to another and preservation of diseases by soil, debris, wind, rain, irrigation water, weeds and insect vectors. Therefore, to mitigate the losses caused by diseases, an integrated disease management approaches becomes necessary for the management of the crop. An integrated disease management strategies does not show single method of disease management, certainly it comprises all the familiar methods of disease management such as cultural, mechanical, physical, chemical, biological as well as varietal resistance in a compatible way from land preparation to harvesting of crops. Therefore, to derive a suitable and effective disease management approach, some facts on various feature like diagnosis, detection, disease assessment, epidemiology and effect of several cultural operation is needed.

To sustain sugarcane farming, based on the past occurrence and present position of disease combination, an integrated disease management approach has been work out. However, it should be well fact that prevalence/popularity and status of various sugarcane diseases in a specific area are not stable but is an active one. The major responsible causes are entry of unrecommended and unreleased susceptible varieties in haphazard way from other zones/states, generation of new virulent strains of pathogen and frequent variation in weather conditions in a specific area of Bihar.

Integrated Disease Management schedule against disease have least care mostly because of the reality that fungicides and antibiotics are comparatively less poisonous than most of the chemicals. The bad effects of several chemicals have also been well established. So to overcome alike circumstances has mandatory to formulate integrated disease management strategies so that sugarcane farming could not bothered.

The Important Segments of Integrated Disease Management

(A) Soil management
- (i) Preparatory cultivation
- (ii) Deep ploughing

(B) Seed management
- (i) Regulative policy
- (ii) Selection of varieties
- (iii) Selection of seed
- (iv) Sett treatment
  - (a) Chemotherapy
  - (b) Thermotherapy
  - (c) Chemo cum thermotherapy

(C) Crop management
- (i) Season of planting
- (ii) Crop rotation
- (iii) Drainage and irrigation
- (iv) Companion cropping
- (v) Monitoring and Roughing
- (vi) Ratoon management/Discontinuation of Roughing
- (vii) Water management
- (viii) Weed management
- (ix) Fertilizer application
- (x) Management through chemical and bio-agent

(A) Soil Management

(i) Preparatory Cultivation

Since inoculum of red rot, smut, wilt, root rot and sett rot diseases remains viable in soil or plant debris, crop residues such as stubbles, dry foliages as well as dead and dried stalks and roots of canes should be destroyed to reduce the inoculums of these diseases from the field. The smut spores also can be covered interior to the soil.

(ii) Deep Summer Ploughing

A good deep tillage of the soil during summer season helps in exposing the inoculum to sun. Due to high temperature during summer, inoculums will destroyed. During this process, crop residues are buried in the soil along with inoculum. Under such conditions red rot and smut spores do not survive in soil and larvae of insect pest present inside the soil also destroyed by birds.

(B) Seed Management

(i) Regulative Policy

For planting of sugarcane, there is need to evolve high quality seed material and such precious materials brought from one place to other place. During the movement, various seed transmitted diseases like wilt, red rot, ratoon stunting, grassy shoot disease, Leaf scald and mosaic *etc.* have been established in locality where the diseases were lacking. Thus, it becomes very essential to follow the vigorously restriction, so that such materials having affected with diseases may be destructed before it reach to the field.

(ii) Selection of Varieties

It is the cheapest way to the cane growers for managing the sugarcane diseases. Although, replacing susceptible varieties with resistant one is the most common way to control the diseases but, breeding and selection of varieties with high resistance against the pathogens keep on changing their behaviour in nature. Minnatullah *et al.* (2016) observed that six genotypes were resistant, sixteen genotypes were moderately resistant and they were found susceptible to red rot disease. Ranjan *et al.* (2018) and Wang *et al.* (2017) evaluated many varieties against pokkah boeng and observed several varieties having resistant to moderately resistant.

(iii) Selection of Seed

Since sugarcane is vegetatively propagated crop, the chances of the pathogen carry on feasible in or on setts. It is also difficult to detect incipient infection of seed piece transmissible disease in seed pieces. It is therefore, advisable to use the seed material from a crop, free from diseases as it bear the primary source of spread. The best way to insure freedom from the diseases is to grow small seed crop nurseries preferably raised from heat treated seed materials. Before planting, both the cut ends of setts with red and black spotted areas should be discarded. In such process, most of the seed piece borne infection can be kept out. In a particular area, at least 5-6 varieties should be planted. The planting material from tissue culture should be prefer to check yellow leaf disease of sugarcane. Soil borne pathogen can be escaped by utilizing three/four budded setts for planting. The collateral host like sorghum should be eliminated to control the sugarcane mosaic virus which is transmitted by aphids (Anwar *et al.*, 2010). The setts cutter should be properly disinfected while cutting the setts for planting (Anwar *et al.*, 2010). The Yellow leaf disease can be managed by planting of healthy setts as disease spreads mainly from the infected setts (El Sayed *et al.*, 2015).

(iv) Sett Treatment

Chemical treatment is used for protection of cut ends to avoid entry of soil borne pathogens (Sett rot, red rot, smut and wilt) through cut ends. Sett treatment with dipping of setts for 30 minutes is recommended. This will not only protect the cut ends from soil borne pathogens but also enhance sett germinability by 10 per cent. Setts are also treated with heat to inactivate or eliminate the pathogens. In thermotherapy, the differential effect of heat on the pathogen and the host is taken advantage of freeing the infected stalks of grassy shoot, ratoon stunting, chlorotic streak, mosaic, spike, leaf scald, downy mildew and smut diseases. In this process, pathogens present in or on setts are inactivated or eliminated at temperature which is not lethal to host tissue. In India 4 types of heat treatment plants have been fabricated for treating the sugarcane seed setts. Impact of sett treatment for the management of primary source of red rot pathogen have been also established previously (Malathi *et al.*, 2016). According to Minnatullah *et al.* (2010) Bavistin was found more effective than *Trichoderma viride* and *Trichoderma harzianum* enhancing sett germination and in reducing red rot disease development.Sett treatment as well as spraying not only reduced the pokkah boeng incidence but also enhance the quantitative and qualitative parameter (Poletine *et al.*, 2012). Ranjan *et al.* (2018) observed that sett treatment as well as spraying with carbendazim was found more effective than other tested fungicides in respect of quantitative as well as qualitative parameters. Sett treatment and three foliar spray at 15 days intervals with carbendazim showed the maximum germination and low pokkah boeng disease incidence in sugarcane (Anonymous, 2016). The wilt can be controlled by sett treatment with aretan @ 0.1 per cent followed by 0.1 per cent Carbendazim along with soil treatment with boric acid @ 15 kg/ha (Singh *et al.*, 1985). The setts should be treated with hot water at 50°C for 3-4 hrs to manage the leaf scald (Govindaraju *et al.*, 2019).

Heat therapy was not only found effective against certain seed borne diseases but also against scale insects, mealy bug and larvals of borers. Recently hot water plus fungicides are now being used in treating the sugarcane setts because this treatment not only provide protection of setts from soil inhibiting pathogens but also stimulate the germination. For protecting the seed materials before planting, the following practices are commonly used. The hot water treatment of setts is reported to reduce the smut infection and it also led to thicker and heavier canes production (Joshi, 1954; Gupta *et al.*, 1978). The moist hot air treatment is also reported to control the smut disease (Misra *et al.*, 1978). The hot water treatment of setts at 55°C for 30 minutes combined with the chemical treatments with bayleton @ 0.1 per cent completely eradicate the setts from smut infection (Mameghmay, 1984). The sugarcane smut can be managed effectively with endophytic bacteria (Jayakumar *et al.*, 2018). The disease free setts should be selected and it can be obtained by heat therapy or tissue culture for the management of ratoon stunting disease (Hoy 2017; Young 2018). The setts can be treated with the hot water at 52°C for 18 minutes to manage the ratoon stunting disease (Anwar *et al.*, 2006). The grassy shoots can be controlled by hot water treatment of the setts at 52°C for 18 minutes (Anwar *et al.*, 2006). The grassy shoot disease incidence can be eliminated by aerated steam therapy of setts at 50°C for 1 hour (Viswanathan, 2001).

Sl.No.	Nature of Heat Therapy	Temp.	Duration	Effects
1	Hot-air	54°C	8 hours	Desiccation of buds and poor germination
2	Hot-water	50°C	2 hours	Softening of buds observed. control of GSD, RSD and Smut diseases.
3	Moist-hot-air	54° C	4 hours	No adverse effect. Most of the pathogens are destroyed
4	Aerated steam	50° C	1 hours	No adverse effect.It is effective against GSD, RSD, Leaf Scald and Smut diseases of sugarcane.

C. Crop Management

(i) Season of Planting

Season and date of planting have a significant influence on disease development. Autumn planted cane crop showed resistance against diseases because its susceptible phase passes at time (March-June) when climatic conditions are unfavourable for pathogen. Spring planted crop is vulnerable to attack as the crop passes through the susceptible phase in (July-October). This time climatic conditions are favourable for pathogen development. Therefore, planting time should be alike that it will not match with the desirable conditions for the development of pathogen.

(ii) Crop Rotation

It is cheapest and best old preventive method. Monoculture of sugarcane over large areas provides congenial conditions for the rapid buildup of inoculums. To break this, crop rotation with green manure crop should be followed because green crop not only increase the population of antagonistic micro flora but also helps in the production of vigorous root system and also helps in controlling soil borne pathogen. Growing of non host crop for a certain period after diseased crop leads towards famine of pathogen and finally its death. Various crops may discharge certain bio chemical compounds that leads to deadly effect to the pathogen. Crop rotation with rice is effective in managing wilt disease, and also red rot affected field should be exchange with rice or green manure crop to arrest the expansion of the disease. Smut disease can be minimized by showing of Arhar as intercropped between rows of sugarcane. So 2-3 years crop alternation should be practiced with non host crop. The crop rotation of sugarcane with paddy or companion cropping of sugarcane with coriander is helpful in suppressing wilt disease (Singh *et al.*, 1985).

(iii) Drainage and Irrigation

Flooding and ill drained conditions help in the spread of inoculums of red rot, smut, wilt and sett rot diseases. To avoid this upland should be selected as far as possible, failing which good drainage facilities should be ensured for mitigating the diseases. Irrigation water should not flow from diseased field to disease free field and avoid preparation of irrigation channel from diseased to healthy field to restrict the spread of diseases. Good drainage facilities and planting of seed setts in 1-2 cm depth can down the risk of various seed as well as soil borne diseases.

(iv) Companion Cropping

Sugarcane when grown with coriander, garlic, linseed and mustard as a companion crop, they substantially reduce wilt incidence. The incidence of red rot was also reduced when sugarcane was grown after the harvest of garlic, coriander, linseed, mustard and onion.

Impact of Different Companion Crops on Red Rot and Wilt Diseases of Sugarcane

Sl.No.	*Treatment*	*Red Rot Incidence (Per cent)*	*Wilt Infection (Per cent)*
1	Garlic + Sugarcane	6.5	8.2
2	Onion + Sugarcane	4.6	7.2
3	Coriander + Sugarcane	4.2	8.6
4	Linseed + Sugarcane	2.8	7.4
5	Mustard + Sugarcane	3.2	5.2
6	Sugarcane alone	20.2	25.6

(v) Frequent Monitoring and Rouging

Frequent and critical survey of cane field should be carried out carefully and diseased clumps should be removed from the field at an early stage to remove the foci of infection and thus preventing wide dissemination of inoculums. In un-roughed field, the incidence of smut disease has been increased considerably year after year. The smut and grassy shoot disease can be managed by destroying the diseased clumps by roughing. After rouging there is a considerable reduction in smut disease infection was observed.

Development of Smut Disease in Rouged and Un-rouged Crop in different Varieties

Sl.No.	*Varieties*	*Incidence of Smut (Per cent)*		
		Plant Crop	*Ratoon Crop Un-rouged*	*Rouged*
1	Bo 141	22.2	45.6	8.0
2	Bo 154	15.0	32.0	6.6
3	CoSe 95422	18.6	34.5	7.0
4	Bo 146	12.4	20.3	4.2
5	Co 1158	30.4	62.2	12.6

(vi) Ratoon Management

After the harvest of sugarcane crop the underground portion of the stubble gives rise to a succeeding crop known as ratoon crop. The number of ratoons in a sugarcane production cycle varies from 1-8. In Bihar raising one or two ratoons is most common though there are instances of multiple ratoons also. Poor ratoon management is the most important limitation in boosting sugarcane productivity. Ratoon cropping of sugarcane is being followed about 40-45 per cent of the total cane areas in subtropical region. However, the yields of ratoon cane are comparatively much poor (30-35 t/ha) as comparison to plant cane (65-70 t/ha). The vital cause in higher expenditure and weak occupancy with advocated practices by the farmers in the field. There is need to give higher priority of management tactices like varietal selection, method and time of harvesting, stubble shaving, dismantling of ridges, gap filling, soil and ratoon treatment, fertilizer application, water management, earthing up, propping, weed management *etc.* for productive cultivation of ratoon crop.

In ratoon crop number of tillers are formed from the sprouting of buds, the inoculums existing in the bud. To restrict further spread of pathogens affected ratoon crop should be discontinued. Ratooning ability is an also essential for good yield, however, it may vary from different location to location. In Bihar situation, BO 110, BO 91, CoP 9301, BO 154, BO 153, CoP 2061, CoP 112 and Rajendra Ganna-1 are having potential ratooning varieties. Harvesting of standing crop should be done adjacent to ground level, in roughly harvested field stubble shaving is necessarily required.

(vii) Water Management

Ratoons require more frequent irrigation than plant crop because of shallower root system. Therefore, irrigation at 15-20 days interval are required in sub-tropics. Total 4-6 irrigations are needed before monsoon. For sugarcane cultivation water requirements differs during various seasons. The climate and soil variations also play an important role in water demand in sub-tropical conditions and it varies from 130-175 cm.

(viii) Weed Management

Sugarcane is a long duration crop, it is affected by several types of weeds. In plant and ratoon crop, mulching with trash fallowed by post emergence utilization of weedicides has been observed effective. Combined effect of weedicide (Atrazine) and intercultural operations like hoeing at 60 and 90 days after planting was found effective. In case of sugarcane ratoon, three hoeing 30, 60 and 90 days after harvest or Atrazine 2.0 kg/ha pre emergence along with hoeing of 60 and 90 days after harvesting was effective in boosting the yield.

(ix) Fertilizer Management

Ratoons are less efficient utilizes of N than plant crop. Therefore, comparatively higher dose of nitrogen is required. Response to phosphate and potash is more consistent. Therefore, 20 per cent more nitrogen than plant crop is advocated. The dose of NPK-125:30:75 kg/acre inorganic fertilizer should be applied in split doses, first split dose should be at ridge breaking, 45 kg N, entire P and 25 kg K should be applied in ratoon crops. Two more splits each with 40 kg N, 25 kg K, after 30 and 60 days of ridge breaking as top dressing. $ZnSO_4$ or $FeSO_4$ @10 kg/acre and foliar application of 0.5 per cent $ZnSO_4$ or $FeSO_4$ should be applied for improving the cane growth and yield.

(x) Management through Chemical and Bio-agent

Several fungicides have been observed in reducing the disease in vitro condition but could not give sufficient result in field conditions. Impervious nature of rind, presence of fibrous nodes in cane restricts the flow of chemicals and scarcities of efficient systemic fungicides are the major factors of the fungicides to give adequate result in preventing the diseases.

Sett treatment before planting and three sprays with carbendazim @ 0.1 per cent at 15 days interval was found effective against Pokkah boeing, red rot, wilt and sett rot disease of sugarcane. Sett treatment with carbendazim fungicide @ 0.1 per cent before planting for 30 minutes has been found effective against the diseases. After removing the whip smut two prophylactic sprays with propiconazole @ 0.1 per cent at 20 days interval has been recommended. Foliar Spray of copper oxycloride and dithiocarbamate fungicide against foliar diseases has been found effective for higher germination and cane yield. Spraying with plantomycin @ 100 gm/Ac in 200 lt. after two months of planting at 15 days interval is affective against bacterial diseases of sugarcane. Two sprays of imidacloprid @ 0.5 ml/lt after detrashing of dry leaves of monthly interval to check transmission by insect vectors was recommended. The soil borne inocula can be managed by the application *Trichoderma* culture @ 10 kg/ ha mixed with compost.The sugarcane sett treated with *Trichoderma* suspension (Conidia 10^{6}/ml) for 15-30 minutes before planting has been found effective against sugarcane diseases.Singh *et al.*, 2008, also reported that the sugarcane setts can be treated with *Trichoderma* suspension (conidia 10^6/ml) for 30 minutes to control red rot disease. The bioagents like *Trichoderma* spp. can be effectively employed for wilt disease management (Viswanathan *et al.*, 2008).Malathi *et al.* (2012) observed that under in vitro conditions complete inhibition of red rot pathogen at 10 ppm of Thiophanate methyl showed satisfactory results in the management of the red rot disease. Bhardwaj and Sahu (2014) observed that out of four tested fungicides against the red rot pathogen, Bavistin showed complete inhibition of mycelia growth of the fungus. The field infection of leaf scald can be effectively managed by spraying streptomycin + tetracycline @ 60 g/500 l water for 1 hectare area (Govindaraju *et al.*, 2019). The pineapple disease of sugarcane can be controlled by application of *Trichoderma harzianum* as it forms lysed zones and inhibits growth of *Ceratocystis paradoxa* (Talukder *et al.*, 2007; Ajayi *et al.*, 2016).

REFERENCES

Ajayi, A.M., Olufolaji, D.B. and Lajide, L. (2016). Pineapple disease of sugarcane: An assessment of three Trichoderma spp. for use as possible biocontrol agent. *International Journal of Advanced Academic Research*, 2(8): 23-33.

Anonymous, (2016). Report of the principle investigator (Plant Pathology). All India Coordinated Research Project on Sugarcane, *Sugarcane Breeding Institute, Coimbatore, (T.N).*

Anuradha, Sanghera, G.S., Kashyap, L., Kumar, R., Pal, R. and Singh, P. (2018). Incidence of pokkah boeng disease on sugarcane clones and varieties in Punjab. *Agriculture Research*, 55 (4): 754-756.

Anwar, M.S., H.W.A. Khan, M. Umer and A.A. Chattha. (2006). Variabilties of red rot (*Collectotrichum falcatum*, Went) in sugarcane germplasm. Abst. *41st Ann. Conv. PSST. 21-22. Aug. 2006 Rawalpindi*, pp. 28.

Anwar, M.S., Khan, H.W.A., Munir, M., Chattha, A.A. and Zia, A. (2010). Integrated control strategies for sugarcane disease. *Pak Sugar Journal*, pp. 8-10.

Bhardwaj, N. and Sahu, R.K. (2014). Evaluation of some fungicides, botanicals and essential oils against the fungus *Colletotrichum falcatum* causing red rot of sugarcane. *The bioscan.* **9** (1): 175-178.

Duttamajumadar, S.K. (2008). Red rot of sugarcane. *Indian Institute of Sugarcane Research, Lucknow,* India.

El Sayed, A.I., Komor, E., Boulila, M., Viswanathan, R. and Odero, D.C. (2015). Biology and management of sugarcane yellow leaf virus: an historical overview. *Arch Virol,* 160(12): 2921-2934.

Gawade, D. B., Pawar, H., Gawande, J. and Vasekar, V. C. (2012). Antagonistic Effect of *Trichoderma* against *Fusarium moniliforme* the Causal agent of Sugarcane Wilt. *American-Eurasian Journal of Agriculture* and *Environment Science.* **12 (9):** 1236-1241.

Gonender, P., Mcfarlane, S.A. and Rutherford, R.S. (2010). Fusarium species causing pokkah boeng and their effect on *Eldana saccharina* walker (Lepidoptera: pyralidae). *Proc. S. Afr. Sug. Technol. Ass.* 83: 267-270.

Govindaraju, M., Li, Y. and Zhang, M. (2019). Emerging Bacterial Disease (Leaf Scald) of Sugarcane in China: Pathogenesis, Diagnosis and Management. *In: Plant Diseases-Current Threats and Management Trends. DOI: http: //dx.doi.org/10.5772/intechopen.88333.*

Gupta, S.C.; Verma, K.P., Singh, M.P. and Misra, S.C. (1978). Control of diseases by hot water treatment of sugarcane seed material. *Indian Sug. Crops J.,* Vol. 5, pp. 28.

Hoy, J. W. (2017). "Is it possible to beat RSD?" *Sugar Journal* 11: 8–9.

ISMA (2019). Indian Sugar Mills Association (ISMA). *Indian Sugar.*

Jayakumar, V., Sundar, R.A. and Viswanathan, R. (2018). Biological Suppression of Sugarcane Smut with Endophytic Bacteria. *Sugar Tech,* https: //doi.org/10.1007/s12355-018-0684-1.

Joshi, N.C. (1954). Hot water treatments of setts for the control of red rot and smut diseases of sugarcane. *Indian sugar,* N.S., 4, pp. 2.

Kalaimani, T. (2000). Effect of chemical sett dips and foliar spray in the control of sugarcane red rot caused by *Collectortichum falcatum* Went. *Indian sugar* **49**: pp. 973-976.

Malathi, P. and Viswanathan, R. (2012). Variation in *Colletotrichum falcatum*. Red rot pathogen of sugarcane in relation to host resistance, *Sugar Tech,* 14(2): 181-188.

Malathi, P., Viswanathan, R. and Amalraj, R. (2016). Mechanized means of sett treatment: An effective way of delivering fungicides for the management of red rot in sugarcane. *Sugar Tech,* 18: 1-7.

Minnatullah, MD., Kumar, A and Chand, H. (2016). Screening of sugarcane genotypes against major diseases of sugarcane and borer pests, *The Bio Scan*, 11(3): 1891-1894.

Minnatullah, MD., Kumar, S., Akhtar, R. and Dohare, S. (2010). Management of red rot disease through sett treatment with bioagaents. *Environment Ecology*, 28(4A): 2660-2661.

Minnatullah, MD. and Kamat, D.N. (2018). Losses due to red rot pathogen in cane juice quality. *International Journal of Current Microbiology Applied Science*.**7** (02): 13-16.

Misra, S.R., Prasad, L., and Singh, K. (1978). Heat therapy against seed piece transmissible disease. Effect of moist hot air treatment on smut disease. *Ann. Rept. Indian Nst. of Sugarcane Research*, Lucknow

Poletine, J.P., Maciel, C.D., Zanato, M.D. and Silva, T.R.B. (2012). Efficiency of seed treatment with fungicides in castot bean crop genotypes. *Journal of Food, Agriculture and Environment*, 10(2): 512-516.

Ranjan, A., Minnatullah, MD., Chand, H., Singh, S.P. and Dubey, S. (2018). Evaluation of varieties against pokkah boeng disease and its natural occurrence. *International Journal of Curent Microbiology and Applied Sciences*, 7(10): 44-50.

Rott P, Sood S, Comstock J C, Gilbert R A, Sandhu H S (2018). Sugarcane Ratoon Stunting. *http;//edis.uft. edu/pdffiles/sc/sc00200.pdf (accessed 16 October 2020).*

Saksena, P., Kumar, S.K., Viswakarma, R., Tiwari, A.K., Singh, A. and Kumar, A. (2013). Pathological and molecular variation in *Colletotrichum falcatum* Went isolates causing red rot of sugarcane in the north west zone of India, *Journal of Plant Protection Research*, 53: 37-41.

Shailbala and Amarendra Kumar (2017). Effect of Thiophanate Methyl on Red rot Disease *Collectorichum falcatum* and yield attributes of sugarcane *Journal of Pharmacognosy and Phytochemistry*, SPI: 163-167.

Sharma, R. and Tamta, S. (2015). A Review on Red Rot: The "Cancer" of Sugarcane. *Journal of Plant Pathology and Microbiology*. **S1** (3): 1-8.

Singh, N., Singh, R.P. and Lal, S. (1985). Effect of sett borne inoculums of wilt pathogens and subsequent growth of sugarcane. *Acta Botanica Indica*, 13: 212-217.

Singh, V., Srivastava, S.N., Lal, R.J., Awasthi, S.K. and Joshi, B.B. (2008). Biological control of red rot disease of sugarcane through *Trichoderma harzianum* and *Trichoderma viride*. *Indian Phytopath*. 61(4): 486-491.

Sundar, A.R., Barnabas,L., Malathi, P. and Viswanathan, R. (2012). A Mini-Review on Smut Disease of Sugarcane caused by *Sporisorium scitamineum*. *In*: *Botany*, pp 107-109.

Talukder, M., Begum, F. and Azad, M. (2007). Management of Pineapple Disease of Sugarcane through Biological Means. Journal *of Agriculture and Rural Development*, 5(1): 79-83.

Tiwari, A.K., Bharti, Y.P., Mishra, N., Tripathi, S., Lal, M., Sharma, P.K., Rao, G.P. and Sharma, M.L. (2010). Biotechnological Approaches for Improving Sugarcane Crop with Special Reference to Disease Resistance. *Acta Phytopathologica et Entomologica Hungarica*, **45** (2): 235–249

Tiwari, A.K., Kumar, S., Vishwakarma and Rao, G.P. (2012). Increasing incidence of sugarcane grassy shoot disease in Uttar Pradesh, India and its impact on yield and quality of Sugarcane. *Phytopathogenic Mollicutes*, 2(2): 63-67.

Vishwakarma, S.K., Kumar, P., Nigam, A., Singh, Aand Kumar, A. (2013). Pokkah Boeng: An Emerging Disease of Sugarcane. *Journal of Plant Pathology and Microbiology*, 4(3) : 170

Viswanathan R, and Padmanaban P. (2008). Hand Book on Sugarcane Diseases and their Management.: *Sugarcane Breeding Institute, Coimbatore*. pp 72

Viswanathan, R. (2001). Different aerated steam therapy (AST) regimes on the development of grassy shoot disease symptoms in sugarcane. *Sugar Tech*, 3: 83-91.

Viswanathan, R. (2010). Plant Diseases: Red rot of sugarcane, New Delhi, *Anmol publications.*

Viswanathan, R. and Rao, G. P. (2011). Disease scenario and management of major sugarcane diseases in India. *Sugar Tech* **13***:* PP. 336-353.

Viswanathan, R., Malathi, P. and Sundar, A.R. (2008). Wilt of crops with special reference to cultural, morphological, molecular characterization and pathogenic variability of isolates in India, *Annual Report 2007-2008, Sugarcane breeding institute, Coimbatore,* pp 49-50.

Viswanathan, R., Poongothai, M., Malathi, P. and Sundar A.R. (2012). Sugarcane Wilt: New Insight into Pathogen Identity, Variability and Pathogenicity. *Functional Plant Science and Biotechnology*, 6(2): 30-39.

Viswanathan, R., Sundar, R.A., Malathi, P. and Padmanaban, P. (2011). Red rot of sugarcane, *Sugarcane Breeding Institute, ICAR, Bulletin Coimbatore*, pp. 1-7.

Wang, Z.P., Duan, W.X., Li, Y.J., Liang, Q., Zhou, Z.G., Zhang, M.Q. and Lin, S.H. (2016). Establishment of resistant evaluation system in the field for sugarcane pokkah boeng. *Journal of South China Agricultural University*, 37(3): 67-72.

Wang, Z.P., Sun, H.J., Guo, Q., Xu, S.Q., Wang, J.H. and Lin, S.H. (2017). Artificial inoculation method of pokkah boeng disease of sugarcane and screening of resistant germplasm resources in subtropical China. *Sugar Tech*, 19(3): 283-292.

Young, A. J. (2018). "Turning a blind eye to ratoon stunting disease of sugarcane in Australia." *Plant Disease* 102: 473–482. https: //apsjournals.apsnet.org/doi/pdf/10.1094/PDIS-06-17-0911-FE.

Chapter 16

Sustainable Management of Leaf Hopper (*Amrasca biguttula biguttula*) on Eggplant/Brinjal (*Solanum melongena* (Linn.) and Related Crops

Sunil Kr. Ghosh[1] and Rajib Karmakar[2]*

[1]Discipline of Agricultural Entomology,
[2]Department of Agricultural Chemicals, Faculty of Agriculture,
Bidhan Chandra Krishi Viswavidyalaya, Directorate of Research,
Kalyani, Nadia – 741235, West Bengal

Pest complex of Eggplant/brinjal is very high in West Bengal particularly in North Bengal, India causes heavy yield loss and their control is very difficult. The jassid, Amrasca biguttula biguttla (Ishida) is a serious pest of brinjal throughout the country causes heavy damage to egg plant/brinjal crop and other vegetables. Leafhopper, Amrasca biguttula biguttula (Ishida) (Homoptera: Cicadellidae) is an alarming pest throughout the crop growth. This pest has assumed serious proportions on Bt cotton in many parts of India and has developed resistance to several commonly used insecticides on cotton. Farmers rely frequently on chemical pesticides, for pest control which produce harmful effect. IPM is one of the most important components in the production process containing all suitable methods and techniques of pest management. Microbial toxin, plant extracts and botanical insecticide are biopesticides having less or no hazardous effects on human health and the environment, and therefore, they can be incorporated in IPM programmes and organic farming.

Introduction

Vegetables rank next to cereals and occupy a very important place in human diet and serve as supplementary source of nutrients like carbohydrates, proteins, vitamins, minerals and energy for all. Besides nutrition, vegetables act as an important source of dietary fiber comprised of inassimilable cellulose, hemicelluloses, protein and lignin. The centre of origin for brinjal is India with China as a secondary centre. It is grown year round throughout most part of India including West Bengal (Ghosh, 1999) and Bangladesh (Singh, 1967).

Pest complex of brinjal is very high in West Bengal particularly in North Bengal, India causes heavy yield loss and their control is very difficult (Ghosh and Senapati, 2002, Ghosh, *et al.*, 2003). Pests control by using pesticides causes damage to the natural suppression of the pests by killing parasites and predators like spider (Ghosh *et al.*, 2006a), *Menochilus sexmaculatus* Berliner (Ghosh *et al.*, 2007), *Coccinella septempunctata* (Chakraborty and Ghosh,2010; Ghosh, 2016). Among the various insect pests causing limitations, shoot and fruit borer, *Leucinodes orbonalis* Guen. has become an alarming pest in eggplant/brinjal growing areas including West Bengal (Ghosh and Senapati,2001a; Ghosh and Senapati,2009; Ghosh,2014). The other important Lepidopteran pests in West Bengal are Brinjal leaf roller (*Eublema olivacea* Walk.) and Brinjal stem borer (*Euzophera particella* Rag) (Ghosh, 1999). The important Hemipteran pests causes heavy damage are Aphid (*Aphis gossypii* Glov.) (Ghosh *et al.*, 2004a, Ghosh,2019 a), Jassid (*Amrasca biguttula biguttula* Ishida) (Ghosh, and Senapati 2003), Thrips (*Thrips tabaci* Lin.) (Ghosh *et al.*, 2005) and Whitefly (*Bemisia tabaci* Genn) (Ghosh *et al.*, 2004b). The important Coleopteran pests are Hadda or spotted leaf beetle (*Henosepilachna vigintioctopunctata* Fabr.) (Ghosh, and Senapati 2001b), Flea beetle (*Phyllotreta* spp.) (Ghosh *et al.*, 2006b) and Blister beetle (*Mylabris pustulata*) (Ghosh, 2020). Mite (*Tetranychus cinnabarinus* Boisd.) also causes heavy damage to eggplant (Ghosh and Chakraborty, 2014; Ghosh, 2019 b).

The jassid, *Amrasca biguttula biguttla* (Ishida) is a serious pest of brinjal throughout the country causes heavy damage to brinjal crop and other vegetables. leafhopper, *Amrasca biguttula biguttula* (Ishida) (Homoptera: Cicadellidae) is an alarming pest throughout the crop growth. This pest has assumed serious proportions on Bt cotton in many parts of India and has developed resistance to several commonly used insecticides on cotton. Jassid has become the serious and most destructive pest of all the vegetables as well as other ornamental plants and agronomics crops also. *Amrasca biguttula biguttula* caused damage from young seedling to mature crops results in 50 per cent yield loss (Halder *et al.*, 2016).

Leaf Hopper, *Amrasca biguttula biguttula* (= *Empoasca devastans* Dist.) (Cicadellidae: Homoptera)

Leafhopper/Jassid

Amrasca devastans Distant/The preferred scientific name for this leafhopper is *Amrasca biguttula biguttula* Ishida (CABI 2012).

(Hemiptera: Cicadellidae)

Taxonomic Rank (according to GBIF backbone taxonomy)

Kingdom: Animalia

Phylum: Arthropoda

Subphylum: Atelocerata

Class: Hexapoda (including Insecta)

Infraclass: Neoptera

Subclass: Pterygota

Order: Hemiptera

Suborder: Auchenorrhyncha

Superfamily: Cicadoidea

Family: Cicadellidae

Subfamily: Typhlocybinae

Tribe: Empoascini

Genus: *Amrasca*

Species: *Amrasca biguttula* (Ishida, 1912)

Subspecies (Subject): *Amrasca biguttula biguttula* (Ishida, 1912)

Geographic Distribution

Leaf hopper/okra jassid (*Amrasca biguttula biguttula* Ishida) is a versatile and widely distributed insect. It occurs in several countries including India, Bangladesh, China, Myanmar, North Africa, Pakistan, Philippines, Sri Lanka, Japan (Honshu Island), Afganistan, Thailand, Oriental region, Vietnam and Taiwan. It is distributed widely throughout Eastern, Western, Southern, Central Africa and Australia (Rahman *et al.,* 2014). It is distributed all over India; but particularly serious in Sind, Punjab and Tamil Nadu. In India it is widely distributed in UP, Bihar, Odisha, W.Bengal, Assam, Tripura, Haryana, Punjab, MP, Gujrat, Punjab, Maharastra, Kerala, TN *etc.* Relatively dry (mean temperature around 32°C) and humid (RH around 70 per cent) weather favours population build-up.

Host Plants Range

This is a highly polyphagous pest attack several malvaceous plants. Apart from feeding on okra, the jassid have a very wide range of the host plants, including herbaceous cultivated plants and weeds, chiefly amongst the Malvaceae, Legunmiosae and Solanaceae. There are about 17 host plants species belong to 8 different families infested by jassid (Rahman *et al.,* 2014). It is a polyphagous pest attacking Abelmoschus esculentus (okra), Gossypium (cotton), Solanum melongena (aubergine/brinja), beans, castor, cucurbits, hollyhock, Solanum tuberosum (potato), Helianthus annuus (sunflower) and other malvaceous plants. The other host plants/species affected are as follows: *Amaranthus* (amaranth), *Arachis hypogaea* (groundnut), *Beta vulgaris var. saccharifera* (sugarbeet), *Cajanus cajan* (pigeon pea), *Calendula* (marigolds), *Cassia* (sennas), *Chloris gayana* (rhodes grass), *Corchorus* (jutes), *Glycine max* (soyabean), *Guizotia abyssinica* (niger), *Hibiscus cannabinus* (kenaf), *Hibiscus sabdariffa* (Jamaica sorrel), *Morus alba* (mora), *Phaseolus vulgaris* (common bean), *Raphanus sativus* (radish), *Solanum lycopersicum* (tomato), *Sorghum bicolor* (sorghum), *Vigna radiata* (mung bean), *Vigna unguiculata* (cowpea), *Zea mays* (maize).

Although initially considered to be a minor or occasionally serious pest of tea, *Empoasca flavescens* has now been established as a regular and serious pest in tea plantations of sub-Himalayan North Bengal in India (Saha and Mukhopadhyay, 2013). The leafhopper is also considered as a severe pest of castor, tea, okra, mulberry and many other crops (Anjani, 2012; Jayasimha *et at.*, 2012). It also occurs on various grasses, including lawns of Bermuda grass (Singh *et al.*, 2018).

Appearance/Morphology

They are small insects (1-3 mm long), green in colour with slender tapered bodies. Leafhoppers are very mobile. The adults hop away when disturbed. Nymphs resemble adults, but have no wings, and run sideways when disturbed. The adult cotton jassid is a long and slender insect about 2.6 mm (0.1 in) in length. It is yellowish-green, with a conspicuous black spot on either side of the head and another near the tip of the fore wing. The head is pale green and the membranous wings transparent and iridescent. On leaf surfaces, the insect tends to move about diagonally (Jayasimha *et at.*, 2012) and when disturbed it immediately jumps and flies away (CABI, 2019). Leafhopper adults are elongated, wedge shaped and somewhat triangular in cross-section. They jump and fly off readily. Depending on species, they range in size from 1/8 to 1/2-inch and their bodies are colored yellow, green, gray or they may be marked with color patterns. Nymphs resemble adults but are wingless. They can run rapidly, occasionally sideways, and hop.

Description Body Parts (CSIRO 1991)

Adult

Colour - Usually yellow or green. Gum tree leafhoppers (Eurymelidae family) are often brightly coloured with vivid red, white, blue black and yellows.

Body - Up to 28 mm. Often wedge shaped.

Wings - 2 Pairs, both used in flight. Held in a tent like position when at rest. Coupled. Adults can fly.

Mouthparts - Stylet. Sucking. Near rear of head.

Antennae - 3-10 segments.

Eyes - 2 Compound and 2 simple.

Legs - Adapted for jumping. Feet (Tarsi) have 4 segments.

Head – Broad. Swollen. Wedge shaped

Thorax - 3 segments.

Abdomen - Eight pairs of spiracles on underside.

Egg - ovoid.

Habits - Can't fly when cold. Suck sap from plants. Jump or hop when disturbed and move to the far side of branches or leaves. Some species are tended by ants who feed on the honeydew produced by the Leafhopper. Often gregarious or in colonies.

Nymphs

Similar to adults. Usually 5 stages (instars). Often have no simple eyes (ocelli).

Autogenerated Description (Zhang, 1990)

Anal hook well-developed, but not extended more than half distance to ventral margin of genital capsule, simple, smooth, curved mesad, not crossing midline. Anal tube ventromedial process absent. Subgenital plate not fused, plates free throughout length. Subgenital plate basolateral angle weakly developed. Subgenital plate lateral margin straight or evenly convex in basal half. Subgenital plate mesal margin distinctly concave. Subgenital plate dorsomedial lobe absent. Subgenital plate group C (macrochaetae) with numerous macrochaetae, reaching apex of plate. Subgenital plate group C setae uniseriate. Subgenital plate group A (basal marginal setae) well differentiated (longer than group B setae) but not greatly thickened. Subgenital plate group A with more than 5 setae. Subgenital plate group B (distal marginal setae) well differentiated, continuous from near midlength to apex. Subgenital plate group D (long, fine setae) with distinct distal and basal groups of elongate setae. Subgenital plate dorsal margin of distal half in lateral view straight. Subgenital plate dorsomedial surface without conspicuous sensilla. Subgenital plate apex compressed. Subgenital plate length in ventral view very long and slender (>5X longer than maximum width). Connective free, well sclerotized, wider than long. Connective lateral margin distinctly constricted. Connective anterior margin straight or weakly concave, without median lobe. Connective corsal keel present on midline posteriorly. Style apophysis (section anterad of articulation to connective) much longer than apodeme. Style preapical lobe absent. Style dentifer not abruptly narrower than basal section. Style distal teeth numerous, well distributed, not densely grouped at apex. Style apex in ventral view curved laterad. Style preapical setae sparse. Styleapex tapered. Aedeagus preatrium longer than half height of genital capsule. Aedeagus dorsal apodeme very short or absent (pair of weakly sclerotized dorsolateral arms may be present). Aedeagus postatrium absent, little if any development of atrial complex beyond base of shaft. Aedeagus shaft orientation in profile > 60° divergence from line of preatrium. Aedeagus shaft shape in lateral view tapering distally - dorsal and ventral margins distinctly convergent through most of length. Aedeagus shaft(not including processes) moderately long, 0.3-0.5X pygofer height. Aedeagus shaft basal half in crossection tubular. Aedeagus shaftin caudoventral view tapering through most of length (broadest in proximal half). Aedeagus atrium in caudoventral view not greatly expanded laterally, less than 2x average width of preatrium or shaft. Aedeagus basal processes absent. Aedeagus distal paired processes absent. Aedeagus distal unpaired process absent. Aedeagus lateral flange absent. Aedeagus dorsal teeth absent. Aedeagus ventral surface smooth. Predominant color of dorsum pale green or yellow with few symmetrical cream-colored markings on head and thorax. Brown or tan mottling absent. Crown{ color pattern without round spots. Anteclypeus concolorous with rest of face, not contrastingly dark. Face without paired dark dorsal spots, without median spot, without median dark brown or black longitudinal stripe. Frontotclypeus without orange ellipse. Pronotum midline concolorous with rest of sclerite. Pronotum without dark band along posterior margin. Forewing inner apical cell unmarked. Forewing brachial cell with black spot distally. Forewing membrane without numerous irregular dark spots. Aedeagus preatrium in lateral view straight. Segment X base with well-sclerotized dorsal arch. Style base linear or slightly sinuate, without distinct bend at point of articulation to subgenital plate. Aedeagal shaft in lateral view (not including processes) curved dorsad. Aedeagal shaft well sclerotized throughout. Segment X apex densely microtrichiate ventrally. Forewing costal vein concolorous with rest of wing. Mesosternum concolorous with rest of thoracic venter. Subgenital plate setal group A encompassing midlength of plate. Male abdominal sternite VIII without conspicuous pair of lateral setae. Mesonotal triangles pale. Mesonotum without dark median spot. Base of pygofer without protuberance on anterolateral margin. Lorum conspicuous, delimited by suture.

Head

Crown in dorsal view distinctly longer medially than next to eye. Crown equal to or longer than width between eyes. Head distinctly wider than maximum width of pronotum. Coronal suture not extended beyond midlength of crown. Lateral frontal sutures extended to or near ocelli, obsolete dorsomedially. Ocelli present, separated from eyes by > ocellar diameter. Eyes in dorsal view as wide as or wider than distance between basomedial angles. Face with maximum length 1-1.5X width across eyes. Male anteclypeus flat or slightly convex. Rostrum distinctly overlapping base of hind coxae.

Wings

Forewing venation with apical cells 2 and 3 both quadrate basally, veins RP and MA free, connected by crossvein. Forewing veins ScR and M free preapically. Forewing apical cell 4 strongly tapered through most of length, MCu ending near apex of anal margin. Forewing CuP distal segment 2X or more longer than segment of CuA between Cu stem and MP. Forewing vein MCu distinctly curved. Punctations along veins absent or indistinct. Hind wing CuA completely confluent with MP (only one apical cell present). Hind wing first apical cell more than 2X longer than wide. Hind wing base of first apical cell considerably distad of apex of CuP.

Legs

Front femur row AV with 1 basal seta distinctly enlarged. Front femur AM1 distinctly enlarged. Middle femur with 1 dorsoapical macrochaeta. Hind tibia row AV with 4 preapical macrochaetae.

Abdomen

2S apodemes extended well past posterior margin of sternite III but little or no beyond posterior margin of sternite IV, parallel or nearly so. Second tergal structure if developed at all, with only a ventrally directed low wall, absent or less developed medially. Third tergal apodemes (and perhaps fourth and beyond) absent. Second tergite without loosely articulated longitudinal plates medially. Posterior margin of 8th sternite angulately emarginate medially.

Male Genitalia

Pygofer paired dorsolateral lobes present. Pygofer distal lobe partly separated from base by distinct dorsal membranous cleft. Pygofer distal lobe apex in lateral view angulate. Pygofer ventral appendage present and free of pygofer wall for at least half its length, elongate, extended well beyond apex of pygofer. Pygofer ventral appendage in lateral view angled upward near base, thence relatively straight. Pygofer ventral appendage in caudoventral view relatively straight through most of length, but curved mediad near apex. Pygofer ventral appendage simple, acuminate. Pygofer ventral appendage vestiture smooth. Pygofer basal portion of ventral appendage not or only slightly broadened. Pygofer ventral appendage without stout spurs apically. Pygofer macrochaetae absent. Pygofer small rigid distal setae scattered over distal portion of lobe. Pygofer mesal surface of lobe without setae. Pygofer mesal sensilla field well developed, conspicuous. Pygofer basolateral setal group absent. Pygofer long, fine setae sparse, inconspicuous.

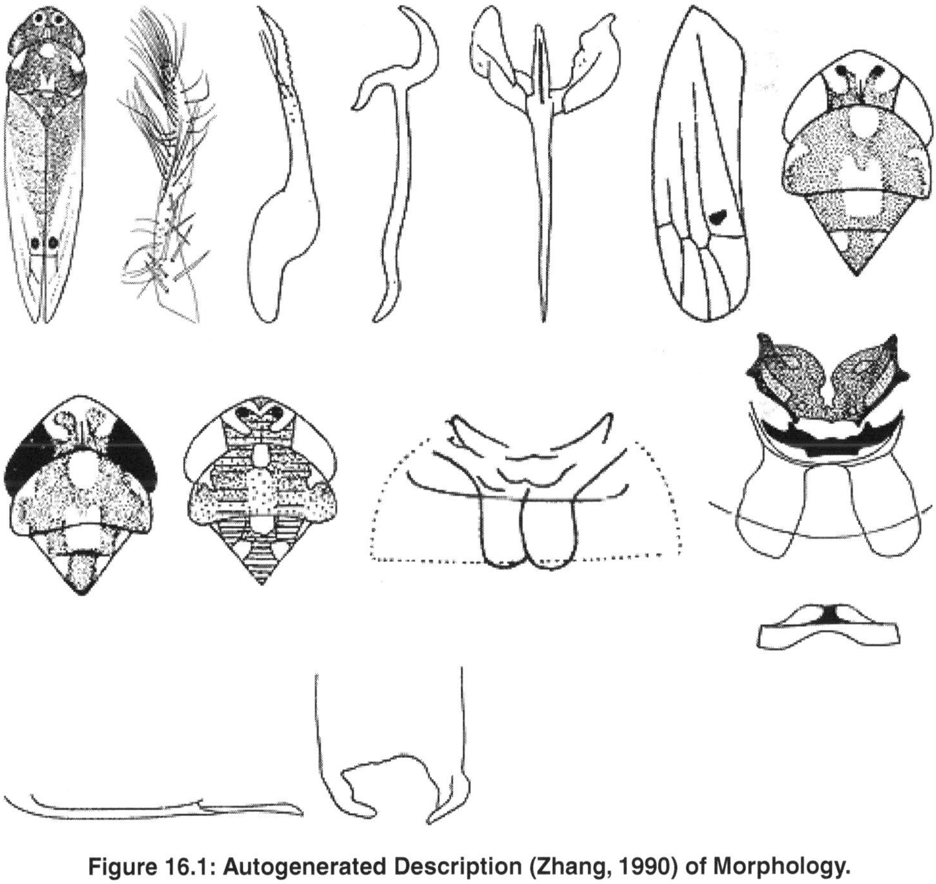

Figure 16.1: Autogenerated Description (Zhang, 1990) of Morphology.

Marks of Identification

- ☆ **Nymph -** Light green, translucent, wingless found between the veins of leaves on the under surface
- ☆ **Adult -** Green, wedge shaped leafhopper.
- ☆ **Movement** -They have a characteristic way of walking diagonally in relation to their body.

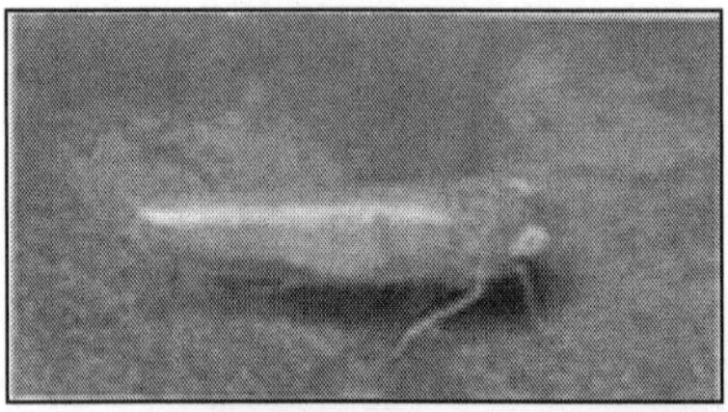

- ☆ The adult is wedge-shaped about 2 mm long and pale green in colour.
- ☆ The forewings have a black spot on their posterior parts.
- ☆ The nymphs are wingless, pale green in colour and are found in large numbers on the lower surface of the leaves.

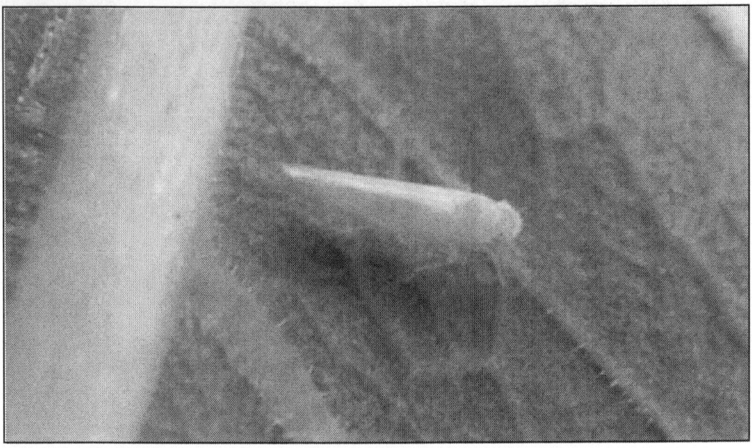

Figure 16.2: An Image of the Leafhopper (*Empoasca flavescens*).

Biology and Life Cycle of Jassid

Egg

The females scratch the leaf tissues and young stems to lay the eggs inside the veins and midribs of the leaves. Eggs are curved, greenish-yellow eggs (0.7-0.9 X 0.15-0.2 mm) (sometimes whitish) are oviposited deeply implanted in the midrib or a leaves vein on either surface of leaf or in a petiole or young stem but never in the leaf lamina (Vennila *et al.*, 2007). On okra, eggs are mainly oviposited inside the tissue of leaf blades, but may also be laid in leaf stalks or in soft twigs. Depending on species, 29-60 eggs can be laid singly and they hatch in 4-11 days (generally one week). The fecundity of females is about fifteen eggs (Jayasimha *et al.*, 2012).

Nymph

Nymphs are flattened, pale yellowish green (Vennila *et al.*, 2007), wedge-shaped, 0.5-2.0 mm long, moves diagonally, confined underside of leaves in daytime, however, during night it can be found anywhere on the leaves. Wingless nymphs hatch from eggs in about 10 days and begin feeding on the tender new growth of their host plant. They develop through five stages (instars) over a period of 12 to 30 days depending on food availability and temperature, leaving shed skins in the feeding area. Nymphal have five instar with the duration of i1.5, i1.1, i1.2, 1.5 and i2.0 days. Mean pre-mating, pre-observation, ovipositionl and post-oviposition period were 2.55, 3.45, 16.57 and 3.90 days respectively. As nymphs grow larger, they develop wing pads. The nymphs resemble the adults, but lack wings. They are pale green in color. They tend to move sideways when disturbed.

Adult

Leafhoppers undergo direct development from nymph to adult without undergoing metamorphosis. They are about 1-3 mm length, body is elongate and wedge shaped with pale green colour. They have fully developed wings. The black spots are present in forewings as well as vertex. The adults may live for one to two months with females living a little longer than males.

The insect is not known to hibernate and if the conditions are favorable rapid multiplication takes place. Most species overwinter as eggs or adults. A generation is completed in a period of one month. The life cycle is completed in 14 to 30 days. Duration of life cycle of jassid was recorded minimum (25.11 days) in June-July and being maximum (36.36 days) in October-November (Ghosh and Senapati, 2003). About 11 generation have been observed in a year, the duration of each varying from 15 to 46 days, but there is considerable overlapping. The adults are generally long-lived and can tide over during adverse climatic conditions. A generation likes 3-4 weeks in the summer jassid is estimated to have 11 generations in a year in India.

The life cycle of the leafhopper is described below in the Figure 16.2 from http://www.yourarticlelibrary.com/zoology/tea-green-fly-*Empoasca*-flavescens-distribution-life-cycle-and-control/24048.

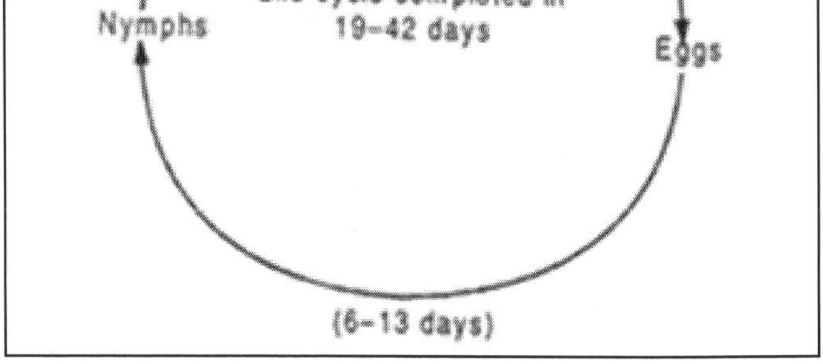

Figure 16.3: The Life Cycle of the Leafhopper.

Nature of Symptoms and Damage

Amongst the important sucking insects pest jassid attacks and causes heavy loss to okra crop. They sucks the plant sap leads to discolouration of the leaves, loses in yield up to 35-40 per cent can increase up to 60-70 per cent during optimal environment (Sultana *et al.,* 2016). Maximum damage was reported during mid season reducing yield (Sahita *et al.,* 2017). Both nymphs and adults suck the sap from the leaves and cause phytotoxic symptoms known as hopper burn/drying of leaves which results in stunted growth and complete desiccation of plants and has become one of the limiting factors in economic productivity of the crop. Fruit setting may be very low and the number of flowers, fruits goes down. Affected leaves become yellow, crinkled, curled and show marginal browning. Under heavy attack, leaves turn yellow and then brown and dry, giving a burned appearance. When several insects suck the sap from the same leaf, yellow spots appear on the leaves, followed by crinkling, curling, bronzing, and drying, or "hopper burn"/rim blight. Both nymphs and adults suck the sap from the lower leaf surfaces through their piercing and sucking mouthparts. While sucking the plant sap, they also inject toxic saliva into the plant tissues, which leads to yellowing. Leafhoppers also damage in okra, cotton, and potato. The nymphs and adults are pale green and are found near the main veins on the under surface and move sideways when disturbed. The eggs are inserted in the leaf tissue on the underside of leaflets. They feed mainly on the underside of eggplant leaves, causing small yellow patches on leaves. Infested leaves sometimes curl upwards along the margin. Leafhoppers multiply rapidly during dry spells and can cause extensive damage. *A. biguttula biguttula* and *Hishimonus* (Dist.) are vectors of brinjal mosaic virus and brinjal little leaf disease. The edges of the leaves become turn pale-green, then yellow and finally brick-red or brown. Severely affected plants have stunted growth. Egg-laying habits can also cause some plant injury. Singh *et al.* (2018) reported that in many areas, this pest regularly occurs on cotton in epidemic numbers.

List of Symptoms/Signs for Identification of Damage

Tender leaves- become yellow

Leaves - abnormal colours

Leaves- margin shows curling downwards and reddening sets in

Leaves - abnormal leaf fall

Leaves - necrotic areas

Leaves - severe infestation leaves becomes browning, typical "hopper burn" symptom

Leaves - necrotic areas (hopper burn/rim blight)

Leaves- leaf margins broke and crumble into pieces when crushed

Leaves- dried up and shed down and stunted growth

Whole plant - dwarfing

Figure 16.4: Hopper Burn.

Seasonal Incidence

Shukla (1989) studied the population dynamics of the pyralid *Leucinodes orbonalis* and the cicadellid *Amrasca biguttula biguttula* (Ishida) on brinjal in the field in Megalaya, India. The population of these pests was positively correlated with average temperature, mean relative humidity and total rainfall. Field studies in Tamil Nadu to determine the relationships between growth stages of brinjal and the occurrence of insect pests made by The age of brinjal plants did not affect the multiplication of *Amrasca bigttula bigttula, Coccidohystris insolitus* and *Henosepilachna vigintiopunctata*, where plants age did affect the incidence and build up of *Amrasca bigttula bigttula, Aphis gossypii, Urenticus hystricellus* (Richt.) and *Lecucinodes orbonalis*. Higher population level was maintained during 3rd—16th standard week with peak population (2.23/leaf) was recorded on 10th standard week in Coochbehar, West Bengal, India (Subba *et al.*, 2015).

The population dynamics of jassid, on okra cv.Azad bhindi-1 in relation to weather factors, during kharif seasons in 2005 and 2006 at Kanpur, Uttar Pradesh, India (Devi *et al.*, 2018). Jassid actively started from first week of August on 3-week-old crop in 2005. In 2006, jassid infestation on shoots started from the fourth week of July on 7 leaf-stage until the third week of September. The seasonal abundance of jassid, on okra was investigated by Halder *et al.* (2016) in Assam, India in 1998-99. Result signifies that meteorological parameters performed a key role in building up of population of cotton jassid. The population of jassid was highest in the end week of May,1998 (37.53 nymphs per leaf) and mid of April in 1999 (30.00 nymphs per leaf). The invasion of jassid begins from 4th week of July and attained in peak on 2nd as well as 4th weeks of September, respectively. In June pest appear and stayed active throughout crop season. The pest population observe below the economic threshold level for around five weeks after germination. The population crossed the threshold level in early June and remained same till late August. Peak population of the pest is in late July (27.8 individuals/leaf). Sowing done on 15th May and 1st June had a lesser incidence of *Amrasca devastans* with a good outputs of marketable fruits (22.9 q/ha). Cicadellid populations remained below the economic threshold level of one insect/leaf for up to 35 days of plant age in khraif and 65 days of plant age in rabi. Most of the cicadellids were found in 35 to 75 days old plants in kharif and 65 to 130 days old plants in rabi season. Plants growth in the kharip season was more vulnerable to insect attack than plant grown in the rabi season (Yadav *et al.*, 2008). Rana (2006) reported that among the sowing dates, 20 February and 5 march for spring sow in and 2 April and 5 June for rainy season sowing gave the least incidence of jassid nymphs.

In an experiment by Laxman and Maheswari (2017), on the effect of dates of sowing on incidence leafhopper on castor genotypes during kharif season 2013, it was revealed that incidence of leafhoppers was in general comparatively high in late sown crop than early sown crop. When castor was sown early, the hopper burn scored was 2 with 11 to 25 per cent injury on varieties DPC-9, DCS-9, PCH-111, PCH-106 and Kranthi and whereas when the crop was sown late, maximum hopper burn score of 4 with more than 50 per cent hopper injury was observed on DPC-9, DCS-9, PCH- 288, PCH-111, RG-2835, PCH-106, PCH-222 and RG-2928. These observations in the experiment above confirm work done by Akashe *et al.* (2015).

Influence of the Weather Factors, on Jassid Population (Temperature, Rainfall and Relative Humidity)

The weather conditions, prevailing in a season, play a vital role in the incidence and subsequent upsurge of the population of insect-pests. Insects are mostly dependent on the environmental temperature for the maintenance of their activity. Prolonged periods of low or high temperatures or sudden changes in them adversely affect the insect development. They found a positive correlation between the maximum and minimum temperatures with the density count of the leaf hopper, while it is found a significant relationship between *A. biguttula biguttula* population-level and mean air temperature. Optimum temperature for the population build up of pest was 27-34°C. The jassid numbers to be negatively correlated with the maximum temperature (34°C) and positively correlated with the minimum temperature (25°C) on both 'okra' and cotton crops. However, statistically, these relations were non-significant on the cotton crop. Patel *et al.* (1997), in their study, observed significant relationship between *A.biguttula biguttula* population-level and maximum temperature (r = 0.76). The population of *A. biguttula biguttula*, increased during the monsoon season with an increase in the maximum temperature from 30-34°C and the severity was found by them to be more when maximum temperature was greater than 36°C. However, there existed a non significant correlation between the population and temperature. They found a positive association of *A. biguttula biguttula* population with the minimum temperature, while the association was seen to be negative with maximum temperature, during winter 1991. All the regression coefficients were non-significant. However, during summer 1992 the population revealed a significant positive association with the maximum temperature, on the brinjal crop. A high temperature (30-36°C) favored the development of jassid population. The maximum and minimum temperatures showed a positive and non-significant correlation with the jassid-population on 'okra'. The maximum and minimum temperatures to be positively and significantly correlated with the population changes of *A. biguttula biguttula*. Recently Solangi *et al.* (2008) reported that the maximum population of jassids (3.93 plant-1), on cotton, was found at a relative mean temperature of 32.88°C.

Jena and Kuila (1996) also reported that leafhopper infestation had a positive correlation with maximum temperature whereas it was negatively correlated with sunshine. The variable effect of different weather parameters on the pest population might be due to the difference in phenology of the crop and time of appearance of the pest at different localities, where crops have been grown. The activity and multiplication of the leafhopper is enhanced in the cold and humid weather of the winter season. Mounica *et al.* (2018) also experimented in 2018 and indicated that the correlation studies on leafhopper population with abiotic factors showed that maximum temperature, minimum temperature, rainfall, number of rainy days and evaporation were negatively correlated but relative humidity and sunshine hours were positively correlated with leafhopper population. A leafhopper population on castor had a significant positive correlation with the maximum temperature and a significant negative correlation with the morning Relative Humidity (RH) in an experiment conducted by Gaur (2014). Their findings conform to the reports of Jayasimha *et al.* (2012) who reported that leafhopper population on okra exhibited positive correlation with the maximum temperature and negative correlation with RH.

Srinivasan *et al.* (1981) reported that the rainfall reduced the mean density and increased aggregation among *A. biguttula biguttula* on 'okra' crop. Similar results were reported by Lal *et al.* (1990), that a continuous rainfall was unfavorable for the population build-up of jassids. Prasad and Logiswaran (1997), found a negative association of *A. Biguttula biguttula* population with the rainfall, during winter 1991. All the regression coefficients were non-significant, while during summer 1992, the population revealed a significant negative association with the rainfall, on brinjal. Correlation studies between the rainfall and jassid population, was significant and negative (Dhamdhera *et al.* (1984), Sekhon and Singh (1985), Lal *et al.* (1990)). Low rainfall period coupled with bright sunshine hours, favored the development of jassid population (Gogoi and Dutta, 2000). The correlation study of Kumawat *et al.* (2000) revealed that a negative non-significant correlation was found between the rainfall and jassid population on 'okra'. This view was also supported by Mahmood *et al.* (2002), who found rainfall to be negatively and non-significantly correlated with the population fluctuations of this pest. Rainfall showed a significant and positive effect on the jassid population. Too much rain or dry weather is not favorable for the development of the insect (Widayat and Winasa 2006). *Empoasca* is most damaging during the period between the dry and rainy seasons.

Management of the Pest

Field Inspection

Inspection should be done weekly interval after the emergence till the crop has six true leaves and represent as jassid/plant. The adult as well as nymphs should be check under side of the leaves. Monitoring of the insect should be done from the underside of the leaves from 20-30 plants. Checking the symptoms should be done from tip damage which appears as extensive blackening of the small leaves within the terminal. Adults are highly mobile and nymphs are very small so counts may vary with the time of day. The damage potential also varies with crop vigour and temperature. For seedlings to six true i\leaves the threshold is 80 per cent reduction in leaf area and 10 jassids/plant (adults and nymphs).

Cultural Practices

Resistant Varieties and Use of Trap Crop

Choose tolerant or resistant cultivars with hairy leaves, as the length and density of the trichomes (hairs) repel leafhoppers. Indian cultivars such as Manjari Gota, Vaishali, Mukta Kesi, Round Green, and Kalyanipur T3 and Bangladeshi variety Bagun 6 are reported to be less susceptible or tolerant to damage. The brinjal cultivars Purple long, Nepali and Neelum were quite resistant to leaf hopper. The varieties Sigatoka beauty and Sitara as moderately susceptible. Host-plant resistance is the most reliable, economical and eco-friendly measure to control leafhopper (Anjani *et al.*, 2018). It was reported that biophysical resistance of brinjal plant to the pest was imparted by long, more erect hairs on the mid- rib, high density with longer hairs on the – leaf –lamina, and high density of hairs on the sides-veins the thickness of lamina and midrib were positively correlated with infestation by the pest, *Amrasca biguttula biguttula*. Gaikwad *et al.* (1991) observed that the varities Manjari Gota and Vaishali were resistant to the cicadellid *A. biguttula biguttula* and also reported that the morphological characters such as leaf thickness, midrib thicknes and leaf area were positively correlated with infestation of level of A. *biguttula biguttula* while the trichome length and trichome density on leaves negatively correlated with the infestation level. The biochemical factors such as the total leaves of sugars free amino acids and poloyphenols in the leaves of brinjal were negatively correlated with the numbers of infesting *A. biguttula biguttula.* In cotton leafhoppers show less preference for varieties of Gossypium arboreum than G. hirsutum. Among the G. hirsutum types resistance is associated with long hair length and tough leaf lamina on the ventral surface of leaves. long internodes in variety HB 69 contributing to resistance to leafhopper attack. On cotton, it has been found that growing a cultivar with hairs on the undersides of the leaves reduces infestation, and that long hairs are better at deterring the insect from laying than are short hairs; this seems to be due to the hairs preventing the insect from getting close enough to the leaf surface to deposit its eggs (Khan and Agarwal, 1984)

Grow okra as a trap crop along the borders of an eggplant field; if pesticides are required to control an infestation, they can be applied to the trap crop. *A. biguttula*

biguttula incidence was significantly lower on cotton intercropped with sunflower, green gram (*Vigna radiata*) and black gram (*Vigna mungo*) than cotton alone.

Intercropping as a Control Measure

Intercropping is an alternative pest management strategy, which can reduce herbivorous pests by making the host-plants difficult to locate, or by increasing the effectiveness of the natural enemies (Faselt, 2014). Intercropping is an essential cultural practice in pest management which is based on the principle of reducing insect pests by increasing the diversity of an ecosystem (Risch, 2005). Most studies indicated that diversification practices such as intercropping in pigeon pea and other crops are beneficial because these practices reduce pest damage (Songa *et al.*, 2007). Rao *et al.* (2012) found that there was a diversity created by introducing cluster bean, cowpea, black gram, or groundnut as intercrops in castor which resulted in a build up of natural enemies (Microplitis, *Coccinellids* and spiders). As a result of the build up of natural enemies, there was much less pest incidence and damage in castor intercropped with cluster bean, cowpea and groundnut compared to the castor mono crop. Rosmana *et al.* (2015) reported that intercropping can reduce the infestation of green leafhopper (*Nephotettix virescens* Distant) on rice varieties in Indonesia and showed a lower incidence of the pests on rice in the intercropped systems than the mono cropped plants.

Mechanical Control

Yellow Sticky Traps

A number of mechanical devices are used for a specific purpose of destruction, reducing or preventing insect populations or their attack on plant. Several colors of sticky traps are used, to sample and control pests in the greenhouses and field crops. Monitor the insects with yellow (570-580 nm) sticky traps placed at random in the field. Yellow is the color of sticky-traps which are used the most to attract insects. Yellow sticky trap, can be used for monitoring and controlling flying insect and seems to give a reliable indication of the general population trend of test insects. These traps are hung immediately above the plants. Locally available, empty, yellow tin coated with grease/vaseline/castor oil, on the outer surface, may also be used. Iron board also painted in yellow colour can be used. Yellow sticky traps, can be used to observe whether the insect population is rising or falling. This can help to keep a vigilant eye on any problem before it arises. More whiteflies and beetles were captured on yellow sticky trap than on traps of other colors. Similarly, Raja and Arivudainambi (2004) reported that the highest number of *A. biguttula biguttula*, per sticky trap (89.82) was recorded in the yellow, with red border traps while the lowest (18.07) was recorded in red-trap in 'okra' crop. According to Held and Boyd (2008) reported that in comparison of captures of adult *Gynaikothrips uzeli* on blue, yellow, white and green sticky traps found the yellow to be the most effective colour.

Biological Control

Avoid the use of broad-spectrum pesticides to encourage the performance of natural enemies.Generalist predators such as ladybird beetles and green lacewings are highly efficient in preying on leafhopper nymphs and adults. Parasitoids such as

Anagrus flaveolus Waterhouse and *Stethynium triclavatum* Enock are effective against leafhopper. Jassid predators *viz.*, the spider, *Lycosa pseudounnulatai* (Blackwall) in rice ecosystem (Krishnaiah *et al.*, 1984), *Chrysoperla* (Yadav *et al.*, 1990), *C. carnea* and *C. septumpunctata* (Kaethner, 1991), predatory mites *Amblysieus* spp (Guddewar *et al.*, 1994) are effective. The population of different predatory in okra ecosystem, spiders, Chrysopids, *Apantelesi* sp. and Coccinellids were the most dominant (Rosaiahi *et al.*, 2001).

Use of chemical insecticides, induces resistance in insects and destroys the natural pest control agents. Therefore, alternatives to insecticides are receiving much attention. Augmentative release of *Chrysoperla carnea* suppresses the infestation of pests to a sub-economic level. Balasubramani and Swamiappan (1998) observed, in a laboratory experiment, that early-instars nymphs of *A. biguttula biguttula* were preferred as a prey by the *Chrysoperla* spp, and each larva of the predator consumed 4.8, 20.06 and 59.6 first-, second- and third-instar nymphs respectively. Fourth- and fifth-instar nymphs and adults were highly mobile and could not be preyed upon by the Chrysopids. While Pawar (1991) recommended the release of *Chrysoperla carnea*, larvae at the rate of 100,000, per hectare, thrice at fortnightly intervals, for controlling the sucking pests including jassids. Recently, Solangi *et al.* (2008) reported that jassid population was below the economic injury level, at all phonological stages of cotton plant, due to a regular increase of the predator population, including Green lacewing.

Biopesticides with Special Reference of Botanical Control

Spray neem products. Commercial and simple (home-made) neem products give control of leafhoppers on eggplants. Thus, an aqueous neem seed extract (10g/l) sprayed at 10 days interval show repellent effects against leafhoppers (Empoasca) in India. Weekly applications of aqueous neem kernel extracts (6 to 50g/l) and of neem oil (5 to 10 l/ha) control the leafhopper. Products derived from the leaves and kernels of 'neem' (*Azadirachta indica*) are becoming popular, in plant protection programmes, mainly because the synthetic pesticides have shown a negative effect on the environment and also on human health. 'Neem' products are suitable for the modern concept of pest control, *i.e.*, IPM and biological control, owing to their unique mode of action on the target insects, limited side-effects on the natural enemies, absence of toxicity to the warm-blooded organisms and owing to being environmental friendly. Considering the importance of pests as well as the ecological and economical aspects, the utilization of 'neem' extracts seems to be the most promising for the small-scale vegetable production (Dreyer and Hellpap, 1991). Spray neem products. Commercial and simple (home-made) neem products give control of leafhoppers on eggplants. Thus, an aqueous neem seed extract (10g/l) sprayed at 10 days interval show repellent effects against leafhoppers (Empoasca) in India. Weekly applications of aqueous neem kernel extracts (6 to 50g/l) and of neem oil (5 to 10 l/ha) control the leafhopper. Spraying of neem oil, neem seed, neem leaf extract, neem seed kernel extract against jassid, aphids, whiteflies as well as thrips were very successful. Shahid *et al.* (1992) tested the 'neem' samples (20 per cent neem oil + 0.3 per cent Azadirachtin), in different concentrations, against the jassids on brinjal and reported that amongst its various concentrations, 16 per cent neem sample gave the best results. The reduction in population was 45.28, 57.91 and 76.19 per cent, after 7 days of the 1st, 2nd and 3rd spray, respectively. Rao and Rajendran (2002) in a laboratory, studied the joint action potential of methanolic extract of 'neem' seed kernel (NSKE) in combination with the methanolic extract of two other botanicals, *viz.*, *Acorus calamus* and *Pongamia glabra*, against 'okra' leaf hopper, at ratios of 1:1:1, 2:1:1 and 3:1:1 (v/v). These combinations recorded the lowest settling response and highest mortality of leaf hoppers, while under field conditions; these combinations at 0.42 per cent concentration gave a superior control of jassids. Neem, green chilli as well as garlic extract mixed with water in the ratio of 1: 2 for managing jassid and aphid (Kanvarjibhai *et al.*, 1993). Thomas *et al.* (1994) reported the hot water extraction of highly pungent chillii along withi few bits of asafetida were effective for managing leafhopper andi mite pests. For controlling jassid crushed garlic bulbs dipped in kerosene for 24 hours and mixed with ground chilli with soap solution and sprayed the crop for effective control of jassid (Thomas *et al.*, 1995). Vijayalakshmi *et al.* (1996), found an effective against jassid, aphids and whiteflies by spraying neem and garlic extract alone and in combination along with additional plant extracts *viz.*, garlic, ichilli, ginger, ineem, tobacco and even cow urine. It was observed that extracts of *Polygonum* plant and *Pongamia* leaves at a concentration of 5 per cent and the microbial insecticide spinosad gave higher Jassid control, recording more than 50 per cent mortality. (Ghosh and Chakraborty, 2015) Lakshmanan *et al.* (2001) found neem oil and neem seed kernel extract alone or in combination along with garlici bulb extract, kerosene, chillii and other extracts effectively managed ijassid, whiteflies, thripsi and tetranychidi mites which infested a number of crops.

Seed Treatment

Sucking insect pests, invade the crop on its early stage of growth and for their effective control seed-treatment is an easy, economic and feasible method. Moreover, this method can be used effectively, even during the rainy season. Seed treatment reduces the usage of pesticides. Jassids can be effectively controlled by using seed-treatment insecticides. Imidacloprid, having no adverse effect on the germination, belonging to nitroguanidines and its systemic action, after its uptake through root system is especially effective against the sucking pests. It is found imidacloprid (17.8 SL) at 9 ml/kg seed, and 70 WP at 10 g/kg seeds to be promising against the jassid-attack and effectively controlled the jassid-attack up to 35 days, after sowing. The seed-treatment with thiamethoxam (Actara 25WG) @ 12 ml/kg of seed was, at par, with imidacloprid seed treatment in reducing leaf hopper infestation. In imidacloprid (Confidor 70 WS) treated plot, jassids remain below the economic threshold level up to seven week, after sowing of the cotton crop. Kannan *et al.* (2004) reported that the seed-treatment of transgenic cotton, with imidacloprid @ 5 g/kg of seed, was effective in keeping the population of leaf hopper, below the economic threshold level up to 40 days, after sowing. Imidacloprid @ 2ml; thiamethoxam and carbofuran @ 2 g/kg seed, were effective in controlling the jassid, on 'okra'. Tomar (2008) showed that the seed treatment with imidacloprid 70 WS 10 g/kg of seed, was effective in suppressing the jassid-population below ETL up to 60 days, after sowing while imidacloprid 70 WS 7.5 g/kg of seed, was effective up to 50 days, after sowing.

Chemical Control

Use only those systemic pesticides recommended by the local extension service. Do not use the same compound or pesticide group continuously to avoid the development of pesticide resistance in insects. Spray any one of the following insecticides. Apply carbofuran 3 per cent CG 25 kg/ha, imidacloprid 200 SL -100ml/ ha, acetamiprid 20 per cent SP -50 g/ha, azadiractin 0.03 per cent WSP-500-1000 g/ha, buprofezin 25 per cent SC -1000 ml/ha, clothianidin 50 per cent WDG-30-40 Kg/ha, difenthiuron 50 per cent WP 600 g/ha, dimethoate 30 per cent EC- 660 ml/ ha, fipronil 5 per cent SC-1500-2000 ml/ha, profenophos 50 per cent EC 1000 ml.ha, thiacloprid 21.7 per cent SC 100-125 ml/ha, thiamethoxam 30 per cent FS 10 kg/ ha, thiamethoxam 25 per cent WG 100 g/ha. From overall observation clothianidin was found best insecticide to reduce 79.18 per cent jassid population followed by flubendiamide 78.38 per cent reduction of pest population (Das *et al.*, 2010).

Imidacloprid acts as an agonist at the insect nicotinic acetlecholine receptor (nAChR). It has an anti-feedant activity against the adults, and has a limited translaminar and contact activity, as a short residual, foliar spray. Sunitha *et al.* (2005) evaluated the efficacy of different groups of chemicals, alone and in combinations, against the jassid, in Andhra Pradesh, India. They found that two spray of imidacloprid, at 0.02006 per cent, was the most effective treatment in reducing the jassid-population, at 5 and 10 days after spraying. Imidaclopridi 70WS, Admire 200SL, against ijassid, aphid and whitefly *of* okra were effective against jassid and others sucking pests. Skip row treatments with systemic insecticides (as alternate or as paired rows) have been found to promote survival of the green lacewing predator Chrysoperla carnea and the coccinellid Cheilomenes sexmaculata, both important natural enemies in the cotton ecosystem. This practice may therefore be of benefit for leafhopper management.

Integration of Control Methods (IPM)

Farmers rely frequently on chemical pesticides, for pest control which produce harmful effect. IPM is one of the most important components in the production process containing all suitable methods and techniques of pest management. The effectiveness of different biocontrol agents against the major pest of 'okra', found that the application of predator, *Chrysoperla carnea* (25,000 larvae/ha/release) + Econeem 0.3 per cent (0.5 l/ha), three times, at 15 days of interval, starting from 45 days after sowing, was effective in reducing the population of sucking pests, including *A. biguttula biguttula* and *B. tabaci*. Misra and Senapati (2003) a botanical insecticide azadirachtin @ 3.0 g a.i./ha against the jassid mixing with thiamethoxam at rates ranging from 25 to 50 g a.i./ha and imidacloprid @ 25.0 g a.i./ha gave a significant/excellent control of jassids (83.3-100.00 per cent) and increased the marketable fruit yield. lmidacloprid was found the most effective treatment for controlling jassid followed by the microbial insecticide spinosad (Ghosh *et al.*, 2011). Parveen *et al.* (2007) reported that the seed-treatment of 'okra' with imidacloprid @ 12 ml/kg of seed, along with foliar spray with fenavalerate or NSKE attributed to an effective control of leaf hoppers. A rapid degradation of persistency is observed in imidacloprid which had a great importance as fruits of vegetables are plucked at frequent interval and consumed after little cooking (Ghosh *et al.*, 2012).

Conclusion

Both nymphs and adults suck the sap from the leaves and cause phytotoxic symptoms known as hopper burn/drying of leaves which results in stunted growth and complete desiccation of plants and has become one of the limiting factors in economic productivity of the crop. Farmers rely frequently on chemical pesticides, for pest control which produce harmful effect. IPM is one of the most important components in the production process containing all suitable methods and techniques of pest management.

REFERENCES

Akashe, V.B., Indi, D.V., Patil, S.R., Jadhav, J.D. and Pawar, P.B. (2015) Incidence of insect pest damage in castor in relation to meteorological parameters in the scarcity zone of Maharashtra. *Journal of Agrometeorology*, **17**: 139-141.

Anjani, K. (2012) Castor genetic resources: A primary gene pool for exploitation Industrial Crop and Products **35**: 1-14.

Anjani, K., Raoof, M.A., Prasad, M.S.L., Duraimurugan, P., Lucose, C. *et al.* (2018) Trait-specific accessions in global castor (*Ricinus communis* L.) germplasm core set for utilization in castor improvement. *Industrial Crops and Products*, **112**: 766-774.

Balasubramani, V. and Swamiappan, M. (1998) Feeding potential of *Chrysoperla carnea* Stephens on cotton leaf hopper *Amrasca biguttula biguttula* Ishida. *Insect Environment*, **2**(3): 69-70.

CABI (2019) *Plantwise Knowledge Bank. CABI. Retrieved 22 November 2019.*

Chakraborty, K. and Ghosh, S.K. (2010) Incidence of *Coccinella septempunctata* in brinjal with some pesticides. *Current advances in Agricultural Sciences*, **2**(2): 129-130.

CSIRO (1991) The Insects of Australia. Melbourne University Press. (1991) p81, 170,

Das, K.; Biswas, S.; Chakraborty,G. and Ghosh, S.K. (2010) Efficacy of insecticides against Iassid (*Amrasca biguttula biguttuka* Ishida) on okra in terai agro-ecology of West Bengal. *J. Appl. Zool. Res.*, **21** (1): 33-35.

Devi, Y.K., Pal,S. and Seram, D. (2018) Okra Jassid (*Amrasca biguttula biguttula* Ishida) biology, ecology and management in okra cultivation. *Journal of Emerging Technologies and Innovative research (JETIR)*. **10** (5): 333-343.

Dhamdhera, S. V., Bahadur, J. and Misra, U. S. (1984) Studies on the occurrence and succession of pest of okra at Gwalior. *Indian J. Plant Prot.*, **12**: 9-12.

Dreyer, M. and Hellpap, C. (1991) Neem a promising natural insecticide for small scale vegetable production in tropical and subtropical countries. Zeitschrift fuer Pflanzenkrankheiten und Pflanzenschutz, 98(4): 428-437.

Faselt, J. (2014) Effects of intercropping on biological control of potato leafhopper, *Empoasca* fabae. *Entomological Society of America Annual Meeting,* 2014, USA.

Gaikwad, B.P., Darekar, K.S. and Chavan, U.D. (1991) Varietal reaction of eggplant against jassid. *J. Maharashtra Agric. Univ.* **16** (3): 354-356.

Gaur RK (2014) Diversity of insect pests of castor, *Ricinus communis* L and their ecological interaction in south-west Haryana. *International Journal of Farm Sciences,* **4**: 147-152.

Ghosh, S.K. (1999) Studies on the pest constraints of brinjal/eggplant (*Solanum melongena* L.) and their management under terai region of West Bengal, India. Ph.D. Thesis awarded by BCKV-Agriculture University, West Bengal, India, Pp. 43-44.

Ghosh, S.K. (2014) Eco-friendly Management of Shoot and Fruit Borer, *Leucinodes orbonalis* (guenee) Infesting Eggplant (*Solanum melongena* L.). Book Edited by P.K.bharati and Avnish Chauhan, Book title *"Agriculture and Environmental Biotechnology" pp. 85-105, 2014.* (Discovery Publishing House Pvt. Ltd. – publication, New Delhi ISBN: 978-93-5056-479-0).

Ghosh, S.K. (2016) Harmful effect of insecticides against predator, Coccinella sp. (Lady bird beetle) on eggplant (*Solanum melongena* L.). *Uttar Pradesh J. Zool.* 36(1): 17-23.

Ghosh, S.K. (2019 a) Sustainable Management of aphid (*Aphis gossypii* Glover) infesting brinjal (eggplant), *Solanum melongena* L.) Book Edited by Sachin Tyagi, Book title "Hi-Tech Horticulture, Volume 4: Plant Protection and Stress Management" *pp. 77-99, 2018.* (New India Publishing Agency, New Delhi ISBN: 978-93-87973-40-4).

Ghosh, S.K. (2019 b) Climate impact on red spider mite (*Tetranychus* sp. Koch) infesting eggplant (*Solanum melongena* L.) and their management using plant extracts. *J. Ent.Res.* **43** (3): 345-350.

Ghosh, S.K. (2020) Sustainable control of blister beetle (*Mylabris pustulata* Thunberg) infesting eggplant (*Solanum melongena* L.) Book Edited by Bishvajit Bakshi, Book title "New perspectives in Agricultural Sciences-An international peer reviewed book."*pp. 115-142, 2020.* (LAP LAMBERT Academic Publishing, Maurtius, ISBN: 978-620-2-66718-4).

Ghosh, S.K. and Chakraborty, K. (2014) **Bio-Efficacy of plant extracts against red spider mite (*Tetranychus spp.*) infesting brinjal (*Solanum melongena* L.).** *Research journal of Agricultural and Environmental Sciences.* **1** (1): 26-31.

Ghosh, S.K. and Chakraborty, K. (2015) Integrated field management of jassid (*Amrasca biguttula biguttula* Ishida.) infesting ladysfinger *Abelmoschus esculentus* (L.) Moench using biopesticides *International Journal of Science, Environment and Technology.* **4** (2): 459-467.

Ghosh, S.K. and Senapati, S, K (2002) Field evaluation of pesticides from different origin against pest complex of brinjal under terai region of W. B. *Crop Res,* **23**(1): 108-115.

Ghosh, S.K. and Senapati, S.K (2003) Biology and seasonal abundance of jassid infesting brinjal in terai region of West Bengal. *Environment and Ecology.* **21**(3): *716-719.* ISSN-0970-0420,

Ghosh, S.K. and Senapati, S.K (2009) Seasonal fluctuation in the population of *Leucinodes orbonalis* Guen. Under the sub-himalayan resion of West Bengal, India and its control on eggplant (*Solanum melongena* L.) *Precision Agriculture,* **10**: 443-449. Springer

Ghosh, S.K. and Senapati, S.K(2001a) Seasonal incidence and biology of brinjal shoot and fruit borer (*Leucinodes orbonalis* Guen.) under terai region of West Bengal, India. *Annals of Ent.* **19**(1): 13-18.

Ghosh, S.K. and Senapati,S.K. (2001 b).Biology and seasonal fluctuation of *Henosepilachna vigintioctopunctata* Fabr. on brinjal under terai region of West Bengal. *Indian J Agric. Res,* **35**(3)': 149-154.

Ghosh, S.K., Laskar, N. and Senapati, S.K,(2004 a) Seasonal fluctuation of *Aphis gossypii* Glov. on brinjal,md field evaluation of pesticides from different origin against *A. gossypii* under tcrai region. *Indian J Agric. Res.* **38**(3): 171-177.

Ghosh, S.K., Laskar, N. and senapati, S.K. (2005) Seasonal fluctuation of thrips on brinjal and field evaluation of pesticides. *Pest Management and Economic Zoology.* **13**(2): 185-190.

Ghosh, S.K., Laskar, N. and Senapati, S.K. (2003) Estimation of loss in yield of brinjal due to

pest complex under terai region of West Bengal. *Environment and Ecology.* **21**(4): 764-769.

Ghosh, S.K. and Senapati, S.K (2003) Biology and seasonal abundance of jassid infesting brinjal in terai region of West Bengal. *Environment and Ecology.* **21**(3): 716-719.

Ghosh, S.K., Laskar, N. and senapati, S.K. (2006 b) Seasonal fluctuation of flea beetle (*Phyllotreta* sp.) on brinjal and field evaluation of some pesticides against flee beetle under terai region of West Bengal. *Juurnal of Entumological Research.***30**(1): 83-87.

Ghosh, S.K., Laskar, N. and senapati, S.K. (2007) Seasonal incidence of predator *Menochilus sexmaculatus* Berliner on brinjal and harmful effect of insecticides on the predator. *Indian J. Agric. Res.***41**(2): 102-106.

Ghosh, S.K., Laskar, N., Basak, S. N. and senapati, S.K. (2004 b) Seasonal fluctuation of *Bemisia tabaci* Genn. on brinjal and field evaluation of some pesticides against *B. tabaci* under terai region of west *Bengal. Environment and Ecology.* **22**(4): 758-762.

Ghosh, S.K., Laskar, N., Basak,S.N. and senapati, S.K. (2006 a) Seasonal fluctuation of spider on brinjal and efficacy of pesticides under terai region of West Bengal. *Orissa Iournal of Horticulture.***34**(1): 86-91.

Ghosh, S.K.; Mahapatra, G.S.S. and Chakraborty, G (2011) Efficacy of biopesticides against jassid (*Amrasca biguttula bighttula* Ishida). *Proceedings* of *International conference- 3rd ISOFAR and 17th IFOAM Organic World Congress pp. 514-517, held in Republic of Korea on 28 September- 1st October, 2011 Publisher:* International Society of Organic Agricultural Research (ISOFAR)

Ghosh, S.K.; Mandal, T.; Biswas, S. and Chakraborty, K. (2012) field evaluation of cultivars and bio-efficacy of insecticides against pest complex of ladysfinger (*Abelmoschus esculentus* L.). *Journal of applied Zoological research* **23**(2): 121-128.

Guddewar, M. B., Shukla, A., Chander, R., Pandey, S., and Saini, M. L. (1994) *Tabenaemomama coronaria* R. Br (Apocynaceae) A potential source of Botanical insecticides. *Plant Protection Bulletin,* **46**: 1-5.

Halder, J., Sanwal, S. K., Deb, D., Rai, A. B. and Singh, B. (2016). Mechanisms of physical and biochemical basis of resistance against leaf-hopper (Amrasca biguttula biguttula) in different okra (Abelmoschus esculentus) genotypes. *Indian Journal of Agricultural Sciences,* **86**(4): 481-484.

Held, D. W. and Boyd, J. D. W. (2008) Evaluation of sticky and insecticides to prevent gall induction by *Gynaikothrips uzeli Zimmerman* (Thysanoptera: Phlaeothripidae) on *Ficus benjamina. Pest Management Sci.,* **64**(2): 133-140.

http: //www.yourarticlelibrary.com/zoology/tea-green-fly-*Empoasca*-flavescens-distribution-life-cycle-and-control/24048.

Jayasimha, G.T., Rachana, R.R., Manjunatha, M. and Rajkumar, V.B. (2012) "Biology and seasonal incidence of leafhopper, Amrasca biguttula biguttula (Ishida) Hemitpera: Cicadellidae) on okra". Pest Management in Horticultural Ecosystems. **18** (2): 149–53. S2CID 85972344.

Jayasimha, G.T., Rachana, R.R., Manjunatha, M. and Rajkumar, V.B. (2012) "Biology and seasonal incidence of leafhopper, *Amrasca biguttula biguttula* (Ishida) (Hemitpera: Cicadellidae) on okra". *Pest Management in Horticultural Ecosystems.* **18** (2): 149–153. S2CID 85972344.

Jayasimha, G.T., Rachana, R.R., Manjunatha. M. and Rajkumar, V.B. (2012) Biology and seasonal incidence of leafhopper, Amrasca biguttula (Ishida) (*Hemiptera*: Cicadellidae) on okra. *Pest Management in Horticultural Ecosystem,* **18**: 149-153.

Jena, B.C. and Kuila, B. (1996) Jassid infestation in groundnut and its chemical control. *Indian J Ent,* **58**: 378-381.

Kaethner, M. (1991) No side effects of neem extracts on Chrysoperla carnea (Steph.) and Coccinella septempunctata L. *Anzeiger fuer Schaedlingskunde Pflanzenschutz Umweltschutz (Germany, FR).* **64** : 97-99.

Kannan, M., Uthamasamy, S. and Mohan, S. (2004) Impact of insecticides on sucking pests and natural enemy complex of transgenic cotton. *Current Sci.,* **86**(5): 726-729.

Kanvarjibhai, N. T. (1993) Insect pest control: Aphid Control. *Honey Bee, 4,* 16.

Khan, Z.R., Agarwal, R.A. (1984) "Ovipositional preference of jassid, *Amrasca biguttula biguttula* Ishida on cotton". *Journal of Entomological Research*. **8** (1): 78–80.

Khattak, M. K., Rashid, M., Hussain, S. A. S. and Islam, T. (2006) Comparative efficacy of neem (*Azadirachta indica* A. juss) oil, neem seed water extract and baythroid TM against whitefly, jassids and thrips on cotton. *Pak. Entomol.*, **28**(1): 31-37.

Krishnaiah, N. V. and Kalode, M. B. (1984). Evaluation of neem oil, neem cake and other non-edible oil cakes against rice pests. *Indian Journal of Plant Protection*, **12**(2), 101-107.

Lakshmanan, K.K., 2001. Garlic a natural pesticide. Hindu March, 1: 8.

Lal, H., Mahal, M. S. Singh, R.and Singh, B. (1990) Influence of rainfall on population build-up of *Amrasca biguttula biguttula* (Ishida) on 'Okra'. *J. Insect Sci.*, 3: 169-171.

Laxman, G. and Maheswari, T.U. (2017) Effect of Dates of Sowing on Incidence of Leaf Hoppers, *Empoasca flavescens* Fab. on Castor Genotypes. *Int J Curr Microbiol App Sci*, **6**: 431-437.

Misra, H. P. and Senapati, B. (2003) Evaluation of new insecticides against okra jassid *Amrasca biguttula biguttula*). *Indian J. Agric. Sci.*, **73**(10): 576-578.

Mounica, B. Venkateswarlu, N.C., Krishna, T.M. Sudhakar, P. and Prasad, K.V.H. (2018) Screening of Castor Genotypes for Resistance Against Green Leafhopper, *Empoasca flavescens* Fabricius. *Int J Pure App Biosci*, **6**: 110-116.

Parveen, K. S., Sajjan, A. S., Patil, R. K. Dharmatti, P. R. and Kurdikeri, M. B. (2007) Influence of Seed Treatment and Foliar Spray with Insecticides and Neem Products on Growth and Seed Yield in okra (*Abelmoschus esculentus* (L.) Moench). *Karnatka J. Agric. Sci.*, **20**(2): 388-390.

Pawar, A. D. (1991) Integrated pest management in cotton in India. *Pak. Cotton.* **5**(3-4): 165-181.

Rahman, A. (2014). *Development of management practices against jassid* (*Amrasca devastans*) *in okra* (Doctoral dissertation, Department of Entomology, Sher-E-Bangla Agricultural University, Dhaka).

Raja, K. M. and Arivudainambi, S. (2004) Efficacy of sticky traps against bhendi leaf hopper, *Amrasca biguttula biguttula* Ishida. *Insect Environment*, **10**(1): 32.

Rana, S. C., Singh, P. B., Pandita, V. K. and Sinha, S. N. (2006) Evaluation of insecticides as seed treatment for control of early sucking pests in seed crop of okra. *Ann. Plant Prot. Sci.*, **14**(2): 112-116.

Rao, N. S. and Rajendran, R. (2002) Joint action potential of neem with other plant extracts against the leaf hopper, *Amrasca devastans* (Distant) on okra. Pest Manag. *Ec. Zool.*, **10**(2): 131-136.

Rao, M.S., Rama Rao, C.A., Srinivas, K., Pratibha, G., Vidya Sekhar, S.M., *et al.* (2012) Intercropping for management of insect pests of castor, *Ricinus communis*, in the semi-arid tropics of India. *J Insect Sci*, **12**: 14.

Risch, S.J. (2005) Intercropping as cultural pest control: Prospects and limitations. *Environ Manage*, **7**: 9-14.

Rosaiah, R. (2001) Performance of different botanicals against the pests complex in bhendi. *Pestology*, **25**, 17-19.

Rosmana, A. Saranga, A.P., Jufri, M. and Rahim, M.D. (2015) Intercropping of Rice Varieties Reduces Green Leafhopper and White Stem Borer Infestation. *Indian Journal of Science and Technology*, **8**: 141-146.

Sahito, H. A., Shah, Z. H., Kousar, T., Mangrio, W. M., Mallah, N. A., Jatoi, F. A. and Kubar, W. A. (2017). Comparative efficacy of novel pesticides against jassid, *Amrasca biguttula biguttula* (Ishida) on cotton crop under field conditions at Khairpur, Sindh–Pakistan. *Singapore Journal of Scientific Research*, 1-8.

Shahid, M. I., Ahmad, F., Ali M. R. and Siddiqui, B. S. (1992) Neem sample efficacy as aninsecticide against *Amrasca devastans* Dist. and *Leucinodes orbonalis* Geun. on brinjalcrop. *Pak. Entomol.*, **14**(1-2): 90-93.

Shukla, R.P. (1989) Population fluctuation of *Leucinodes orbonalis* and *Amrasca biguttula biguttula* on brinjal (*Solanum melongena* L.) in relation to abiotic factors in Meghalaya. *Indian J. Agric Sci.* **59** (4): 260-264.

.Singh, A., Singh, J., Singh, K. and Rani, P. (2018) "Host range and biology of Amrasca biguttula biguttula (Hemitpera: Cicadellidae)".

Singh, H. (1967) Vegetables. Chapter-7. pages 387-406 in cd. C.G. Raghava Kurup. Hand Book of Agriculture, Indian Council Agric. Res., New Delhi, pp.877.

Solangi, G. S., Mahar, G. M. and Oad, F. C. (2008) Presence and Abundance of Different Predators against Sucking Insect Pest of Cotton. *J. Entom.*, **5**(1): 31-37.

Songa, J.M. Jiang, N., Schulthess, F. Omwega, C. (2007) The role of intercropping different cereal species in controlling lepidopteran stemborers on maize in Kenya. *J Appl Entomol*, **131:** 40-49.

Srinivasan, K., Krishnakumar, N. K., Ramachander, P. R. and Krishnamurthy, A. (1981) Spatial distribution pattern of leaf hopper *Amrasca biguttula biguttula* on okra *Abelmoschus esculentus.Entomon.*, **6**(3): 261-264.

Subba, B.; Ghosh, S.K.; Banerjee, D. and Jasudasu, G.S. (2015). Seasonal incidence of Jassid Infesting tomato (*Lycopersicon esculentum* L.) and their sustainable management.*Annals of Plant and Soil Research* **17**: 19-22 Special Issue).

Sultana, P., Alam, M. J., Das, K., Azad, M. A. K. and Islam, M. T. (2016) Eco-friendly control of okra jassid using botanicals from jute (*Corchorus capsularies*). *Bangladesh Journal*, **30**, 65-70.

Sunitha, P., Rao, G. R. and Rao, P. A. (2005) Bio-efficacy of certain eco-friendly insecticides against sucking pests on okra *Abelmoschus esculentus* (Linn.). *J. Applied Zool. Res.*, 1**6**(2): 186-187.

Tomar, S. P. S. (2008) Effect of different insecticides on seed treatment for controlling early season sucking pests in cotton. *Green Farming*, **1**(6): 23-26.

Thomas, M. S. (1995, December). A few traditional practices in rice farmer in Kerala. In *Key Note Paper and Extended abstract of 2nd Congress on Traditional Sciences and Technology of India, Anna University, Annamalinagar*. December 27-31.

Vijayalakshmi, K., Subhashini, B., and Shivani, V. K. (1996) Plant in pest control: garlic and onion. *Centre for Indian Knowledge System, Chennai*, 1-20.

Widayat, W. and Winasa, I.W. (2006) Bioecology of *Empoasca flavescens* and its natural enemy. *Journal of Research of Kina and Tea*, **9**: 12-19.

Yadav, D. N. and Patel R. (1990) Effect of some botanicals on oviposition of *Chrysoperla scelestes* and their ovicidal action. In: *Proceedings of Symposium on Botanical Pesticide in Integrated Pest Management, Rajamundrai*, pp. 166-169.

Yadav, J. B., Singh, R. S. and Tripathi, R. A. (2008) Evaluation of Biopesticides against pest complex of Okra. *Annals of Plant Protection Sciences*, **16**(1): 58-61.

Yadav, P., Singh, P., Yadav, R. L., and Lal, R. (2004). Ameliorative organic manures and nitrogen levels on okra. *Haryana Journal of Horticultural Science*, **33**(1): 124-126.

Zhang, Y.L. (1990). A taxonomic study of Chinese Cicadellidae (Homoptera). Yangling Shaanxi, China: Tianze Eldonejo. Iv+218p (In Chinese) (illustration, records, synonymes).

Chapter 17

Pests and Diseases of Silkworm

P.L. Ambulkar[1], Amit Kumar Sharma[2] and R.K. Jhade[3]

[1]*Department of Entomology,* [3]*Department of Horticulture,*
JNKVV Krishi Vigyan Kendra, Chhindwara – 480003 Madhya Pradesh
[2]*Department of Entomology, JNKVV College of Agriculture,*
Jabalpur – 482004, Madhya Pradesh

Bombyx mori, the domestic silk moth, is an insect from the moth family Bombycidae. It is the closest relative of Bombyx mandarina, the wild silk moth. The ***silkworm*** *is the larva or caterpillar of a silk moth. It is an economically important insect, being a primary producer of silk. A silkworm's preferred food are white mulberry leaves, though they may eat other mulberry species and even the osage orange. Domestic silk moths are closely dependent on humans for reproduction, as a result of millennia of selective breeding. Wild silk moths are different from their domestic cousins as they have not been selectively bred; they are thus not as commercially viable in the production of silk.*

Introduction

Sericulture, the practice of breeding silkworms for the production of raw silk, has been under way for at least 5,000 years in China, whence it spread to India, Korea, Nepal, Japan, and the West. The domestic silk moth was domesticated from the wild silk moth *Bombyx mandarina*, which has a range from northern India to northern China, Korea, Japan, and the far eastern regions of Russia. The domestic silk moth derives from Chinese rather than Japanese or Korean stock.

Silk moths were unlikely to have been domestically bred before the Neolithic Age. Before then, the tools to manufacture quantities of silk thread had not been developed. The domesticated *B. mori* and the wild *B. mandarina* can still breed and sometimes produce hybrids. Domestic silk moths are very different from most members in the genus *Bombyx*; not only have hey lost the ability to fly, but their colour pigments have also been lost. The present production of Muga silk in Assam is 117 MT accounting to 98 per cent of total Muga silk produced in the country while rest is contributed by the other north eastern states of India (Singh, R.N., 2013). The Muga silkworm, Antheraea assamensis (Helfer) the producer of glittering golden yellow silk is endemic to Assam and is a semi domesticated, polyphagous, multivoltine species. This silkworm feeds on two primary food plants namely Som, *Persea bombycina* and Soalu, Litsea polyantha and completes six life cycles in a year. The six crop cycles are Jethua (Apr-May) and Kotia (Oct-Nov) which are considered as commercial crops, Chotua (Feb-Mar), Bhodia (Aug-Sep) as seed and Jarua (Dec-Jan) and Aherua (June-July) as pre seed crops (Chakravorty, 2004). However, Muga rearing being outdoor, its yield is constrained by several natural a biotic and biotic vagary, one of which is predation of silkworm by insect pests that damages crop significantly. Information on pest's complex in a particular agro-climatic condition is a prerequisite, which helps in designing a successful pest's management strategy (Srilaxmi, K. and Paul, R. 2010). Earlier several authors have reported that insect pest infesting a particular crop differs from place to place. The use of chemical insecticides for control of insect pest in sericulture is often impractical and undesirable, there is an urgent need for an alternative approach that can complement and partially replace chemical based pest management in sericulture (Singh and Saratchandra, 2008).

Mulberry silkworms can be categorized into three different but connected groups or types. The major groups of silkworms fall under the univoltine ("uni-"=one, "voltine"=brood frequency) and bivoltine categories. The univoltine type is generally linked with the geographical area within greater Europe. The eggs of this type hibernate during winter due to the cold climate, and cross-fertilize only by spring, generating silk only once annually. The second type is called bivoltine and is normally found in China, Japan, and Korea. The breeding process of this type takes place twice annually, a feat made possible through the slightly warmer climates and the resulting two life cycles. The polyvoltine type of mulberry silkworm can only be found in the tropics. The eggs are laid by female moths and hatch within nine to 12 days, so the resulting type can have up to eight separate life cycles throughout the year.

The major pests and diseases of mulberry silkworm are given below-

Pests and Diseases of Mulberry Silkworm

Major Pests of Mulberry Silkworm

Uzi Fly

Occurrence and Symptom

The uzi fly, *Exorista bombycis* is a serious endo-larval parasitoid of the silkworm, Bombyx mori, inflicting 10-15 per cent damage to the silkworm cocoon crop in the premier silk producing states of Karnataka, Andhra Pradesh and Tamil Nadu. Uzi fly occurs throughout the year, but severe during rainy season. Presences of eggs or black scar on the silkworm body and maggot emergence hole at the tip of the cocoon are the typical symptoms of uzi fly attack.

As soon as the uzi fly enters into rearing house, it lays one or two eggs on each silkworm larva. After 2-3 days, egg hatches, enters inside the larva and feed on internal contents for 5-7 days, after which it comes out by rupturing the larva. The maggot pupates in a dark corner or cracks and crevices in about 18-24 hours. The pupal stage lasts for 10-12 days. If the uzi fly infests at last instar, the uzi maggots come out after cocoon formation by making a circular hole.

Management

Exclusion Method

Provide wire mesh/nylon net on all windows/doors. Provide doors with automatic closing mechanism.Provide anteroom at the entrance of the rearing house. Keep the leaf in the verandah of the rearing house and observe for the uzi fly before shifting leaf into the rearing house.

Physical (Using uzi trap)

Dissolve one table in 1 litre of water and keep the solution in white trays both inside and outside the rearing house at window base from 3rd instar onwards up to spinning.

Place uzi traps inside the rearing house/mounting hall after spinning up to 20 days under close-door condition to trap uzi flies emerging inside.

Biological

Release Nesolynx thymus (a pupal parasitoid of the uzi fly) inside rearing house on 2nd day of V instar.After mounting of all spinning worms transfer the same pouches near the chandrikes.After harvesting of cocoons keep the same pouches near the manure pit.Two pouches are required for 100 dfls.

Proper Disposal of Silkworm Litter after Cocoon Harvest

Separate the silkworm litter from mulberry twigs. Do not throw silkworm litter in open space/litter pit, as it contains hundreds of Uzi fly pupae. Instead, pack it in plastic bags and keep for 15 to 20 days to prevent the emergence of uzi fly from litter. Alternatively, it can be buried in soil or burnt immediately.

Dermestid Beetles

Occurrence and Symptom

Dermestid beetles, are known to attack pierced cocoons in cocoon storage rooms. The female beetles lay about 150-250 eggs in the floss of cocoons. The beetles migrate from cocoon storage room to grainage and attack green cocoons as well as moths also. Generally they attack the abdominal region of the moth. The damage is estimated to be 16.62 per cent on cocoons and 3.57 per cent on moths.

Management

Preventive Measures

Storage of rejected cocoons and perished eggs for long period should be avoided.Rearing house and cocoon storage rooms should be cleaned periodically. Grainage premises should be cleaned before and after moth emergence.Provide wire mesh to door and windows in pierced cocoon (PC) storage rooms.

Wooden articles of storage room and grainage should be dipped in 0.2 per cent malathion solution for 2-3 minutes. Trays *etc.*, should be thoroughly washed and sun dried for 2-3 days before reusing.

Mechanical Control

Collect the grubs and adults by sweeping or by using a vacuum cleaner, destroy by burning or dipping in soap water.

Chemical Control

Store pierced cocoons in Chlopyriphos treated bags i.e., soak the bags in 0.2 per cent Chlorpyriphos solution (1 ltr : 100 ltr water) and dry in shade. Spray 0.2 per cent Chlorpyriphos solution on walls and floor of PC room once in 3 months. Sprinkle bleaching powder (200 gm/sq.mt) all around inner wall of PC room to prevent crawling of grubs from PC room

Major Diseases of Mulberry Silkworm

Grasserie

Causative agent: Bombyx mori Nuclear Polyhedrosis Virus

Occurrence: The disease prevails all through the year but its severity is more during Summer and Rainy seasons.

Source of infection: Silkworm gets infected when it feed on contaminated mulberry leaves. The milky white fluid released by the grasserie larvae, contaminated silkworm rearing house and appliances are the sources of infection.

Predisposing factors: High temperature, low humidity and poor quality mulberry leaves.

Symptoms: The skin of infected larvae becomes shining before moult and fails to moult. Inter segmental swelling appears and the colour of the body becomes yellowish. The infected larvae move restlessly in the rearing bed/along the rim of the trays. Infected larval body ruptures easily and turbid white haemolymph oozes out.

Management

Practice thorough disinfection of rearing house, its surroundings and appliances with any recommended disinfectant. Conduct an optional disinfection with 0.3 per cent slaked lime solution when high incidence of disease noticed in the previous crop. Practice personal and rearing hygiene. Collect the diseased larvae and ensure its proper disposal. Maintain optimum temperature and humidity in the rearing house.Feed quality mulberry leaf and avoid overcrowding. Apply recommended bed disinfectant as per schedule and quantity. Feed Amruth as per schedule to control grasserie disease.

Flacherie

Causative agent: Bombyx mori Infectious flacherie virus/Bombyx mori Densonucleosis virus or different pathogenic bacteria *viz.*, Streptococcus sp./ Staphylococcus sp./Bacillus thuringiensis/Serratia marscesence individually or in combination of bacteria and viruses.

Occurrence: The disease is common during Summer and Rainy seasons.

Source Infection: Silkworm gets infected by eating contaminated mulberry leaf. Dead diseased silkworm, its faecal matter, gut juice, body fluid are the sources of pathogen contamination. The infection can also takes place through injuries/ cuts/wounds.

Predisposing factors: Fluctuation in temperature, high humidity and poor quality of leaves.

Symptoms

The larvae become soft and flaccid.The growth of infected larvae retarded, becomes inactive and vomit gut juice. The faeces become soft with high moisture content. Sometimes chain type excreta and rectal protrusion also observed.Larval head and thorax become translucent.When infected with Bacillus thuringiensis symptoms of toxicity such as paralysis and sudden death are observed. After death, larvae turn black in color and gives foul smell. Sometimes, the dead larvae turn red when infected with Serratia sp.

Management

Disinfect the rearing house, its surroundings and equipment's with recommended disinfectant mentioned above.Pick up diseased larvae and dispose them by burning. Provide good quality leaf grown under good Sunlight and recommended inputs. Do not provide over matured/over stored/dirty leaf to the silkworms.Avoid starvation, overcrowding and accumulation of faeces in the rearing bed.

Rear silkworms under optimum temperature and humidity. Avoid injury to the larvae.Apply recommended bed disinfectant as per schedule and quantity. Feed Amruth as per schedule to control flacherie disease.

Muscardine

Causative agent : Among fungal diseases, White Muscardine is common. The disease is caused by *Beauveria bassiana*.

Occurrence: The disease is common during Rainy and winter seasons.

Source of Infection: The infection starts when conidia come in contact with silkworm body. Mummified silkworms/alternate hosts (most are lepidopteron pests), contaminated rearing house and appliances are sources of infection.

Predisposing factors : Low temperature with high humidity.

Symptoms: The larvae loose appetite and become inactive.Presence of moist specks on the skin.The larva vomits and turns flaccid.After death, larva gradually becomes hard followed by mummification due to growth of aerial mycelia and conidia over the body and body turns chalky white.

Management: Disinfect the rearing house, its surroundings and equipment's with recommended disinfectant as mentioned above. Control mulberry pests in the mulberry garden.Pick up diseased larvae before mummification and dispose them by burning. Avoid Low temperature and high humidity in the rearing house. If required use heater/stove to raise the temperature. Regulate bed humidity during rainy season by dusting slaked lime powder during moult. Apply bed disinfectant, Vijetha and Vijetha supplement/Ankush/any recommended bed disinfectant as per schedule and quantity.

Pebrine

Pebrine is the most serious disease of muga silkworm caused by a protozoan of Nosema sp. It is unique in being transmitted to offspring by transovarial/transovum means from other moths. If infection is primary, more than 50 per cent larvae die before 3rd moult and rarely any larva goes for spinning. When healthy larvae get infected through contamination during rearing, it is called secondary infection. Secondary infection during early 4th larval stage leads to formation of flimsy cocoons, whereas larvae infected during 5th larval stage form well formed cocoons.

Causative agent: *Nosema bombycis*/different strains of microsporidia.

Occurrence: Non-seasonal

Sources of Infection: Silkworm gets infected through eggs (Transovarian/ Transovum transmission) or by eating contaminated mulberry leaf. Infected silkworms, faecal matter, contaminated rearing house and appliances and alternate hosts (mulberry pest) are the sources of infection.

Symptoms: Irregular hatching of silkworm eggs. Irregular size of the larval body and moulting. The infected larva looses its appetite and becomes inactive with wrinkled skin.Black pepper-like spots appear on the body of the infected worms. White postules appear on the silkgland when examined under microscope with presence of shining oval spores.

Management

Disinfect the rearing house, surroundings and with recommended disinfectant as mentioned above.

Conduct strict mother moth examination and surface disinfection of silkworm eggs to produce and rear disease free layings.Follow strict hygiene maintenance during rearing. Control mulberry pests in and around the mulberry garden.Apply recommended bed disinfectant, Vijetha/Ankush as per schedule and quantity. Monitor seed crops constantly to eliminate the microspodian infection.

Disinfection of rearing house, its surroundings and appliances:

Select any recommended disinfectant for disinfection purpose. CSR and TI, Mysore has recommended the following disinfectants:

0.05 per cent Asthra solution (Add 50g Asthra powder in 100 liters of water and stir thoroughly and keep for 2 hours for dissolution of the powder). 2.5 per cent Sanitech/Serichlor in 0.5 per cent Slaked lime solution (To prepare 100 liters of solution, take 250g of activator in to a basin/bucket and add 2.5 liters of Sanitech/ Serichlor solution. Keep it for 10 minutes. Add activated solution to the rest of water. To this solution, add 500 g slaked lime powder and mix thoroughly). 2 per cent Bleaching powder in 0.3 per cent slaked lime solution (To prepare 100 liters of solution, add little water to 2 kg bleaching powder and 300g slaked lime powder and make a paste. Add this paste to the rest of water and stir thoroughly. Keep for 10 minutes and use the supernatant). 0.3 per cent Slaked lime solution (optional disinfection if viral diseases noticed in previous crop – Add 300g of slaked lime to 100 liters of water and stir thoroughly. Keep for 10 minutes and use supernatant). The total requirement of disinfectant solution for disinfection is estimated based on the rearing house floor area (Length × Breadth of floor). The quantity of disinfectant solution required is 1.5 lt./sq. m or 140 ml/sq. ft. floor area of rearing house (height 3 m/10 ft.) + 10 per cent of total quantity of disinfectant solution. Disinfect the rearing house, appliances and surroundings by spraying the solution with power sprayer. Two times disinfection recommended for each crop (once 3days before initiation of rearing and after completion of rearing).

Source

- ☆ Central Sericulture Research and Training Institute, Mysore, Karnataka
- ☆ Package of Practices of Mulberry Sericulture for Eastern and North Eastern Region, Central Sericultural Research and Training Institute, Berhampore, West Bengal

Pests and Diseases of Muga Silkworm

Major Pest of Muga Silkworm

Uzi Fly (*Exorista sorbillans*)

Nature of Damage

It is the major pest of muga silkworm. The fly lays eggs on the integument of the worms in the dorsal and dorso-lateral side. After hatching from the eggs, the maggots of the fly penetrate into the larval body and feed on the tissues of the worms. The mature maggots come out of the larvae/pupae and undergo pupation in the rearing field or grainage hall. The uzi infested muga silkworm dies during larval or pupal stage. Season of incidence: Prevalent throughout the year, peak during December to march.

Management

Rear the silkworm under nylon mosquito net during peak infestation period (December to March), which ensures 80-90 per cent control.During transfer of late stage worms, remove the fly eggs from the integument of the silkworm larvae with the help of forceps. Keep the rearing field clean and dust with bleaching powder during rearing. Mount uzi infested worms in separate Jali. Harvest and stifle the uzi infested cocoons on 4th and 5th day of spinning. Collect and destroy the maggot/ pupae of the fly. Burn heavily infested worms.

Apanteles (*Apanteles stantoni*)

Nature of Damage

Usually infect the early stage silkworms. Apanteles lays eggs inside the body of silkworm larvae by inserting the ovipositor through the tubercles after maturation. The mature maggots form pupae in aggregation outside the body of the silkworm larvae.

Season of Incidence

Prevalent during summer and winter months of the year.

Management

Rearing of silkworm under nylon mosquito net. Keep the rearing field clean and dust with bleaching powder during rearing. Collect and destroy the maggots/ pupae of the fly along with the silkworm larvae.

Spider

Attacks 1st instar worms.

Wasp (*Vespa orientalis*)

It occurs during June-July to August-September months. It attacks early stage worms by lacerating and picking young age worms. It can be controlled by covering silkworm rearing by nylon nets and destroying hives.

Red Ants

The red ants are also serious pest in many muga growing areas. It attacks 1st stage worms. The lost due to red ants are reported to be 5-10 per cent. They can be controlled with the spray of 0.2 per cent Chlorpyriphos (insecticide) before 15 days of rearing or burning down their nest well before the rearing.

Major Diseases of Muga Silkworm

Pebrine Disease

Peak season: Throughout the year.

Symptoms

Early stage of infection: The infected silkworm larvae appear normal. However, microscopic examination of silkworm larvae may indicate the presence of spores of the pathogen.

Later stage of infection: The silkworm larvae loose appetite, vary in size, retarded growth, moult irregularly and the colour of the larvae become light yellowish green instead of deep green colour (normal healthy larvae). The infected late stage larvae show black dots or specks on the surface of the body and hence the disease is known as Phutuka (spotted disease in Assamese).

Sources of Infection

Egg stage: Transovarial. Surface contamination of eggs (transovum). Contaminated grainage appliances.

Larval stage: Contaminated egg laying kharika, Transovarially infected larvae, Faecal materials of infected larvae, Contaminated foliage, Contaminated rearing site and Contaminated rearing appliances.

Moth stage: Infected seed cocoons, Infected moth, Infected grainage appliances, Meconium and moth scales and Grainage dust.

Spread of the disease: Pebrinized larvae extrude faecal material, gut juice and vomit which contaminate the rearing environment, appliances and host plant foliage. Mostly, consumption of contaminated foliage/egg shell results in spread of the disease.

Management

Scientific inspection of individual mother moths for detection of pebrine during egg production. Disinfection of grainage appliances with 2 per cent formalin before and after every grainage operation.Use microscopically tested disease free disinfected eggs only. Surface sterilization of eggs with 2 per cent formalin for 5 minutes. Maintain hygienic condition in egg production room and rearing sites. Follow cellular method of rearing for basic stock maintenance. Disinfection of rearing appliances before use.Check the faecal materials, unequal/lathergic/ unsettled irregular moulters periodically.If pebrine spores are detected, reject the entire infected crop.

Source: Central Muga Eri Research and Training Institute, Lahdoigarh, Jorhat, Assam

Flacherie

Flacherie; commonly known as anal protrution is a syndrome of bacterial diseases in muga silkworm caused by Pseudomonas sp. Sometimes it is caused by an ultra virus, which acts as an exciting agent followed by secondary infection of bacteria. Sudden fluctuation in temperature and humidity, bad weather, poor quality leaves with high water content are predisposing factors. Peak season: Throughout, intensive during rainy summer months (June to August).

Symptoms

The infected larvae become lethargic and motionless.The colour of the haemolymph turns black.

Excreta looks like chain. Sealing of anal lips. Rectal protrusion. Infected larvae die within a short time.

Source of Infection

Larvae get infected upon feeding of contaminated/poor quality leaves of host plants. Diseased larvae, its gut juice, faecal material, body fluid. Contaminated rearing site and appliances.

Spread of the Disease

Secondary infection of larvae due to feeding on contaminated leaves. Infected worms ooz-out body fluid containing pathogen throughout the incubation period and contaminate the rearing environment. Feeding of late-stage worms with very tender succulent leaves. Sudden fluctuation of temperature and humidity during rearing period.

Management

Use disinfected good quality seeds of disease-free zones. Orient the brushing to protect the young larvae form direct sun shine. Disinfection of rearing site with 2 per cent formalin solution before rearing. Dust 3 per cent slaked lime in addition to formalin in case of high incidence of the disease in the preceding year. Inspect rearing field regularly and pick out stunted/sluggish/irregular moulters and destroy. Destruction of diseased/doubtful worms by burying with 5 per cent formalin solution. Wash the hands with formalin solution at the time of transfer of worms. Maintain hygienic condition during rearing. Feed with good quality disease free leaves. Do not feed late-stage worms with tender succulent leaves.

Grasserie

It is a major viral disease of muga silkworm caused by a baculovirus. High temperature clubbed with high humidity, poor quality host plant leaves are predisposing factors.

Peak season: Throughout the year, predominant during rainy summer months of the year.

Symptoms

The silkworm larvae fail to moult. The integument becomes fragile and intersegmental region becomes swollen and that is why the disease is known as Phularog (swelling disease in Assamese). The body tissues and haemolymph of the infected larvae get disintegrated into turbid white fluid and the larvae hang upside down with the anal claspers after dying. The turbid fluid contains large number of hexagonal polyhedral bodies.

Source of Infection

Feeding of contaminated foliage, Disintegrating diseased silkworms, their body fluids and Contaminated rearing sites and appliances.

Spread of the Disease

The diseased silkworm larvae extrude the pathogen along with oozing of body fluid due to injury and breakage of dead and/or diseased larvae. The body fluid and broken body parts of the larvae contaminate the foliage, rearing site and appliances. The disease spreads to healthy worms on feeding of the contaminated leaves and/or use of contaminated appliances during rearing.

Management

Disinfection of rearing site with 2 per cent formalin solution before rearing. Dust 3 per cent slaked lime in addition to usual disinfection in case of high incidence of disease in preceding rearing. Pick out growth retarded/lethargic/irregular moulters and destroy. Ensure destruction of diseased/doubtful worms by burning or burying with 5 per cent formalin solution. Ensure proper hygiene during rearing. Use certified disinfected disease-free laying only. Ensure rearing on good quality leaves.

Mascardine

Mascardine is one of the major diseases of silkworm. However, it is less prevalent in muga silkworm and occurs under certain specific environmental influence only. The disease appears at an interval of 2-3 years. The causal organism is a fungus and it is yet to be identified for muga silkworm. Low temperature with high humidity is predisposing factors. Peak season: Winter months of the year when night temperature falls down and the day temperature remains comparatively high associated with high humidity *i.e.* foggy weather.

Symptoms

Infected larvae loose appetite and become inactive. Colour of the larvae turns pale. Gradual cease of movement within 12-18 hours of infection. The larvae hang on tree twig/trunk and harden.The larvae die within 16-18 hours of infection. A white encrustation appears on the larval body and it covers the whole larval body within 24 hours of death. The dead larval body becomes dry, brittle and mummified.

Source of Infection

Mummified/diseased larvae and Contaminated rearing environment.

Spread of the Disease

The conidia/spores of the pathogenic fungus are dispersed by wind. The conidia on contact with larval integument germinate, penetrate into the larval body and cause infection.

Management

Orientation of brushing towards sun shine during winter. Disinfection of rearing site with 2 per cent formalin solution before rearing. Dust slaked lime in the field to control humidity at the time of rearing. Dust Tasar Kit Oushad developed by CTR and TI, Ranchi on the body of the larvae at the time of transfer. Spray 0.5 per cent sodium hydroxide solution on the worms after 24 hours of each moult as prophylactic measure. Maintain hygienic condition during rearing. Collection and destruction of dead/diseased larvae Pickout sick or dead worms with forceps/chopstick and put in 2 per cent formalin solution. Bury the carcasses in a pit and cover the soil. Wash hands with formalin or dettol solution after handling dead or infected larvae.Do not allow birds, ants or poultry to eat the carcasses.

Conclusion

Sericulture, the practice of breeding silkworms for the production of raw silk. Silk moths were unlikely to have been domestically bred before the Neolithic Age. Before then, the tools to manufacture quantities of silk thread had not been developed. Information on pest's complex in a particular agro-climatic condition is a prerequisite, which helps in designing a successful pest's management strategy. Earlier several authors have reported that insect pest infesting a particular crop differs from place to place. The use of chemical insecticides for control of insect pest in sericulture is often impractical and undesirable, there is an urgent need for an alternative approach that can complement and partially replace chemical based pest and disease management of silkworm in sericulture.

REFERENCES

B.K. Singh, N. Tiken Singh, 2010, Muga Silkworm Seed Organization (MSSO), P-4 Unit, Mendipathar, East Garo Hills, Meghalaya.

Chakrabaty, R. (2004). Diversity of muga and eri sericulture and its prospects I Himalayan states. Proceedings of thye national workshop on potential and strategies for sustainable development of vanya silks in the Himalayan states. Nov.8-9, 9-16 p

Gupta V, Jaiswal K, Gangwar SK, Dwivedi P, Srivastava A, Kumar R (2006) Disease susceptibility in different hybrid breeds of *Bombyx mori* L. in Uttar Pradesh climatic conditions. Proceedings of the *"Regional Seminar on Prospects and problems of sericulture as an economic enterprise in north-west India"*, Dehradun, India, pp 286–289.

Rajesh Kumar Sanp,Harjindra Singh and V.S.Acharya (2015). Major diseases and predators of silkworm and their control. Krishisewa.

Silkworm School Science Project Instruction" (PDF). Archived from the original (PDF) on 14 March 2012 Retrieved 18 October2011.

Singh, R.N and Saratchandra, B. (2008). Push Pull Strategies for pest management in Sericulture. Indian Silk., 47 (7): 4-6.

Srilaxmi, K. and Paul, R. (2010). Diversity of Insect pest of pigeonpea (Cajanus Cajan L. Mill Sp.) and their succession in relation to crop phenology in Gulbarga, Karnataka. The Ecosan 4 (4): 273-276

Subharani, S. and Jayaprakash, P. (2015). Insect Pests of Muga Silkworm, Anthereae Assamensis (Lepidoptera: Saturniidae). G.J.B.A.H.S.,Vol.4(2): 130-134.

Trevisan, Adrian. "Cocoon Silk: A Natural Silk Architecture". Sense of Nature. Archived from the original on 7 May 2012

Chapter 18

Plant Parasitic Nematodes: Hidden and Serious Enemy of Ornamental Crops

Deepak Kumar[1], Vinod Kumar[1], Anil Kumar[1] and Aman[2]

[1]Department of Nematology, CCS Haryana Agricultural University, Hisar – 125004, Haryana
[2]Department of Farm Power and Machinery, IGKV, Raipur – 492012, Chhattisgarh

Floriculture is an important part of horticulture which makes the farmers more profitable for its high aesthetic value, perfume, edible and medicinal oils with highest income per unit area. India has been known for a long tradition history of cultivation of ornamental crops and many records are present like the Vedas, Ramayana, Mahabharatha etc. But the production of flowers crops on commercial scale has assumed from last few years in our country. The floriculture industries are explored as a viable and profitable alternative as income and self employment generation among small and marginal farmers. The practices like production of cut flowers and fresh or dry flower products earn foreign exchange in developing countries. Flowers are recorded to cover an area around 324003 hectares with the production of 1962028 MT and 822750 MT of loose and cut flowers in India respectively. The major flower cultivating states in India are Chhatisgarh, Andhra Pradesh, Jammu and Kashmir, Gujarat, Tamil Nadu, Madhya Pradesh, West Bengal, Uttar Pradesh, Maharashtra, Orissa and Haryana. There are various biotic and abiotic factors that affect the flower production under open field and controlled environmental conditions Ornamentals are mostly grown in monoculture and so become hot beds of infestation by plant parasitic nematodes (PPNs) which are damaging pests of a wide range of ornamental plants. Most PPNs feed on the fine and fibrous roots but few species attack on bulbs, corms, leaves and shoot tips. Various ornamental crops which are grown commercially and subjected to damage by Meloidogyne spp., Radopholus spp., Pratylenchus spp., Heterodera spp., Xiphinema spp., Longidorus spp., Paratrichodorus spp., Criconemoides spp., Belonolaimus spp. and Hoplolaimus spp. resulted into considerable injury to ornamentals. Among foliar nematodes, Aphelenchoides ritzemabosi and Ditylenchus dipsaci are economically important nematode pest of commercially grown ornamental crops. These PPNs are highly diversified microorganisms but not all are harmful and distributed equally in all type of lands.

Importance of Nematodes in Ornamental Crops

Monoculture is the common feature of growing important flower crops which provides congenial environment for one or the other parasitic nematode species. If the nematode population increases unchecked then it will reach an alarming level within a limited time and may finally wipe out the whole infested crop. Worldwide annual yield loss in ornamentals due to plant parasitic nematodes is 11.10 per cent (Trivedi, 2003). In India, an average annual national loss of Rs. 21,068.73 millions has been estimated due to plant parasitic nematodes under protected cultivation (Rao *et al.*, 2005). The loss in yield varies from 27 per cent in carnation to 30 per cent in gerbera due to root-knot nematode. Nagesh and Parvatha Reddy (2000) estimated 31.10 per cent yield loss in gerbera crop while Khan and Mustafa (2005) recorded the loss of 11-22 per cent in yield of gladiolus crop by *M. incognita*. In tuberose (cv. Calcutta Single), the yield loss may occur to the extent of 59 per cent (Pathak and Khan, 2008) and the presence of PPNs in the cut flower stalk is one of the main constraint to export the flower. However, there are not so much losses data available in ornamental crops due to different plant parasitic nematodes but many PPNs are reported in association with different ornamental crops in various regions of India (Table 18.1).

Table 18.1: List of some Important Nematodes and their Ornamental Hosts (Dhaliwal *et al.*, 2004)

Sl.No.	*Nematodes*	*Ornamental Hosts*
1	*Meloidogyne* spp.	Rose, Crossandra, Jasmine, Antirrrhinum, Phlox, Poppy, Naustertium, Carnation, Celosia, Begonia, Pyrogustia, Hedichium, Acalypha, Geranium, China aster, Cacti *etc.*
2	*Ditylenchus dipsaci*	Narcissus, Carnation, Phlox, Begonia, bulb crops *etc.*
3	*Tylenchorhynchus claytoni*	Azalea, Ilex, Sedum *etc.*
4	*Aphelenchoides ritzemabosi* and *A. besseyi*	Tuberose, Zinnia, Chrysanthemum *etc.*
5	*Scutellonema brachyurum*	Begonia, Camellia, Carnation African violet *etc.*
6	*Radopholus similis*	Anthurium, Canna, Celosia, Shoe flower, Ixora, Jasmine *etc.*
7	*Pratylenchus penetrans*	Rose *etc.*

Identification of the Problem

PPNs are the serious problem throughout the country, so proper identification of nematode species is critical to initiate a better management program. Once these tiny organisms enter and establish themselves in the field, it become so cumbersome to complete eradicate them. Proper identification and detailed knowledge about nematode species become important for the grower so that the selection of location and crop can be done timely. Through various methods, one can analyze the nematode species along with population density in soil and plant samples. The population density may be varying depending upon time and host availability (Deimi *et al.*, 2008).

Nematodes Associated with some Important Commercial Flowers

A. Foliar Nematodes

1. *Ditylenchus dipsaci*

Stem and bulb nematode is an obligate parasite of plants and infect various flower crops like daffodils, narcissi, tulips, hyacinth, hydrangea and phlox. Although different biological races of this nematodes has been occur on the different crops. *D. destructor* is another species this nematode which can be observed as a pest of iris and gladiolus. This nematode is commonly distributed in Europe countries and can also be found in cooler regions of the world where the bulb crops are grown. It can be easily disseminated through propagating material of the flowering plants (La Mondia, 1999).

Symptoms

Generally distortion of the leaves can be seen on the infected ornamental plants. Some time symptoms like swellings or yellowed blisters can be easily seen. Feeding on the bulbs produce initially yellow necrotic lesions which later convert into dark-brown colour (La Mondia, 1999). In advanced stages of infestation, nematode infected bulbs may colonized by secondary pathogens like fungi and bacteria which leads to complete decaying of bulbs. In narcissus, twisting and malformation of leaves is caused by *Ditylenchus dipsaci.*

Management

Due to residual impact of chemical nematicides, these are less acceptable in most of the countries. Planting material like bulbs should be nematode free which is effective means of preventing losses. In narcissus, hot water treatment of bulbs can be effective for elimination of PPNs from planting material. Presoaking of bulbs should be done at room temperature and then placed in hot water (44-46°C) for 3 hours. As the symptoms of nematode infestation appear in the field, removal or 'roguing' should be done which reduce the level of infected bulbs (Windrich, 1985).

2. *Aphelenchoides ritzemabosi*

The chrysanthemum foliar nematode is an important nematode pest of chrysanthemum crop in temperate regions of the world but has been reported from high altitudes of tropical countries such as South America and Africa. Some other closely related species are *A. fragariae*, *A. besseyi*, *A. subtenuis*, and *A. blastophorus* which also occur as pests of flower crops (Oliveira and Kubo, 2006). *A. besseyi* become a major limiting factor for cultivation of tuberose in various regions of West Bengal. Specially 'Calcutta single' variety of tuberose is more susceptible to *A. besseyi* than other varieties.

Symptoms

This foliar nematode is above ground migratory parasite and generally feed on mesophyll tissues of leaves. Initially, minute yellow spots can be seen between the large veins which later turn into brown colour. On severity, these spots coalesce and a large V-shaped angular area of necrotic tissue is formed on leaves. It produces characteristically angular necrotic patch in chrysanthemum leaves and *A. ritzemabosi* remains inside tissues while it is not strictly an endo-parasitic and survive in leaves, but not in the soil. Secretions from nematodes cause shortening of internodes (rosetting) giving a bushy appearance to the affected plants. These secretions by nematode also leads to production of low, premature side shoots and distorted leaves. Blindness can be seen on infested plants which is characterized by shoot browning and failure to grow. The disease tends to progress from the lower to the upper leaves as the nematode moves up the plant. *A. fragariae* cause similar symptoms to *A. ritzemabosi* on plnats like begonia, lilies and violets *etc.* Some other nematodes like *A. subtenuis* has been observed infectious on narcissus and *A. blastophorus* in *Scabiosa* plants produces symptoms like twisting and distortion of leaves (Chalanska *et al.*, 2017).

Diseases Caused

Yellow and brown inter-veinal necrosis is caused by nematode feeding on chrysanthemum leaves.

Management (Chalanska et al., 2017)

Sanitation : Removal of infected plants and diseased debris from the field.

Healthy planting material : Cuttings should be taken from the upper portion of the plant and disinfected with a hot water treatment (50°C for 5 min or 44°C for 30 min).

Cultural Practices : Mulching over fallen leaves in ornamental plantings should be done to prevent upward migration of the nematode. Space plants to allow good air flow and rapid drying.

B. Below Ground Nematodes

3. *Meloidogyne* spp.

Flower crops are hosts for various *Meloidogyne* spp. which are the chief limiting factors for flower crop productions. There are four species under *Meloidogyne* genus which are found associated with most of floral crops (Agudelo *et al.*, 2006). *M. hapla* and *M. ardenensis* can be found associated with flower crops in temperate regions but this is mostly a pest of the crops in tropical regions. *M. incognita* and *M. javanica* are the most frequently recorded species on different flowers including carnations (*Dianthus*), rose, gerbera, *Coleus*, *Heliconia*, *Alpinia*, and *Proteus* under field and glass conditions (Borgohain, 2016). A study was conducted to test the influence of *M. incognita* and *Rotylenchulus reniformis* infection on the growth of ten ornamental plant species which revealed that four out of ten (Chinese amaranth, coleus, carnation and mountain thyme) has been found to be affected by these nematodes in terms of growth parameters with different degrees.

Symptoms

It is a sedentary endoparasites in which second stage juvenile (J_2) is infective stage which preferably invade the roots near root cap. The parasitic nematodes suck and drain the fine hair like roots and create knots and limiting the development of the root system. These tiny swellings (galls) appear all over the mass of the roots on heavy infection. The damaged roots can no longer take up water or nutrition into the upper parts of the plant ultimately leads to weak and stunted plants which don't have normal healthy green coloration. Flowering is also sparse upon nematode infestation (McSorley and Frederick, 1994).

Management

Nematodes can be controlled by planting nematode-free plants into parts of the garden where nematodes haven't been a problem previously. Marigolds or *Alliums* as under plantings repel nematodes. Select type of crop and cultivar based on host response to nematode (Jones, 1995) (Table 18.2).

Table 18.2: Relation of Root-knot Nematode, *M. incognita* with some Woody Ornamentals (Bernard and White, 1987)

Host Plant	*Response to M. incognita*
Camellia sasanqua	Intolerant
Cedrus deodara	Tolerant
Chamaecyparis pisifera	Tolerant
Cotoneaster horizontalis	Tolerant
Elaeagnus pungens	Tolerant
Hedera helix	Intolerant -Tolerant
Ilex cornuta cv. Burfordii	Highly Intolerant
Ilex creneta cv. Hetzii	Intolerant
Jasmenum nidiflorum	Highly Intolerant
Juniperus horizontalis cv. Douglasii	Intolerant -Tolerant
Loropetalum chinese	Intolerant -Tolerant
Osmanthus x fortunei	Highly Intolerant
Poncirus trifolilata	Intolerant
Prunus laurocerasus cv. Zabeliana	Tolerant
Syringa persica	Highly Intolerant
Thuja occidentalis cv. Woodwardii	Tolerant
Thuja orientalis cv. Berkmanns	Tolerant
Vitex angus-castus	Highly Intolerant

Note :

- ☆ Plants highly intolerant (severe stunting, branch dieback and death, heavy nematode reproduction)
- ☆ Plants intolerant (some stunting, but plants will grow satisfactorily, some nematode reproduction)
- ☆ Plants tolerant, will grow satisfactorily

Sensitivity of Some Annuals to Root Knot Nematode (McSorley and Fredrick, 1994) :

A. *Immune*

- ☆ *Argemone* sp.
- ☆ *Tagetes* sp.
- ☆ *Ageratum* sp.
- ☆ *Gaillardia* sp.
- ☆ *Rudbeckia* sp.
- ☆ *Coreopsis lanceolata*

B. *Highly Resistant*

- ☆ *Aster tradescanti*
- ☆ *Lupine* sp.
- ☆ *Coreopsis tinctoria*
- ☆ *Cosmos bipinnatus*
- ☆ *Gomphrena globosa*
- ☆ *Catharanthus rosea*
- ☆ *Matthiola* sp.
- ☆ *Lobularia maritima*
- ☆ *Salvia splendens*
- ☆ *Arctotis stoechadifolia*
- ☆ *Mirabilis jalapa*
- ☆ *Zinnia elegans*
- ☆ *Torenia fournieri*
- ☆ *Torenia* sp.
- ☆ *Thunbergia* sp.
- ☆ *Gerbera jamesonii*

C. *Moderately Resistant*

- ☆ *Godetia* sp.
- ☆ *Penstemon* sp.
- ☆ *Dianthus* sp.
- ☆ *Portulaca* sp.
- ☆ *Verbena* sp.
- ☆ *Reseda odorata*
- ☆ *Lathyrus latifolius*
- ☆ *Liatris spicata*
- ☆ *Clarkia* sp.
- ☆ *Chrysanthemum maximum*
- ☆ *Iberis umbellata*
- ☆ *Callistephus chinensis*

D. *Susceptible*

- ☆ *Helipterum roseum*
- ☆ *Linaria* sp.
- ☆ *Ipomoea* sp.
- ☆ *Chrysanthemum* sp.
- ☆ *Eschscholtzia californica*
- ☆ *Hunnemania fumariaefolia*
- ☆ *Dimorphotheca sinuata*
- ☆ *Ibellis perennis*
- ☆ *Nicotiana alata*
- ☆ *Papaver* sp.

E. *Highly Susceptible*

- ☆ *Coleus* sp.
- ☆ *Aquilegia* sp.
- ☆ *Helianthus annuus*
- ☆ *Cynoglossum* sp.
- ☆ *Gypsophila* sp.
- ☆ *Schizanthus* sp.
- ☆ *Ipomoea* sp.
- ☆ *Delphinium* sp.
- ☆ *Lobelia erinus*
- ☆ *Helichrysum* sp.

- ☆ *Gilia* sp.
- ☆ *Calendula officinalis*
- ☆ *Antirrhinum majus*
- ☆ *Althea rosea*
- ☆ *Salpiglossis sinuata*
- ☆ *Viola tricolor*
- ☆ *Centaurea cyanus*
- ☆ *Amaranthus* sp.
- ☆ *Impatiens balsamina*
- ☆ *Trachymene caerulea*
- ☆ *Lathyrus odoratus*
- ☆ *Celosia argentea*
- ☆ *Dolichos* sp.
- ☆ *Cucurbita* sp.

4. Lesion Nematode (*Pratylenchus* spp.)

It is a migratory endoparasites and all the juvenile and adult stages are infective stage. Eggs are laid inside the roots. Juveniles on hatching move inside the root tissue in cortex both inter as well as intracellular causing extensive damage to the root tissue. The nematode takes about 46-55 days to convert from egg to adult. Nematode can survive up to eight months in moist soil in absence of the host crop. Survival period is reduced to three months in dry soil (Chalanska *et al.*, 2016). Important species are *P. vulnus, P. pratensis* and *P. penetrans.*

Symptoms

Infested plants are weak and stunted with pinkish to purple and yellow coloured leaves. Inflorescence and flower size is reduced along with root system and roots bear brown to black lesions (Chalanska *et al.*, 2016).

Management

Application of FYM and inter-culturing of african marigold reduces the nematode population significantly. Interculturing with marigold and post-planting drenches with DBCP (nemagon) controls the nematodes (Davis and MacGuidwin, 2014). Carbofuran 3G @ 1g per plant is effective.

5. Dagger Nematode (*Xiphenema diversicaudatum*)

Life cycle of this nematode may take three years to complete. Females survive for more than five years and mate several times during this period with males which are as abundant as females. It can survive in the soil devoid of higher plants for up to three years. Eggs are killed at 5°C, but juveniles and adults survive this temperature. It is a major pest of various ornamental crops in countries like USA, Russia, Canada, Australia, New Zealand, India *etc.* It has wide host range like rose, carnations, gerbera, coleus and strawberry *etc.* (Archidona-Yuste *et al.*, 2016).

Symptoms

It is a migratory ectoparasitic nematode that mainly infects the root tips and causes galling of root tips in roses. Malformation, chlorosis and dwarfing occur in infested plants (Palomares-Rius *et al.*, 2012).

Management

Fallowing and crop rotation are practically not feasible against this nematode due to its vast host range. This nematode poses more threat as a virus transmitting agent. Application of organic amendments like neem cake or mustard cake @ 25q/ha (Moseley *et al.*, 2010).

Other Nematodes of Flower Crops

Despite the frequent occurrence of harmful PPNs, several other species are observed on ornamental crops which causes serious damage to most of economic crops. Many of the PPN species show their distinct attitude towards varied hosts (Table 18.3) based upon their host selection and feeding. Some commonly occurring nematodes, such as *Rotylenchulus reniformis*, *Radopholus* sp., *Heterodera* sp., *Longidorus* sp., *Paratrichodorus* sp., *Criconemoides* sp., *Belonolaimus* sp. and *Hoplolaimus* sp. which affect quantity and quality of produce (Borgohain, 2016). That's why their management become very important so that nematode's direct and indirect impacts can be minimized in high valued ornamental crops.

Some examples of diseases caused by nematodes alone or in complex with other microorganisms:

- ☆ *Heterodera trifoli* cause disease complex with association of wilt causing fungus (*Fusarium oxysporum* f.sp. *dianthi*) on *Dianthus*
- ☆ Strawberry latent ring spot NEPO virus on lilies is vectored by nematodes

Table 18.3: Host Response of some Ornamental Plant Species to *Meloidogyne incognita* and *Rotylenchulus reniformis*

Plant Species	*Meloidogyne incognita*	*Rotylenchulus reniformis*
A. tricolor	Susceptible	Tolerant
C. roseus	Resistant	Resistant
C. blumei	Susceptible	Tolerant
D. caryophyllus	Highly Susceptible	Susceptible
G. globosa	Resistant	Highly Resistant
O. basilicum	Tolerant	Highly Susceptible
P. grandiflorum	Resistant	Susceptible
P. odoratissimum	Resistant	Resistant
P. amboinicus	Highly Susceptible	Highly Susceptible
T. erecta	Highly Resistant	Highly Resistant

Integrated Nematode Management Approaches

a) For producing healthy seedlings under protected conditions

Soil mixing with following mixtures for producing healthy seedlings of crossandra and China asters.

- ☆ Neem cake @ 50 kg + carbofuran or phorate @ 5 kg
- ☆ *Trichoderma harzianum* + *Pseudomonas fluorescens* and Neem cake @ 50 kg each at the rate of 2 kg/ton can be used.

b) For producing healthy the seedlings of ornamental crops in the raised nursery beds in open field conditions

- ☆ Apply Neem cake/Pongamia cake/Castor cake @ 500 gm per sq. m. followed by solarization of the nursery beds and leave the raised nursery beds for a period of 15-20 days.
- ☆ When the nematode densities are higher, apply carbofuran or phorate @ 50 gm per sq. m.
- ☆ Along with above mentioned botanicals or chemicals, apply formulations of *Trichoderma harzianum* or *Pseudomonas fluorescens* and *Paecilomyces lilacinus* (each 50g/sq.m) to the nursery beds.

c) For producing nematodes free crop plants in the main field

Recommended dose of well rotten organic amendments and fertilizers should be added before brining soil to fine tilth. Also add 200g neem/pongamia/mahua cake + carbofuran or phorate @ 20–25 kg per acre. Maintain proper moisture content in the beds for decomposition of neem/pongamia/mahua cake. For eco-friendly management of PPNs, add 2 tons of well rotten farm yard manure or 500 kg of cake (neem/pongamia) or 1ton of vermicompost enriched with biological control agents (*Pseudomonas fluorescens* + *Trichoderma harzianum* + *Paecilomyces lilacinus*) during land prearation.

Conclusion

Plant-parasitic nematodes (PPNs) are one of the major constraints for the cultivation of ornamental crops in India and presently this problem is gaining greater attention to the commercial flower growers particularly for production of quality flowers in protected systems. So, the management of these tiny organisms become very important for sustainable production of ornamental crops to meet out increasing demand. Available strategies and methods for managing nematodes consist resistant cultivars, cultural practices, biological control agents and physical methods. Further, there is an urgent need to study the biology, host pathogen relationship and epidemiology of diseases to arrive at sound integrated strategy. Popularization of importance of nematodes and the damage caused by them among the farming community is greatly needed to boost the productivity as well as quality of ornamental plants in India.

REFERENCES

Agudelo, P. A., Lewis, S. A. and Abril, M. A. (2006). First report of root-knot nematode, *Meloidogyne javanica* on chrysanthemum in Colombia. *Plant Diseases* 90: 828.

Archidona-Yuste, A., Navas-Cortes, J. A., Cantalapiedra-Navarrete, C., Palomares-Rius, J. E., and Castillo, P. (2016). Remarkable diversity and prevalence of dagger nematodes of the genus *Xiphinema* Cobb, 1913 (Nematoda: Longidoridae) in olives revealed by integrative approaches. *Plos One* 1-54.

Bernard, E. C. and White, W. T. (1987). Parasitism of Woody Ornamentals by *Meloidogyne hapla*. *Annals of Applied Nematology* **1**: 41-45.

Borgohain, N. (2016). Plant parasitic nematodes associated with some important commercial flowers. *Journal of Global Biosciences* **5**(8): 4541-4549.

Chalanska, A., Bogumi³ A. and Labanowski G. (2017). Management of foliar nematode, *Aphelenchoides ritzemabosi* on Anemone hupehensis using plant extracts and pesticides. *Journal of Plant Diseases and Protection* **124** (2).

Chalanska, A., Labanowski, G. and Sas, D. (2016). Root-lesion nematodes (*Pratylenchus* spp.) in ornamental plant nurseries - influence of soil texture, acidity, salinity and organic matter content. *Communications in Biometry and Crop Science* **11** (2): 98–104.

Davis, E. L. and MacGuidwin A. E. (2014). "Lesion Nematode Disease". The plant health instructor (2000). The American Phytopathological Society. 1030-032.

Deimi, A. M., Chitambar, J. J. and Maafi, Z. T. (2008). Nematodes associated with flowering ornamental plants in mahallat, Iran. *Nematologia Mediterranea* **36**: 115-123.

Dhaliwal, G. S., Arora, R. and Dhawan, A. K. (2004). Crop losses due to insect pests in Indian agriculture. *Indian Journal of Ecology*. **31**: 1-7.

Jones, R. K. (1995). Nematode management in bedding plants in the landscape. Plant Pathology Information Note. Ornamental No. 31. North Carolina Cooperative Extension Service.

Khan, M. R. and Mustafa, U. (2005). Corm rot and yellows of Gladeoli and its bio-management. *Phytopathologica Mediterrranaea*, **44:** 208-215.

La Mondia, J. A. (1999). Efficacy of insecticides for control of *Aphelenchoides fragariae* and *Ditylenchus dipsaci* in flowering perennial ornamentals. *Journal of Nematology* **31**(4S): 644-649.

McSorley, R. and Frederick, J. (1994). response of some common annual bedding plants to three species of *Meloidogyne*. *Journal of Nematology*. **26** (4S): 773-777.

Moseley, D., Patton, A., Bateman, R. and Kirkpatrick, T. (2010). Controlling nematodes on golf courses. Cooperative Extension Service, University of Arkansas, MP 481, 12 pp.

Nagesh, M. and Reddy, P. P. (2000). Crop loss estimation in carnation and gerbera due to root-knot nematode, *Meloidogyne incognita* (Kofoid and White) Chitwood. *Pest Management in Horticultural Ecosystems* **6**: 157-159.

Oliveira, C. M. G. and Kubo, R. K. (2006). Foliar nematodes (*Aphelenchoides* spp.) on begonia in Brazil. *Revista Brasileira de Horticultura Ornamental* **12**: 134-137.

Palomares-Rius J. E., Gutiérrez- Gutiérrez, C., Cantalapiedra Navarrete, C. and Castillo, P. (2012). Prevalence and diversity of grapevine fan leaf virus in southern Spain. *Plant Pathology*. **61**: 1032-1042.

Pathak, B. and Khan, M. R. (2008). Yield loss potential of foliar nematode, *Aphelenchoides besseyi* in tuberose. *Indian Journal of Nematology* **39**(1): *In press.*

Rao, M. S., Umamaheswari, R. and Chakravarty, A. K. (2005). Plant parasitic nematodes: a major stumbling block for successful crop cultivation under protected conditions in India. *Current Science*. **108**: 1.

Sigariova, D. D. and Karpliyk, V. G. (2015). Parasitic nematodes in flowering and ornamental plants: effect of parasites on the plants and response of the plants to the presence of nematodes. *Vestnik zoologii*, **49**(2): 427-432.

Tivedi, P. C. (2003). Advances in Nematology. *Scientific Publisher*. pp 294.

Windrich, W.A. (1985). Control of stem nematode, *Ditylenchus dipsaci*, in tulip bulbs with aldicarb and oxamyl. *Crop Protection*, **4**(4) : 458-463.

Chapter 19

Pesticide Regulation in India

Neeru Dumra[1], Krishan Rolania[1], Neha Sharma[2] and Sachin Kumar[3]

[1]Department of Agricultural Entomology, [3]Department of Horticulture Fruit Science, CCS, HAU, Hisar – 125001, Haryana
[4]Department of Agricultural Entomology, CSA University of Agriculture and Technology, Kanpur – 208002, Uttar Pradesh

A pesticide is any substance or mixture of substances intended for preventing, destroying, repelling, or mitigating any pest (insects, mites, nematodes, weeds, rats, *etc.*) including insecticide, herbicide, fungicide, and various other substances used to control pests (EPA, 2009). Definition of pesticide varied with times and countries. However, the essence of pesticide remains basically constant, *i.e.*, it is a (mixed) substance that is poisonous and efficient to target organisms and is safe to non-target organisms and environments.

The setup of guidelines and rules which followed by concerned authorities and users to run the process in efficient, accountable and transparent way. Legal particulars at every step from manufacture (Industry) to application (Farmers) of insecticide is known as regulation.

Need for Regulation

Worldwide approximately 9,000 species of insects and mites, 50,000 species of plant pathogens, and 8,000 species of weeds damage crops. Insect pests cause an estimated 14 per cent of loss, plant pathogens cause a 13 per cent loss, and weeds a 13 per cent loss (Pimentel, 2009). Pesticide is so indispensable in agricultural production. About one-third of the agricultural products are produced by using pesticides (Liu *et al*,2002). Without pesticide application the loss of fruits, vegetables and cereals from pest injury would reach 78 per cent, 54 per cent and 32 per cent respectively (Cai, 2008). Crop loss from pests declines to 35 per cent to 42 per cent when pesticides are used. Varieties and consumption of pesticides worldwide have been increasing dramatically as increased human population and crop production (Liu and Liu, 1999).

The pesticide consumption in India during the period 1954-2000 shows that it has raised from 434 MT to 46,195.16 MT (Anonymous, 2015h). Cotton (36 per cent) is the highest pesticide consuming crop followed by rice (20 per cent). Andhra Pradesh is the highest pesticides consuming state (23 per cent) followed by Punjab and Maharashtra (Bhardwaj and Sharma, 2013). India is now the second largest manufacturer of pesticides in Asia after China and ranks twelfth globally (Mathur, 2010). Researchers at the New Delhi based Centre for Science and Environment have found alarmingly high levels of pesticides in blood samples of villagers in Punjab, the showpiece state of India's green revolution. India needs to urgently take a tough look at the indiscriminate and careless use of pesticides (Menon, 2006).

Agricultural development continues to remain the most important objective of Indian planning and policy. In the process of development of agriculture, pesticides have become an important tool as a plant protection agent for boosting food production. Further, pesticides play a significant role by keeping many dreadful diseases. However, exposure to pesticides both occupationally and environmentally causes a range of human health problems. A vast majority of the population in India is engaged in agriculture and is therefore exposed to the pesticides used in agriculture. Although Indian average consumption of pesticide is far lower than many other developed economies, the problem of pesticide residue is very high in India. Pesticide residue in several crops has also affected the export of agricultural commodities in the last few years. In this context, pesticide safety, regulation of pesticide use, proper application technologies, and integrated pest management are some of the key strategies for minimizing human exposure to pesticides (Abhilash and Singh, 2008).

In this process pesticide misuses become more and more serious, which has resulted in heavy environmental pollution and health risk of humans. Majority of pesticides are not specifically targeting the pest only and during their application they also affect non-target plants and animals. Over time, repeated application leads to loss of biodiversity and increases in pest resistance, while its effects on other species can facilitate the pest's resurgence. Many pesticides are not easily degradable, they persist in soil, leach to groundwater and runoff can carry pesticides into aquatic environment and contaminate wide environment. Overall, intensive pesticide application results in several negative effects in the environment that cannot be ignored (Sitaramaraju *et al.*, 2014).

In India there is a trade-off between agricultural production and increasing soil, air and water pollution and associated health hazards (Gupta, 2004). Many of the chemicals used in pesticides are persistent soil contaminants, whose impact may endure for decades and adversely affect soil conservation. Pesticides enter the soil via spray drift during foliage treatment, wash-off from treated foliage, release from granulates or from treated seeds in soil. The transport, persistence or degradation of pesticides in soil depend on their chemical properties as well as physical, chemical and biological properties of the soil. All these factors affect sorption/desorption, volatilisation, degradation, uptake by plants, run-off, and leaching of pesticides. Persistence of pesticides in soil can vary from few hours to many years in case of OC pesticides. Despite OC pesticides were banned or restricted in many countries, they are still detecting in soils (Shegunova *et al.*, 2007; Jiang *et al.*, 2009 and Anonymous, 2015d).

Unfortunately, India is one of the few remaining countries still producing and using some of the chlorinated pesticides such as DDT and lindane (Abhilash and Singh, 2009).

In 1962, Rachel Carson, an American courageous woman and scientist, wrote down her nature observation and pointed out sudden dying of birds caused by indiscriminated spraying of pesticides (DDT). Her book, Silent Spring, became a landmark. It changed the existing view on pesticides and has stimulated public concern on pesticides and their impact on health and the environment. Silent Spring facilitated the ban of the DDT in 1972 in the United States (Zilberman and Millock, 1997).

Modern agricultural practices have been proved a great promise for economic development of nation. Farm productivity is directly proportional to use of agrochemicals as observed from the first green revolution. Improper and unsafe use of these agrochemicals, especially pesticides is not only harmful to environment but also human health. Pesticides cause 14 per cent of all known occupational injuries in agriculture and 10 per cent of all fatal injuries. Total 100 agricultural labourers were interviewed with pre-tested interview schedules by using accidental sampling procedure. It is found that more than 75 per cent of labourers used either "moderately hazardous" or "highly hazardous" pesticides as classified by World Health Organisation (WHO). However, 88 per cent did not use any form of protection, while handling pesticides. Poverty and illiteracy are greatly responsible for improper handling of pesticides.The study also found that there is ample scope for reducing pesticide exposure through training, agricultural extension and community mobilisation (Dhanraj and Ravasaheb, 2012).

In this process pesticide misuses become more and more serious, which has resulted in heavy environmental pollution and health risk of humans. The FQPA directs the U.S. Environmental Protection Agency (EPA) to make more rigorous and conservative evaluation of risks and hazards and mandates a special emphasis on the safety of infants and children. The EPA provides incentives for the industry to register materials that are designated "reduced risk". The future for the registrant industry will include continued reduction in numbers of registrants through mergers and acquisitions. Conventional chemicals will remain as important pest management components, and the processes of combinatorial chemistry and high-throughput bioassays will allow the rapid synthesis and testing of large numbers of candidate compounds. Biopesticides will become more important tools in pest management, with microbial pesticides and transgenic crops being likely to play important crop protection roles. There will be a continuing need for research-based approaches to pest control. (Wheeler, 2002)

Pesticides have contaminated almost every part of our environment and pesticide residues are found in soil, air and in surface and groundwater. Pesticide contamination poses significant risks to the environment and non-target organisms ranging from beneficial soil microorganisms to insects, plants, fish, and birds. Recent studies have indicated that our environment is chronically polluted by pesticides and levels of biocidal contamination have increased tremendously. The environmental deterioration due to pesticides is endangering the situation of future.

International Level

The International Code of Conduct on the Distribution and Use of Pesticides was one of the first voluntary Codes of Conduct in support of increased food security, while at the same time protecting human health and the environment. It was adopted in 1985 by the FAO Conference at its Twenty-third Session, and was subsequently amended to include provisions for the Prior Informed Consent (PIC) procedure at the Twenty-fifth Session of the FAO Conference in 1989. The Code established voluntary standards of conduct for all public and private entities engaged in, or associated with, the distribution and use of pesticides, and since its adoption has served as the globally accepted standard for pesticide management. Experience over the last 15 years has shown that the Code, in conjunction with its supplementary technical guidelines, has been instrumental in assisting countries to put in place or strengthen pesticide management systems. Surveys show that the number of countries without legislation to regulate the distribution and use of pesticides has greatly decreased; awareness of the potential problems associated with pesticide use has grown significantly; involvement in various aspects of pesticide management by NGOs and the pesticide industry has been strengthened; and further successful Integrated Pest Management (IPM) programmes are being implemented in developing countries.

The objectives of this Code are to establish voluntary standards of conduct for all public and private entities engaged in or associated with the distribution and use of pesticides.

1. The Code is designed for use within the context of national legislation as a basis whereby government authorities, pesticide manufacturers.
2. The Code describes the shared responsibility of many sectors of society to work together so that the benefits to be derived from the necessary and acceptable use of pesticides are achieved without significant adverse effects on human health or the environment.
3. The Code addresses the need for a cooperative effort between governments of pesticide exporting and importing countries to promote practices that minimize potential health and environmental risks.
4. The entities which are addressed by this Code include international organizations, governments of exporting and importing countries, pesticide industry, application equipment industry, traders, food industry.

The Federal Insecticide, Fungicide, and Rodenticide Act, 1972

The regulation of pesticides by the federal government began in 1910 with the passage of the Federal Insecticide Act. This act was passed in response to concerns from the United States Department of Agriculture (USDA) and farm groups about the sale of fraudulent or substandard pesticide products, which was common at the time. Thus, the first federal pesticide legislation was to ensure the quality of pesticide chemicals purchased by consumers. Specifically, the act set standards for the manufacture of Paris green, lead arsenate, insecticides, and fungicides, and also provided for inspections, seizure of adulterated or misbranded products, and prosecution of violators. Further amendments to the FIFRA occurred in 1975, 1978, 1980, and 1981. These amendments were designed to improve the pesticide registration process and provide for consideration of agricultural benefits of pesticides in regulatory decisions made by the EPA. The 1975 amendment required the EPA to notify the Secretary of Agriculture in advance of regulatory decisions affecting pesticides, consider the impact of pesticide cancellations on the production and costs of agricultural commodities, and establish a Scientific Advisory Panel to review proposed regulations. The amendment to the FIFRA in 1978 allowed the EPA, under certain circumstances, to grant conditional registrations of pesticides before all supporting data had been submitted by the registrant. This provision was intended to reduce the backlog of registration applications.

Codex Alimentarius Commission

The Codex Alimentarius is a collection of internationally recognized standards, codes of practice, guidelines and other recommendations relating to foods, food production and food safety (Anonymous, 2015g). Its texts are developed and maintained by the Codex Alimentarius Commission, a body that was established in early November 1961 by the Food and Agriculture Organization of the United Nations (FAO), was joined by the World Health Organization (WHO) in June 1962. (Anonymous 2015)

Objectives

- Protect consumers adequately from illness or injury caused by food; policies need to consider the vulnerability of the population, or of different groups within the population
- Provide assurance that food is suitable for human consumption
- Maintain confidence in internationally traded food
- Provide health education programmes which effectively communicate the principles of food hygiene to industry and consumers.

Industry should apply the hygienic practices set out in this document to:

- Provide food which is safe and suitable for consumption
- Ensure that consumers have clear and easily-understood information, by way of labelling and other appropriate means, to enable them to protect their food from contamination
- Maintain confidence in internationally traded food.

National Level

The Indian Parliament sets up the Joint Parliamentary Committees (JPC) only on critical issues of public interest. Only five such committees have been formed in the history of independent India. The JPC formed on Pesticide Residues in and Safety Standards for Soft Drinks, Fruit Juices and Other Beverages, in 2003, was the fourth and the only committee to have been set up on public health issue. The rest have been to investigate scams. The committee found loopholes in the pesticides regulations as well as their implementation.

In 2005, the JPC set out a clear agenda for governments to ensure the safe use of pesticides. The committee recommended to make mandatory the setting of maximum residue limits (MRL) for pesticides before registering it, setting MRLs for deemed registered pesticides, reviewing the set MRLs for compliance with the Acceptable Daily Intake (ADI) of pesticides and monitoring pesticide residues regularly. Pesticide use in India is regulated by the Central Insecticides Board and Registration Committee (CIBRC) and the Food Safety and Standards Authority of India (FSSAI). The CIBRC registers pesticides for crops while the FSSAI sets the maximum residue limits of pesticides for the crops it has been registered for.

It was found that recommendations of JPC have not been followed properly. Of the 234 pesticides registered in the country, the FSSAI has not set MRLs for 59 pesticides. A review of MRL status of 20 commonly used and recommended pesticides showed that the MRLs set for 18 pesticides are not complete. MRLs have not been set for all the crops these pesticides have been registered for. A few MRLs have been set for crops for which the corresponding pesticide is not registered. MRLs have been set for broad groups like fruits, vegetables and food grains rather than specific crops while the pesticides have been registered for specific crops.

The agriculture universities, departments and boards have recommended many pesticides that have not been registered for some crops. Recommendations of waiting periods for pesticides are not complete. The farmers were found to be unaware of the registered uses of pesticides. They mostly followed the pesticides as the dealers recommended them. The outreach of state agriculture universities and departments to the farmers was minimal. The State Agriculture Universities (SAUs), State Agriculture Departments (SADs) and other institutions related to specific crops like National Horticultural Board (NHB) and Spices Board of India make another set of recommendations for agricultural practices including use of pesticides. These recommendations are important considering the local needs of the states and research about specific crops, their diseases and insects.

Farmers use pesticides to kill pests in crops as it helps to increase agricultural productivity. But rampant abuse of pesticides results in Pesticide residues in some component of the environment after its application, spillage or dumping. Pesticide residues may pose significant risks to human beings and environment if, it exceeds the maximum residues levels (MRLs).There is efficient monitoring system which regularly tests food items for pesticide residues to ensure that it does not exceed (MRLs) and there is no potentially adverse effect on humans and environment before they can be licensed and sold. Some of the government agencies involved in the regulation are The Central Insecticides Board and Registration Committee (CIBRC), The Codex Committee on Pesticide Residues (CCPR) and Environmental Protection Agency (EPA). The objective of this paper is to give an overview of work that has been done on pesticide residue and how the overuse of pesticide can be controlled by obeying regulations (Kochar and Urkude, 2014).

Pesticides Regulations

Pesticides regulations are governed in India under following Acts/Rules:

1. The Insecticides Act, 1968 and Rules, 1971
2. The Environment (Protection) Act, 1986
3. Hazardous Waste (Management and Handling) Rules, 1989
4. Water (Prevention and Control of Pollution) Act, 1974
5. Air (Prevention and Control of Pollution) Act, 1981
6. Prevention of Food Adulteration Act, 1954
7. The Factories Act, 1948
8. Bureau of Indian Standards Act, 1986

The Committee have been informed that pesticides Consumption in some of the major countries, is as follows:

USA : 7.0 Kg/ha

Europe : 2.5 Kg/ha

Taiwan : 17 Kg/ha

Japan : 12 Kg/ha

Korea : 6.6 Kg/ha

India : 0.5 Kg/ha

(Anonymous, 2013)

Irrational use may because of following factors:

(i) Indiscriminate use of chemical pesticides
(ii) Non-observance of prescribed waiting periods
(iii) Use of sub-standard pesticides
(iv) Wrong advice and supply of pesticides to the farmers by pesticide dealers
(v) Continuance of DDT and other uses of pesticides in Public Health Programmes
(vi) Effluents from pesticides manufacturing units
(vii) Wrong disposal of left over pesticides and cleaning of plant protection equipments
(viii) Pre-marketing pesticides
(ix) treatment of fruits and vegetables

Environmental (Protection) Act, 1986

This Act was promulgated in the year 1986 as a sequel to Bhopal gas tragedy. The Act has inter-alia entrusted the responsibility to lay down the procedures and safeguards for handling the hazardous substances and prevention of accidents. The Ministry of Environment and Forests has been designated as the Nodal Ministry for "Management of hazardous substances and prevention of chemical disasters". Following sets of Rules had been notified to regulate handling and use of hazardous chemicals including pesticides:

(i) Manufacture, Storage and Import of Hazardous Chemicals Rules, 1989 and amendments made.

(ii) Chemical Accidents (Emergency Planning, Preparedness and Response) Rules, 1996

(ii) Hazardous Wastes (Management and Handling) Rules, 1989 and amendments made therein

(iv) Environmental Protection Rules, 1986

The principal objectives of the Rules are the prevention of major accidents arising from industrial activities, limiting the effects of such accidents both on human and environment and the harmonisation of various control measures to prevent and limit the major accidents. These Rules have been amended in 1994 and in 2000.

Prevention of Food Adulteration (PFA) Act, 1954

Prevention of Food Adulteration is a subject in the Concurrent list of the Constitution. Under the "Prevention of Food Adulteration Act, (PFA) 1954" and the "Prevention of Food Adulteration Rules, 1955" standards for different food articles as well as the provisions for their distribution, storage and sale are prescribed. The permissible limit for food additives and the maximum residue limits for different pesticides are also regulated through the PFA Rules. Noncompliance of these standards and provisions are punishable under the provisions of the Act.

The Act has been amended thrice, in 1964, 1976 and in 1986 with the objective of plugging the loopholes and making the punishments more stringent and for empowering consumers and voluntary organisations to play a more effective role in its implementation. The Central Government primarily plays an advisory role in its implementation, besides carrying out various statutory functions/duties assigned to it under the various provisions of the Act.

The Central Committee for Food Standards (CCFS) constituted under Section 3 of the Prevention of Food Adulteration Act, 1954 advises Central/State governments on all matters arising out of implementation of the Prevention of Food Adulteration Act, 1954 and the PFA Rules, 1955, including review and formulation of rules, regulations and standards of food articles. The Central Committee for Food Standards (CCFS) has representatives from different departments in the Central Government, State Governments, trade, industry, technical experts and consumers, organisations besides representatives from the National Institute of Nutrition, Central Food Technology and Research Institute and the Central Food Laboratories.

The Ministry of Agriculture regulates the manufacture, sale, import, export and use of pesticides through the 'Insecticides Act, 1968' and the rules framed thereunder. Central Insecticides Board (CIB) constituted under Section 4 of the Act advises Central and State Governments on technical matters. The Registration Committee (RC) constituted under Section 5 of the Act approves the use of pesticides and new formulations to tackle the pest problem in various crops. The monitoring of pesticides residue levels in food comes under the preview of Union Ministry of Health and Family Welfare.

Insecticides Act, 1968

The Insecticides Act, 1968 regulates import, manufacture, sale, transport and distribution and use of insecticide, with a view to prevent risk to human beings or animals and the matters connected therewith. This Act was passed by the Parliament in the Nineteenth year of Republic of India and came into force on 01.03.1971 (Anonymous, 2015a).

Presently 290 pesticides stand approved for use in the country (Anonymous, 2015h). Out of these 71 pesticides belonging to deemed registered category are used in the country. Waiting periods for some pesticides are not mentioned on the leaflets due to non-availability of data on residues on the crops against which the products are approved. Prior to insecticide act 1968, already there were some pesticides being used in the country. So, that is why, they were given what is called the deemed registration status. Many of them are already phased out. To overcome the gaps, the Registration Committee has constituted an Expert Group to examine total data available with the pesticide industry and the Registration Committee.

Insecticides Rules, 1971

Under the Act, Central Government has the power to frame rules. Under this Rule various functions are prescribed (Anonymous, 2015b).

The Board shall consist of the following primary members, namely:

a. The Director-General of Health Services, ex officio, who shall be Chairman;
b. The Drugs Controller, India, ex officio;
c. The Plant Protection Adviser to the Government of India, ex officio;

Functions of the Board:

(a) Advise the Central Government on the manufacture of insecticides under the Industries (Development and Regulation) Act. 1951 (65 of 1951).
(b) Specify the uses of classification of insecticides on the basis of their toxicity as well as their being suitable for aerial application.
(c) Advice tolerance limits for insecticide residues and on establishment of minimum intervals between the application of insecticides and harvest in respect of various commodities.
(d) Specify the shelf-life of insecticides.

(e) Suggest colourisation, including colouring matter which may be mixed with concentrates of insecticides particularly those of highly toxic nature.

(f) Carry out such other functions as are supplemental, incidental or consequential to any of the functions conferred by the Act ot these rules.

Registration Committee

Established under Section 5 of the Insecticides Act, 1968.

Objectives

1. To register insecticide after scrutinizing their formulae and verifying claims made by the importer or the manufacturer, as the case may be, as regards their efficacy and safety to human being and animals; and
2. To perform such other functions as are assigned to it by or under this Act.

Functions

The Registration Committee shall, in addition to the functions assigned to it by the Act, perform the following functions, namely

1. Specify the precautions to be taken against poisoning through the use or handling of insecticides
2. Carry out such other incidental or consequential matters necessary for carrying out the functions assigned to it under the Act or these rules.

Registration Committee meets normally once in a month to transact its business. It meets fortnightly exclusively for registration for export, if required.

Registration of Insecticides

1. Fill Form I including the verification portion, shall be signed in case of an individual by the individual himself or a person duly authorized by him, in case of partnership firm by the managing partner, in case of a company, by any person duly authorized in that behalf by the Board of Directors. The Registration Committee may, if necessary direct inspection of the testing facility for establishing the authenticity of the data.
2. The application form duly filled together with a treasury challan evidencing the payment of registration fees for rupees one hundred shall be sent to the Registration Committee, Insecticides Act, Department of Agriculture, Government of India, New Delhi.
3. The registration fee payable shall be paid by a demand draft drawn on the State Bank of India, Faridabad, in favour of the Accounts Officer, Directorate of Plant Protection, Quarantine and Storage, Faridabad, Haryana.
4. The certificate of registration shall be in Form II or Form II-A. (Anonymous, 2015c).

Computerized Registration of Pesticides (CROP)

It is a web-based application for Registration of Pesticides developed for the Central Insecticides Board and Registration Committee. (Anonymous, 2015e) The application is supposed to automate the entire registration procedure. The highlights of the application are summarized below:

- Online Application for Registration of Pesticides under 9(3) and 9(3b)
- On line status checking.
- Frequent email notification to the applicant at certain milestone of the registration process till Deficiency Reporting and Approval.
- Online application for the grievances.

Objectives

The objective of this manual is to aid the Applicant (Manufacturers, which happen to be the intended audience of this document) in understanding the features of the CROP application. The user manual provides a step-by-step description of the various functionalities.

Grant of Licenses

I. Fill Form III or Form IV, as the case may be to the licensing officer and shall be accompanied by a fee of rupees fifty for every insecticide for which the license is applied, subject to a maximum of rupees five hundred.

II. If an insecticide is proposed to be manufactured at more than one place, separate applications shall be made and separate licenses shall be issued in respect of every such place.

III. Licensing officer may after giving reasonable opportunity of being heard, to the applicant, refuse to grant any license.

IV. A fee of rupees five shall be paid for a duplicate copy of a license.

V. Applications for the grant or renewal of a license to sell, stock or exhibit for sale or distribute insecticides shall be made in Form VI or Form VII.

VI. Any license issued or renewed under this chapter shall, unless sooner suspended or cancelled, be in force for a period of two calendar years. Provided that the license to manufacture insecticides, if any, issued on the basis of provisional registration granted under sub-section (3-B) of Sec. 9 of rule 10 or under sub-rule (3) of rule 10-A shall expire or be renewable along with the main license.

VII. The license granted by endorsement on main license under Clause (iii) of sub rule(8) of Rule 9 under Clause (iii) of Sub rule 4 of Rule 10 or under sub rule 3 of rule 10A shall expire or be renewable along with main license.

Labeling

The following particulars shall be either printed or written in indelible ink on the label of the innermost container of any insecticide

i. Name of the manufacturer
ii. Name of insecticide
iii. Registration number
iv. Kind and name of active and other ingredients
v. Net content of
vi. Batch number.
vii. Expiry date
viii. Antidote statement.

The label shall be so affixed to the container that it cannot be ordinarily removed (Anonymous, 2015f).

Classification of the Insecticides	Medium Lethal dose by the oral route (acute toxicity) LD 50mg/kg. body weight of test animals	Medium lethal dose by the dermal route (derma toxicity) LD 50 mg/kg. body weight of test animals	Colour of identification band on the label.
1	2	3	4
1. Extremely Toxic	1-50	1-200	Bright red
2. Highly Toxic	51-500	201-2000	Bright yellow
3. Moderately Toxic	501-5000	2001-20000	Bright blue
4. Slightly Toxic	More than 5000	More than 20000	Bright green

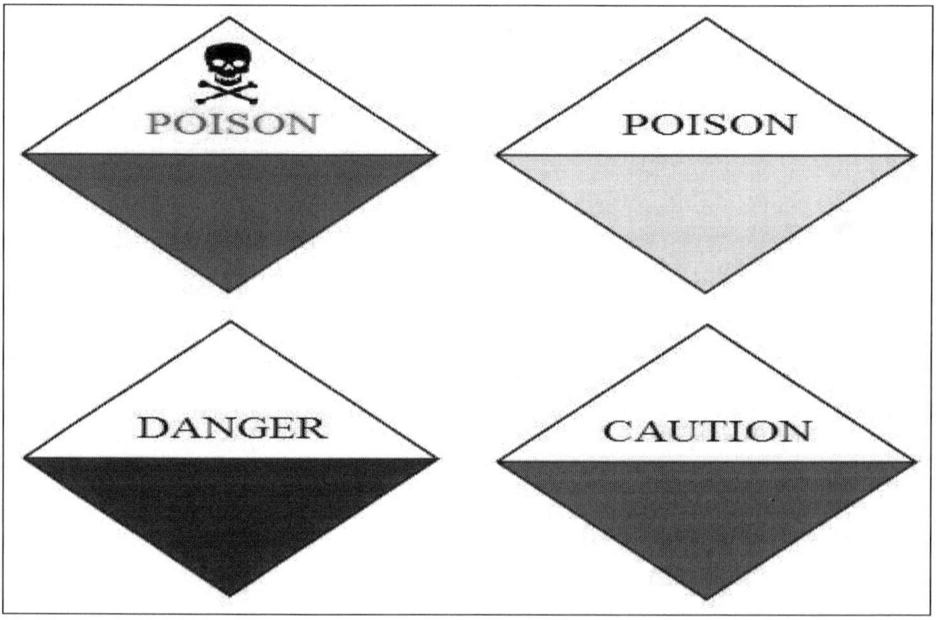

The Insecticides Analyst

The Insecticides Analyst shall have the power to call for such information of particulars or do anything as may be necessary for the proper examination of the samples sent to him either from the Insecticide Inspector or the person from whom the sample was obtained.

I. He shall analyze or cause to be analyzed or test or cause to be tested such samples of insecticides as may be sent to him by the Insecticide Inspector under the provisions of the Act and shall furnish report or results of such tests or analysis.

II. He shall, from time to time, forward to the State Government reports giving the result of analytical work and investigation with a view to their publication at the discretion of the Government.

Insecticide Inspector

The Insecticide Inspector shall have the following duties namely:

1. To inspect not less than three times in a year all establishment selling insecticides within the area of his jurisdiction
2. To satisfy himself that the conditions of license are being complied with
3. To procure and send for test and analysis, samples, of insecticide which he has reason to suspect are being sold, stocked or accepted for sale in contravention of the provisions of the act or rules.
4. To investigate any complaint in writing which may be made to him
5. To institute prosecution in respect of breaches of the Act and the rules made there under
6. To maintain a record of all inspections made and action taken by him in the performance of his duties including the taking of samples and seizure of stocks and to submit copies of such record to the licensing officer
7. To make such inquiries and inspections as may be necessary to detect the sale and use of insecticides in contravention of the Act.

Central Insecticides Lab

By virtue of the Section 16 of the Act, the Central Government has established a Central Insecticides Laboratory.

Functions of CIL

The functions of the Laboratory as envisaged under Rule 5 of Insecticides Rules 1971 are as follows:

- ☆ To analyze such samples of insecticides sent to it under the Act by any officer or authority and submission of certificates of analysis to the concerned authority.
- ☆ To carry out such investigations as may be necessary for the purpose of ensuring the conditions of Registration of Insecticides.

- To determine the efficacy and toxicity of insecticides;
- To carry out such other functions as may be entrusted to it by the Central Government or by a State Government with the permission of the Central Government and after consultation with the Central Insecticides Board.

Activities of CIL

i. To verify Quality of samples of Insecticides, submitted by any Officer or Authority Authorized by the Central or State Government.
ii. To investigate Insecticides with a view to verify conditions of registration
iii. To determine efficacy and Toxicity of Insecticides
iv. Any other function as may be entrusted by the Central Government or by a State Government with the permission of the Government and after consultation with the Central Insecticides Board.

Offences and Punishment

The Person

- Imports, manufactures, sells, stocks or exhibits for sale or distributes any insecticide deemed to be misbranded under sub-clause (i)
- Imports or manufactures any insecticide without a certificate of registration
- Manufactures, sells, stocks or exhibits for sale or distributes an insecticide without a license
- Obstructs an Insecticide Inspector in the exercise of his powers or discharge of his duties under this Act, or the rules made there under, shall be punishable
- For the first offence, with imprisonment for a term which may extend to six months, or with fine or with both
- For the second and a subsequent offence, with imprisonment for a term which may extend to one year, or with fine or with both.
- If any person convicted of an offence under this Act commits a like offence afterwards it shall be lawful for the Court before which the second or subsequent conviction takes place to cause the offender's name and place of residence, the offence and the penalty imposed to be published in such newspapers or in such other manner as the Court may direct.
- When an offence has been committed by a company, every person who at the time the offence was committed was in charge of, or was responsible to the company for the conduct of the business of the company, as well as the company, shall be deemed to be guilty of the offence and shall be liable to be proceeded against and punished accordingly.

Pesticides Management Bill 2008

We need a strong bill that will regulate and control the pesticides and its use so as to ensure that they do not affect the environment and the health of people.

The Pesticides Management Bill 2008 is now tabled in the parliament (Anonymous, 2015i). The bill was introduced in October 2008, after which it was sent to the Standing Committee on Agriculture. The committee submitted its report in 2009. The Standing Committee recognized a few problems but not all. The proposed Pesticides Management Bill 2008 incorporates a few changes but they are incremental. The larger loopholes remain. For the effective management of pesticides a lot more needs to be done.

Proposed Changes

i. Precautionary Principle should be enshrined in the object of the Bill. In case of any uncertainity human and environmental health should be given priority.

ii. Pesticides cannot be regulated by Ministry of Agriculture as it promotes the use of pesticides. Either the Ministry of Environment and Forests or the Ministry of Health should be given the power to regulate pesticides.

iii. New pesticides safer than older one should be allowed to be used. Any pesticide that is being registered for the first time, should be registered on the principle of Comparative Risk Assessment which means that existing data on pesticides should be used and considered to avoid any future hazards.

iv. Considering that pesticides useful in one agro-ecological zone can be detrimental to another, states must have powers to regulate the use of pesticides.

v. Liability of environmental contamination and health impacts should be on the pesticides manufacturers as per the Polluter Pays Principle.

Comparison of Insecticide Act and Pesticide Management Bill

Topic	*The Insecticides Act, 1968*	*The Bill*
Coverage	Covers insecticides which are defined as any substance included in the Schedule to the Act.	Covers pesticides which are defined as any substance of chemical or biological origin used to control the spread of pests in agricultural commodities or animal feed.
Powers of registration committee	Power to cancel the registration of an insecticide lies with the central government.	Committee can suspend or cancel registration in case of violations of the Act or due to adverse impact on crops, humans or animals.
Conditions and Process of Registration	The Act did not require tolerance limits for pesticide residues to be specified. Tolerance limits are specified under the Prevention of Food Adulteration Act, 1954. There is no protection of data submitted for registration. Same data could be used by multiple applicants.	Bill requires tolerance limits to be specified as a condition for registration. Such limits are specified by the Food Safety and Standards Authority, set up under the Food Safety and Standards Act, 2006. Data submitted for registration by one applicant cannot be used by others for three years, without permission.

vi. Compensation should be provided under the Pesticides Management Bill, 2008 instead of the Consumer Protection Act, 1986.

vii. The Bill should ensure complete transparency and data disclosure.

viii. The fixation of tolerance limits of pesticides as pre-requisite to registration

ix. Requirement of minimum qualification of licensees

x. An elaborate procedure for drawl of pesticide samples and inspection of pesticides

xi. Make punishments more stringent to check production and sale of misbranded, Sub-standard and spurious pesticides.

Conclusion

Although our nation has a defined regulatory system regarding insecticide use and application but still there are loopholes in our system. We need a more defined, strict, more accountable system. The government seems to be in flip-flop mode. It is noteworthy that even if the authorities did adopt such measures, they would merely be treating symptoms and not the root cause. Significantly, the legislation does not explicitly recognise environmental hazards of pesticides or the threat they pose to biodiversity. Neither does it specifically mention how the adverse effects of pesticides can be minimised on water, soil, air and non-target organisms. Apart from such fundamental flaws, there are lacunae at every stage of the law right from the registration process to the review procedure. A company seeking to get a new molecule registered has to provide various types of stipulated data. There are essentially two different kinds of data: environment independent (involving chemical characteristics) and environment dependent (pertaining to field data). Information that falls into the former category remains constant, irrespective of the region where the molecule is sought to be registered. But environment dependent data varies according to different agroclimatic zones in the country. It needs to be generated independently by the company which has applied to get the new molecule registered. Field level data includes details about different crops and pests, dosage of the pesticide, levels of its residues and its persistence. No one actually follows procedures. Mostly, the industry has its way. Producers can get a molecule registered despite its failing certain tests. There is an unholy nexus between private labs and the industry. At times, private laboratories don't even bother to collect samples. They are more than willing to toe the industry's line. If the law is followed, registration would be possible only after the new molecule has undergone rigorous tests. The truth, however, is that pesticide testing in India is a scientific sham. Another data-related drawback is that the information furnished by the manufacturer is verified only prior to registration. Once the molecule is introduced into the market, virtually no tabs are kept on it. This leads to its unrestrained use. Any case of pesticide poisoning in the country needs to be notified under section 26 of the Insecticides Act. Our regulatory bodies turn a blind eye to hundreds of reported cases. Even in instances like Kerala's endosulfan tragedy, which have hit the headlines, the government has failed to act with conviction. It is ironical that though persistence data and residue analysis of pesticides are a part of registration, there is no way of controlling these poisons from entering the food chain. The act does not wholly address the issue of pesticide residues in the environment. Though scientific analyses to ascertain contamination levels are conducted periodically, the law is silent on remedial measures. Surprisingly, in India manufacturers are expected to educate farmers on the use of pesticides. This includes awareness about its dosage, spraying frequency and precautions to be taken. In the event, a major fraud is perpetrated by the industry. Manufacturers do comply with the law, but as a mere formality. They insert leaflets containing information in pesticide packets. The catch is that these primers are illegible. Salespersons educate farmers on how much of the pesticide to use. Obviously, they tell them to use more so that they can sell more. Indian farmers are gullible. They spray pesticides, perceive an immediate benefit and resort to indiscriminate use of the chemicals.Overuse has another fallout. The more pesticides are used, the more resistant the insects become. The farmer then has to spray even more. It's a vicious circle. From time to time, the Union government sets up its own expert committees to review various pesticides being used in the country. This appears to be an exercise in futility because even when review panels suggest steps to curb the use of dangerous pesticides, their recommendations are not taken seriously. "Regulation is either non-existent or painfully slow. India has taken too long to ban molecules which are highly hazardous to human health and the environment. We should not ban chemicals just because their use has been prohibited in other countries. This should be done only when the molecule has been phased out universally. The pesticide lobby has always had a say in policy matters and so far it has thwarted all moves by the government to make the law more stringent. This time, too, the industry threw a spanner in the works. And now in the packaged drinking water case, it seems that the pesticide manufacturers are up to their nefarious games again.

REFERENCES

Anonymous (2009) EPA http: //www.epa.gov/opp00001/about.

Anonumous (2015h) indiastat.com.

Anonymous (2013) parliamentofindia.nic.in/ls/jpc/chapter3.pdf.

Anonymous (2015a) http: //www.cibrc.nic.in/insecticides_act.htm.

Anonymous (2015b) http: //www.cibrc.nic.in/insecticides_rules.htm.

Anonymous (2015c)http: //www.cibrc.nic.in/reg_procedure.htm.

Anonymous (2015d) http: //www.cibrc.nic.in/list_pest_bann.htm.

Anonymous (2015e)https: //pesticides-registrationindia.nic.in.

Anonymous (2015f)http: //www.cibrc.nic.in/guidelines.htm.

Anonumous (2015g) www.codexalimentarius.org/standards/list of standards

Anonymous (2015i) prs.org.

Abhilash P C and Singh N (2008) Pesticide use and application: An Indian scenario *J hazard mater* **65**: 1-12

Abhilash P C, Singh N (2009) Pesticides use and application: An Indian scenario. *J hazard mater*. **165**: 1-12

Bhardwaj T and Sharma J P (2013) Impact of Pesticides Application in Agricultural Industry: An Indian Scenario. *Inter J Agri Food Sci Tech* **4**: 817-22.

Cai D W (2008) Understand the role of chemical pesticides and prevent misuses of. *Bulletin of Agricultural Science and Technology*. **1**: 36-38.

Dhanraj A Patil and Ravasaheb J Katti (2012) modern agriculture, pesticides and human health : a case of agricultural labourers in Western Maharashtra *J Rur Dev*. **31**: 305 – 18.

Gupta P K (2004) Pesticide exposure-Indian scene. *Toxicol*. **198**: 83-90.

Jiang Y F, Wang X T, Jia Y, Wang F, Wu M H, Sheng G Y, Fu J M (2009) Occurrence, distribution and possible sources of organochlorine pesticides in agricultural soil of Shanghai, China. *J hazard mater* **170**: 989-997.

Kochar R and Urkude R (2014) Perspective on pesticide residues in food and environment and its Regulation in Crop Protection. *J App Chem* **43**: 40-42.

Liu C J, Men W J and Liu Y J (2002) The pollution of pesticides in soils and its bioremediation. *System Sciences and Comprehensive Studies in Agriculture*. **18**: 295-97.

Liu Z J and Liu Z B (1999) The Status of the world agrochemicals. *Chem Tech Mrkt*. **22**: 14-17.

Mathur S C (2010) Future of Indian pesticides industry in next millennium. *Pest Info*. **24**: 9–23.

Menon R (2006) Pesticide Poisoning in Punjab; India Together.

Pimentel D (2009) Pesticides and pest control. In: Integrated Pest Management: Innovation Development Process *Springer* **1**: 83-87.

Shegunova P, Klanova J, Holoubek I (2007) Residues of organochlorinated pesticides in soils from the

Czech Republic. *Env Pol* **146**: 257-261.

Sitaramaraju S, Prasad N V V S D, V. Reddy C and Narayana E (2014) impact of pesticides used for crop production on the environment. *J Chem Pharma Sci* **54**: 75-79.

Wheeler B Wills (2002) Role of Research and Regulation in 50 Years of Pest Management in Agriculture. *J Agric Food Chem*. **50**: 4151-55.

Zilberman D and Millock K (1997) Pesticide Use and Regulation: Making Economic Sense Out of an Externality and Regulation Nightmare. *J of Agric Res Econ* **22**: 321-32.

Chapter 20

Red Spider Mite (T*etranychus urticae* Koch) Infestation on Brinjal/Eggplant (*Solanum melongena* L.) and Environmental Sustainability

Sunil Kumar Ghosh[1] *and Wajid Hasan*[2]

[1]*Department of Agricultural Entomology, BCKV (Ag. University), AINP on Acarology, Directorate of Research, Kalyani, Nadia – 741235, West Bengal*
[2]*Krishi Vigyan Kendra, Jehanabad, BAU, Bihar*

Eggplant or brinjal (Solanum melongena L.) crop is susceptible to various insect and mite pests of which red spider mite, Tetranychus urticae (Tetranychidae: Acarina) is the most predominant. Tetranychus urticae was most active during May i.e., 22-24 Standard Meteorological Week (SMW) and September-October i.e., 40-43 SMW. Highest mite population (22.87/leaf) was recorded on 42nd SMW (first week of October). Sudden fall of population was found in last week of June because of heavy rains. The mite population always recorded higher on the upper canopy (52.75 per cent population) of the plant as compared with the middle (30.64 per cent population) and lower canopy (16.61 per cent population). This result implies that mites were most densely populated in the young and new leaves of eggplant. The mite population had significantly positive correlation with temperature, minimum and average relative humidity where as non-significant positive correlation with maximum relative humidity and weekly total rainfall. Among the seven treatments evaluated microbial toxin- avermectin resulted in the best suppression of mite population (87.10 per cent suppression), closely followed by chemical insecticide, fenazaquin and mixed formulation of botanical pesticide, azadirachtin with botanical extract, Spilenthes (79.24 per cent and 70.66 per cent suppression). Spectophotometric scanning of crude methanolic extract of Spilanthes flower showed strong absorbance wave length between 645-675 nm. Considering the level of peaks of wave length the flower extract contain some important chemicals of which polysulphide compounds are important and responsible of pest control. azadirachtin and botanical extract individually did not produce good results (moderate mite suppression) but when azadirachtin is used as a mixture with botanical extracts provided better results recording more than 65 per cent suppression. Microbial toxin, plant extracts and botanical insecticide are biopesticides having less or no hazardous effects on human health and the environment, and therefore, they can be incorporated in IPM programmes and organic farming.

Introduction

Eggplant or brinjal (*Solanum melongena* L.) is one of the most common vegetable crops grown in India and other parts of the world. In the sub-Himalayan region of north east India eggplant is cultivated at a commercial scale but insect and mite pest damage constitutes a limiting factor for its successful production (Ghosh, 2020). Pests control by using pesticides causes damage to the natural suppression of the pests by killing parasites and predators like spider (Ghosh *et al.*, 2006a), *Menochilus sexmaculatus* Berliner (Ghosh *et al.*, 2007), *Coccinella septempunctata* (Chakraborty and Ghosh,2010; Ghosh, 2016). Among the various insect pests causing limitations, shoot and fruit borer, *Leucinodes orbonalis* Guen. has become an alarming pest in eggplant/ brinjal growing areas including West Bengal (Ghosh and Senapati,2001a; Ghosh and Senapati,2009; Ghosh,2014). The other important Lepidopteran pests in West Bengal are Brinjal leaf roller (*Eublema olivacea* Walk.) and Brinjal stem borer (*Euzophera particella* Rag). The important Hemipteran pests causes heavy damage are Aphid (*Aphis gossypii* Glov.) (Ghosh *et al.*, 2004a, Ghosh,2019 a), Jassid (*Amrasca biguttula biguttula* Ishida) (Ghosh, and Senapati 2003), Thrips (*Thrips tabaci* Lin.) (Ghosh *et al.*, 2005) and Whitefly (*Bemisia tabaci* Genn) (Ghosh *et al.*, 2004b). The important Coleopteran pests are Hadda or spotted leaf beetle (*Henosepilachna vigintioctopunctata* Fabr.) (Ghosh, and Senapati 2001b), Flea beetle (*Phyllotreta* spp.) (Ghosh *et al.*, 2006b) and Blister beetle (*Mylabris pustulata*) (Ghosh, 2020). Mite (*Tetranychus cinnabarinus* Boisd.) also causes heavy damage to eggplant (Ghosh and Chakraborty, 2014; Ghosh, 2019 b). The crop is susceptible to various insect and mite pests of which red spider mite, *Tetranychus sp.* (Tetranychidae : Acarina) is most predominant. In Bihar, mites are reported to causes losses to the extent from 36.8 per cent to 83.2 per cent in different vegetable crop. Losses of vegetable production was found from 2 to 35 per cent in eastern part of the India due to mites infestation. The outbreak of this pest is assumed to be the consequences of frequent and indiscriminate use of toxic chemicals, especially pyrethroid insecticides by the vegetable growers. Moreover, warm and dry weather is favourable for the multiplication and spread of this pest. In India, it has been reported as one of the important pests of vegetable crops (Gupta, 1985). Anitha and Nandihali (2008) reported that incidence of mite on summer crop was commenced from 16th standard week (2.12 mites/3leaves) and peak infestation on 2nd week of May (14.61 mites/3leaves), and on kharif crop peak during 4th week of October (29.25 mites/3leaves). Weekly population counts on mites showed non-significant positive correlation (p=0.05) with temperature, maximum relative humidity, total rainfall and significant positive correlation with minimum and average relative humidity (Ghosh, 2013). The first sight of infestation by red spider mite is usually chlorotic, stippled appearance on the leaves. As the mites feed on the underside of the leaves, they remove leaf cell contents, including the chlorophyll. Without the chlorophyll, those empty cells appear whitish or bronze. Heavily infested leaves turn completely pale, dry up, and fall off.

The control of this pest through the use of synthetic pesticides during the fruit bearing stage is rather difficult as the fruits are harvested at frequent intervals, and there is every possibility that, if spray applications are made, harvested fruit would contain toxic residues that may cause health hazards. Biopesticides are often preferred and over synthetic one to overcome this problem. These pesticides are safe to animals and plant life and are biodegradable with sufficient action. Kumar *et al.* (2007) reported that methanolic extract of neem and karanj at 1 per cent concentration proved 78.6 and 71.9 per cent control of *Tetranychus sp.* in laboratory condition respectively. Direct contact toxicity of neem product has been demonstrated against termites and aphids (Srivastava, 2003). The mite mortality per cent on chilli crop was found high from Propergite 57 EC @ 1.5 ml/lit which was at par with the treatments Prophenophos 50 EC and Fenazaquin 10 EC (Bala and Ghosh, 2016). Ten days after spraying dicofol was found to be the most effective acaricide against *Aceris tulipae* on garlic and very low population (1.40 mites per sq. cm.) was found in this treatment closely followed by ethion (3.0 mites per sq. cm.) and fenazaquin (5.08 mites per sq. cm) treated plot as compared to untreated control (6.78 mites per sq. cm.). (Bala *et al.*, 2015). Azadiractin-rich neem formulations, *viz.*, Neem Azal (10,000 ppm) and Neem Gold (10,000 ppm) were effective against cotton whitefly, aphids, jassids and spotted boll worm. Dicofol treatment resulted in the best suppression of mite population on brinjal/eggplant crop (83.16 per cent suppression), closely followed by mixed formulation of botanical pesticide, azadirachtin and chemical pesticide, dicofol (71.41 per cent suppression) (Ghosh and Chakraborty, 2014).). Four days after spraying, dicofol treatment was found most effective (86.35 per cent suppression) very closely followed by mixed formulation of botanical pesticide, azadirachtin and chemical pesticide, dicofol (73.42 per cent suppression) against this mite (Ghosh and Chakraborty, 2014). Acharya *et al.* (2002) studied the efficacy of new insecticides imidacloprid, abamectin and reported the new molecules evaluated were safer to lady bird beetles. Azadiractin individually did not produce good results (moderate mite suppression) but when used as a mixture with low dose of (1.5 ml/L) of chemical insecticide, propergite provided better results recording more than 85 per cent suppression (Ghosh, 2013). Documentation of the population abundance and spatial distribution of this pest in vegetables and selection of newer acaricides would open up new scope for our farmers to combat spider mites in vegetable cultivation. The objective of this study was to determine the efficacy of microbial toxin, the plant extracts against red spider mite, *Tetranychus sp.* infesting eggplant and to formulate suitable control measure.

Host Range of Mite

Large populations are often accompanied by fine webbing. Host plants are many and include strawberries, melons, beans, tomatoes, eggplant, ornamental flowers, trees and most houseplants. *T. telarius* is a polyphagous pest and is known to feed on 183 species of plants including cucurbits, brinjal, ladysfinger on which it is occasionally it is very serious.

Biology and Life Cycle of Mite

The mite is active from March to October and passes the winter as a gravid female. Most mite species overwinter as eggs on the leaves and bark of host plants. As the season warms up in March, it spins webs on the undersurface of the leaves of various host plants and lays 60-80 eggs. After mating, females continuously produce as many as 300 eggs over a couple of weeks. Parthenogenesis (reproduction without fertilization) is common. The eggs are spherical attached to fine silk webbing and hatch in 2-6 days. The emerging larvae are light brown and have three pairs of legs. They feed underneath the webs and, within 3-4 days, change into nymphs which have four pairs of legs. The nympgs grow to maturity in two stages within 4-9 days and the adult follows the second nymphal stage. The adults live for 9 to 11 days with individual females living an average of two months. The life cycle is composed of the egg, the larva, two nymphal stages (protonymph and deutonymph) and the adult. The length of time from egg to adult varies greatly depending on temperature. The life cycle during the active period is completed in 9-19 days. There are several overlapping generations per year. In Punjab, India, the mite is believed to complete 32 generations in a year. Hot, dry weather favors rapid development of these pests.

Nature of Symptoms and Damage

Many species of the spider mite (family: Tetranychidae), attack both indoor and outdoor plants. They can be especially destructive in greenhouses. These are minute, polyphagous insects found in large colonies on the underside of the leaves covered with fine silky webs and feed by piercing leaf tissue and sucking up the plant fluids. Feeding marks show up as light dots on the leaves. White streaks appear on the leaves as a result of their feeding. These streaks coalesce and appear as white patches. The entire affected leaf becomes discoloured and gets withered. As feeding continues *i.e.* in case of heavy infestation the leaves turn yellow, and may dry up and drop off. Large populations can severely defoliate plants. The damage is more during warm and dry season. In case of severe infestation it may be found all over the leaf surface. Characteristic blotching of leaves may be seen due to removal of plant sap and green matters on account of feeding. The damage is caused by both the nymphs and adults. A large number of webs are formed on the leaves giving an unhealthy appearance. All the active stages usually feed on the underside of the leaves by sucking cell sap. The webbing interferes with plant growth. There is a poor setting of the fruits and the yield is considerably reduced as the plants become weak and photosynthetic activity is seriously hampered.

Seasonal Incidence of Mite

The two spotted spider mite prefers the hot, dry weather of the summer and fall months, but may occur anytime during the year. Overwintering females hibernate in ground litter or under the bark of trees or shrubs. The outbreak of this pest is assumed to be the consequences of frequent and indiscriminate use of toxic chemicals, especially pyrethroid insecticides by the vegetable growers. Moreover, warm and dry weather is favourable for the multiplication and spread of this pest.

Studies were conducted in the Instructional Farm of Uttar Banga Krishi Viswavidyalaya (State Agricultural University) at Pundibari, Coochbehar, West Bengal, India for two years (2010-11). Analysis of pooled mean data for the two years on mite infestation on brinjal revealed that the pest was active throughout the growing period (Figure 20.1). However, population appeared during January-February and remained very low up to middle of April and thereafter increased gradually with the rise of temperature. Pest population reached high (19.19 mites/leaf) during 23rd SMW (last week of May) and thereafter started decline with the onset of monsoon and heavy rainfall, and this tendency was continued up to 36 SMW (last week of August). After rainy season, again pest population increased and reached highest population (22.87/leaf) on the 42nd SMW (first week of October) when the average temperature, average relative humidity and weekly total rainfall were 28.65°c, 79.88 per cent and 23.80 mm. respectively. However, mite was most active during May *i.e.*, 22-24 SMW and September-October *i.e.*, 40-43 SMW. There was a sudden fall of population was found with the heavy rains (weekly total 201.95 mm) during monsoon in 25th SMW (last week of June) and continued up to end of August.

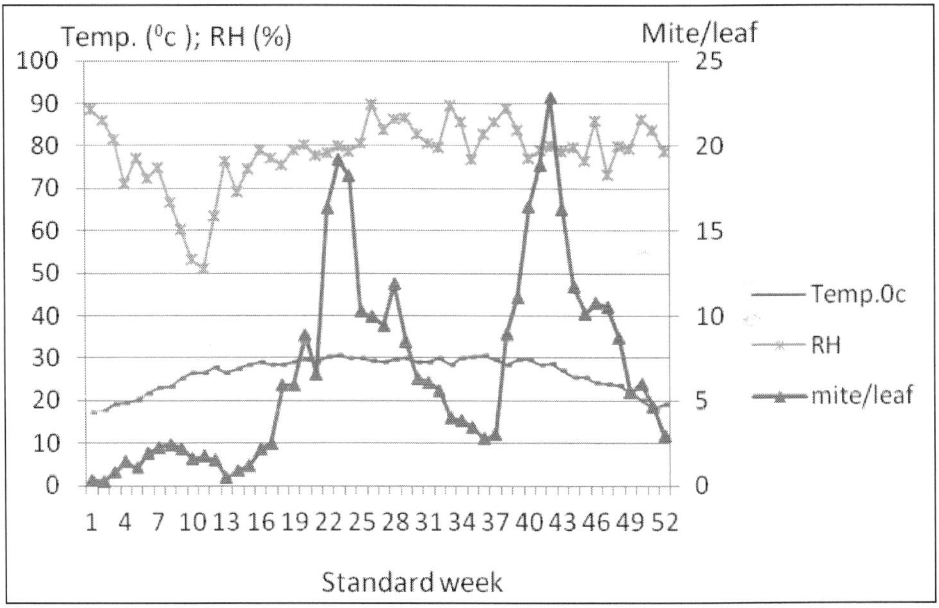

Figure 20.1: Incidence of Mite as Influenced by Temperature and RH.

The mite population always recorded higher on the upper canopy (52.75 per cent population) of the plant as compared with the middle (30.64 per cent population) and lower canopy (16.61 per cent population). This result implies that mites were most densely populated in the young and new leaves of eggplant. The incidence of mite population always remained higher on the upper canopy of the plant (Ghosh, 2013). The mite population had significantly positive correlation with temperature, minimum and average relative humidity where as non-significant positive correlation with maximum relative humidity and weekly total rainfall.

Table 20.1: Correlation Co-efficient between Mite (*Tetranychus* sp.) and Weather Parameters

Environmental Parameter		*Correlation Co-efficient (r)*	*Co-efficient of Determination (R^2)*	*Regression Equation*
Temperature °C	**Maximum**	0.408**	0.166	Y = 0.206x + 29.66
	Minimum	0.424**	(-)0.179	Y = 0.390x + 18.96
	Difference	(-)0.338*	0.114	Y = (-)0.184x + 10.69
	Average	0.434**	0.186	Y = 0.298x + 24.30
Relative Humidity (per cent)	**Maximum**	0.045	0.002	Y = 0.055x + 82.83
	Minimum	0.382**	0.146	Y = 0.685x + 68.74
	Average	0.264*	0.069	Y = 0.371x + 75.78
Weekly rainfall	**Total**	0.229	0.052	Y=1.566x+54.75

*: Significant at 5 per cent level of significance; **: Significant at 1 per cent level of significance.

Management of Mite

Safe control of mite in agricultural field is the burning topic of the days. We should follow the Integrated Pest Management (IPM) which include cultural method, mechanical method, biological control including biopesticides, genetic method and lastly pesticide control. We should always try to avoid much use of pesticides for environmental safety.

Cultural Control

Proper irrigation and clean cultivation are essential to keep the pest population under control. Water stress makes both trees and garden plants more susceptible to mite infestations. Make sure your plants are properly watered. Monitor fields weekly from June through August. Look for plants with yellowish leaves or defoliated plants. Examine plant leaves for webbing, especially on older leaves in the middle of the plant with symptoms. For spring-planted fields that are pruned in early July, treat after pruning for better coverage. Dust on leaves, branches, and fruit encourages mites. A mid-season hosing (or two!) to remove dust from trees is a worthwhile preventative. Spider mites are wind surfers. They disperse over wide areas riding their webbing on the breezes. Careful containment and disposal of infested plants is crucial.

Mechanical Control

Collection and burning of severely infested plant parts reduce further multiplication of mites. Prune leaves, stems, and other infested parts of plants well past any webbing and discard in trash (and not in compost piles). Don't be hesitant to pull entire plants to prevent the mites spreading to its neighbors. Overwintering mites may be reduced in numbers by the destruction of weeds such as pokeweed, Jerusalem oak, Jimson weed, wild blackberry, wild geranium and others. Use the Bug Blaster to wash plants with a strong stream of water and reduce pest numbers.

Biological Control and Predatory Mites

Though predaceous mites are not as effective in controlling pest mites in crops such as eggplant with hairy leaves. spider mites are commonly attacked by predator mites. Five species are commercially available, *Phytoseiulus persimilis, Mesoseiulus longipes, Neoseiulus californicus, Galendromus occidentalis* and*Amblyseius fallicus*. Predatory mites can be distinguished from spider mites because of their longer legs. The front pair of legs is often extended forward. They are more active and move about at a fast pace. They are often red or orange in color. Predatory mites such as *Phytoseiulus persimilis* Athias-Henriot and several species of *Amblyseius*, especially *A. womersleyi* Schicha and *A. fallacies* Garman can be used to control spider mites. They are more effective under protective structures and in high humidity conditions.

Biopesticide and Pesticidal Control

Insecticidal soap, neem oil, or botanical insecticides can be used to spot treat heavily infested areas. On fruit trees, horticultural oil should be applied early in the season or late in the fall to destroy overwintering eggs.

Horticultural oil: A self-emulsifying concentrate, Bonide All Seasons Horticultural Spray Oil kills insect eggs and soft bodied adults by smothering them. Works great indoors or out on aphids, whitefly, mites and scale, and can be used as a dormant spray (no leaves) or delayed dormant (green tip) spray. A superior type parafinic oil, Bonide All Seasons Spray Oil degrades rapidly and has almost no toxicity. Highly recommended for use on fruit trees, shade trees, shrubs, ornamentals, roses and vegetables. Includes a spreader/sticker.

Management strategies must take into account the fast development time of this pest, especially during warm weather when eggs are laid continuously. Just targeting the adults will do little good if eggs and larvae survive. Repeat treatments arc almost always necessary.

Experiment for Bio-control

Studies were conducted in the Instructional Farm of Uttar Banga Krishi Viswavidyalaya (State Agricultural University) at Pundibari, Coochbehar, West Bengal, India for two years (2010-11).

Among the seven treatments evaluated microbial toxin- avermectin resulted in the best suppression of mite population (87.10 per cent suppression), closely followed by chemical insecticide, fenazaquin and mixed formulation of botanical pesticide, azadirachtin with botanical extract, *Spilenthes* (79.24 per cent and 70.66 per cent suppression). Spectophotometric scanning of crude methanolic extract of *Spilanthes* flower showed strong absorbance wave length between 645-675 nm. Considering the level of peaks of wave length the flower extract contain some important chemicals of which polysulphide compounds are important and responsible of pest control. azadirachtin and botanical extract individually did not produce good results (moderate mite suppression) but when azadirachtin is used as a mixture with botanical extracts provided better results recording more than 65 per cent suppression. Microbial toxin, plant extracts and botanical insecticide are biopesticides having less or no hazardous effects on human health and the environment, and therefore, they can be incorporated in IPM programmes and organic farming.

Harmful Effect of Chemical Insecticides

Since the discovery of DDT as an insecticide in 1939 by Dr. Paul Muller, there has been a great expansion in the use of chemicals for pest control. Pesticides during manufacturing, transportation, storage and actual use enter in the abiotic and biotic components of the environment through air, water and soil and disturb the ecosystem, causing great disaster sometimes. Miss Rachel Carson published an epoch-making book "Silent Spring" in the year 1962 awakened the people referring the forceful account of the danger of pesticides. These days the pollution of the environment is a problem of great importance and is of everybody's concern. The pesticides may accumulate in the environment and contaminate all the systems *i.e.* air, water, soil, plants animals *etc.* by being transported from one system to another. Although they are protecting the crops from pests for boosting up the agricultural produce but bring out ecological disturbance and environmental pollution. A variety of insecticides like BHC, aldrin, dieldrin, heptachlor, chlordane, toxaphene, methyl parathion, phorate *etc.* are being used in our country for the control of termite, white grub, cutworm, root borer *etc.* It is found that the pesticides disturb the microbial activity of the soil, adverse effect on the earth worm and may harm the predatory mites and carabid beetles. They may have adverse effect on some invertebrates that were responsible soil fertility. The uptake of insecticide residues by some crops adversely affects our health. The people around vicinity of pesticide factories even upto 5-7 km area badly feel off flavor and such suffocated environment ultimately results in different kind of diseases among the residents. The contamination of air during application of pesticides may also take place which could pose serious health hazards, if the concentration in air raised above the thresh hold values. Different doses of pesticides at different dose levels are being used on the different types of crops against noxious pest all over the country for the last four decades. Saxena *et al.* (1990) reported that the residue of malathion was found more than the permissible limits in tomato, okra, cauliflower, brinjal and beans. Gupta *et al.* (1985) reported that above 60 per cent samples of potato, brinjal, cabbage, cauliflower okra and cucurbits were containing the residues of organophosphatic insecticides more than the permissible level. Water has been found contaminated with pesticides by different ways. In order to destroy unwanted plants, insects and fishes *etc.* the deliberate use of pesticide is being done due to which water is contaminated. Water has also been found to be contaminated with pesticides through run off from fields, through sewage disposal, through the effluents of industries using pesticides, through dead and decayed plants treated with pesticides. In UK the presence of insecticides was reported in rain water. Fresh rain water on the mountain top of the

Himalaya was found possessing pesticide residues. Commonly cultivated crops, vegetables and fruit plants have been found to be affected with injudicious use of pesticides right from germination through growth to harvest. The application of pesticides may bring the resurgence of the target pests against which the chemicals are applied, the outbreak of some unimportant pests. The pesticides are also harmful to beneficial fauna such as honey bees and natural enemies like parasites, predators and pathogens of the pests *etc.* As a result of pesticide pollution several diseases may develop in human beings and domestic animals.

Importance of Biopesticides for Pest Control in Vegetable Field

In view of the harmful effects of chemical pest control, several alternatives have been explored to control the insect pests. In more recent times, there is a renewed interest in eco-friendly approaches to pest management and the rationale for biological approaches is irrefutable. In a broader sense, biopesticides include pesticides of biological origin.

Botanical Extracts

Botanical pesticides are either naturally occurring plant materials or products derived rather simply from such plant materials. They are water extracts or organic solvent extracts of insecticide components of plants. Plant substances are produced as byproducts of major biochemical pathways and chemically, they include alkaloids, terpenoids and phenols as well as a number of other compounds. These chemicals repel approaching insects, deter feeding and oviposition on the plants, disrupt behaviour and physiology of insects in various ways and even prove toxic to different developmental stages of many insects (Dhaliwal and Arora, 2006). The use of plants as pesticides has been practiced since times immemorial. The Hindu book, the 'Rig Veda' written in India in 2000 BC, makes a mention of the use of poisonous plants for pest control. Crude botanical pesticides have been used for several countries and were known in tribal or traditional indigenous technologies around the world before being introduced to Europe. Those with long histories of traditional use include neem in India, rotenone in East Asia and South America, pyrethrum in Persia, and sabadilla in Central and South America. Neem (*Azadirachta indica*) tree is indigenous to India and is a storehouse of large number of pesticidally active tetranortriterpenes, commonly called limonoids, among which azadirachtin, salanin and nimbin are the most active. The most abundant neem constituent, azadirachtin is considered an excellent botanical pesticide because of its biodegradability, demonstrated low toxicity to vertebrates, environmental safety, and safety to non-target. Its multi-pronged effects against insects including repellent, antifeedant, oviposition deterrent, molting or growth disruption, sterilant, ovicidal and oviposition deterrent, *etc.* help to an effective control of a variety of farm and household insect pests and pathogens infesting agricultural plantation and cash crops. Several commercial formulations of neem like Neem Gold, Neemazal, Econeem, Neemark, Neemcure and Azatin among now available in many countries including the United States, India, Germany and several Latin American countries. Inadequate stability, problems in large-scale production, lack of effective delivery system, poor self life, diminished residual toxicity under field condition, non-availability of reliable standards and inadequate availability of the raw material are some impediments in successful commercialization of neem pesticides. *Polygonum* and *Pongamia* are such type of plants having insecticidal properties needed to be conservation. *Polygonum hydropiper*, a well known weed in northeast India especially in the foot hill of the Himalaya, locally known as "biskanthali" is widely used indigenously in different tribal areas for pest control in vegetable field. *Pongamia pinnata*, a medium sized tree is well known in terai region of West Bengal having humid and subtropical climate. It is one of the few nitrogen fixing trees often planted as an ornamental and shade tree. *Pongamia* oil and leaf extract is useful for pest control in vegetable field.

More than 1000 species of plants have been reported to have chemicals in leaves, stems, flowers, seeds and roots which have insecticidal properties. Various scientists direct their attention to discover such type of plant toxicant which are eco-friendly and play a better role in the control of pest. It is an important component of IPM. Plant origin insecticides offer a great advantage by being compatible with other low risk options such as pheromones, biopesticides and bioagents such as entomopathogenic fungi and nematodes, predators and parasitoids, *etc.* which can be integrated in IPM programmes.

Conclusion

The control of this pest through the use of synthetic pesticides during the fruit bearing stage is rather difficult as the fruits are harvested at frequent intervals, and there is every possibility that, if spray applications are made, harvested fruit would contain toxic residues that may cause health hazards. Biopesticides are often preferred and over synthetic one to overcome this problem. These pesticides are safe to animals and plant life and are biodegradable with sufficient action. Plant origin insecticides offer a great advantage by being compatible with other low risk options such as pheromones, biopesticides and bioagents such as entomopathogenic fungi and nematodes, predators and parasitoids, *etc.* which can be integrated in IPM programmes.

REFERENCES

Acharya, S., Mishra, H.P. and Dash, D. (2002) Efficacy of insecticides against okra jassid (*Amrasca biguttula biguttula* Ishida). *Annals of Plant prot. Sci.* **10** (2): 230-232.

Anitha, K.R. and Nandihalli. B.S. (2008) Seasonal incidence of sucking pests in okra ecosystem. *Karnataka J. Agric. Sci.* **21** (1): 137-138.

Bala, S.C. and Ghosh, S.K. (2016) Host plant resistance-cum-chemical control approach for the sustainable management of yellow mite, *Polyphagotarsonemus latus* (Bank) *J. Ent.Res.* **40** (4): 373-377.

Bala, S.C.; Karmakar, K. and Ghosh, S.K. (2015) Population dynamics of mite, Aceria tulipae Keif. on garlic (Allium sativum L.) and its management under Bengal basin. *International Journal of Science, Environment and Technology.* **4** (5): 1365-1372.

Chakraborty, K. and Ghosh, S.K. (2010) Incidence of *Coccinella septempunctata* in brinjal with some pesticides. *Current advances in Agricultural Sciences,* **2**(2): 129-130.

Dhaliwal, G.S. and Arora. (R.) 2006. Integrated Pest Management: Concepts and Approaches. *Kalyani Publishers, New Delhi, India,* pp 235-245.

Dutta, N.K., Alam, S.N., Uddin, M. K., Mahumudunnabi, M. and Khatun, M.F. (2011). Incidence and spatial distribution of red spider mite, *Tetranychus Urticae* in brinjal and its chemical control. *Bangladesh J. Entomol.* **21**(1): 45-52.

Fitzgerald, J., Xu, X., Pepper, N., Easterbrook, M. and Solomon, M. (2008). The spatial and temporal distribution of predatory and phytophagous mites in field-grown strawberry in the UK. Exp. *Appl. Acarol.* **44**(4): 293-306.

Ghosh, J., Ghosh,S.K., Chatterjee, H. and S.K. Senapati. S.K. (1999) Pest constraints of Okra under terai region of West Bengal. *Indian J. Entomol.* **61** (1): 362-71.

Ghosh, S.K. (2013). Incidence of red spider mite (*Tetranychus urticae* Koch) on okra (*Abelmoschus esculentus* (L.) Moench) and their sustainable management. *Current Biotica* **7**(1 and 2): 40-50.

Ghosh, S.K. and Chakraborty, K. (2014) Bio-Efficacy of plant extracts against red spider mite (*Tetranychus* spp.) infesting brinjal (*Solanum melongena* L.). *Research Journal of Agricultural and Environmental Sciences.* **1** (1): 26-31.

Ghosh, S.K. (2013) Sustainable management of red spider mite (*Tetranychus sp.*) infesting eggplant (*Solanum melongena* L.) at field level. *Uttar Pradesh J. Zool.* **33**(2): 175-180.

Ghosh, S.K. and Senapati, S.K (2003) Biology and seasonal abundance of jassid infesting brinjal in terai region of West Bengal. *Environment and Ecology.* **21**(3): 716-719.

Ghosh, S.K., Laskar, N. and Senapati, S.K. (2003) Estimation of loss in yield of brinjal due to pest complex under terai region of West Bengal. *Environment and Ecology.* **21**(4): 764-769.

Ghosh, S.K. (2020) Biopesticides and environmental Sustainability for controlling Epilachna beetle (*Henosepilachna vigintiocto punctata* Fabricous) infesting eggplant (*Solanum melongena* L.) Book Edited by Wajid Hasan, Book title "Integrated Pest Management –an Applied perspective."*pp. 239-253, 2020.* (Biotech Book Publishing, New Delhi, ISBN: 978-81-7622-472-7).

Ghosh, S.K. (2020) Sustainable control of blister beetle (*Mylabris pustulata* Thunberg) infesting eggplant (*Solanum melongena* L.) Book Edited by Bishvajit Bakshi, Book title "New perspectives in Agricultural Sciences-An international peer reviewed book."*pp. 115-142, 2020.* (LAP LAMBERT Academic Publishing, Maurtius, ISBN: 978-620-2-66718-4).

Ghosh, S.K. and Chakraborty, K. (2014) Bio-Efficacy of plant extracts against red spider mite (*Tetranychus* spp.) infesting brinjal (*Solanum melongena* L.). *Research journal of Agricultural and Environmental Sciences.* **1** (1): 26-31.

Ghosh, S.K., Laskar, N. and senapati, S.K. (2006 b) Seasonal fluctuation of flea beetle (*Phyllotreta* sp.) on brinjal and field evaluation of some pesticides against flee beetle under terai region of West Bengal. *Juurnal of Entumological Research*.**30**(1): 83-87.

Ghosh, S.K. (2019 b) Climate impact on red spider mite (*Tetranychus* sp. Koch) infesting eggplant (*Solanum melongena* L.) and their management using plant extracts. *J. Ent.Res.* **43** (3): 345-350.

Ghosh, S.K. and Senapati,S.K. (2001 b).Biology and seasonal fluctuation of *Henosepilachna vigintioctopunctata* Fabr. on brinjal under terai region of West Bengal. *Indian J Agric. Res*, **35**(3)': 149-154.

Ghosh, S.K., Laskar, N., Basak, S. N. and senapati, S.K. (2004 b) Seasonal fluctuation of *Bemisia tabaci* Genn. on brinjal and field evaluation of some pesticides against *B. tabaci* under terai region of west *Bengal. Environment and Ecology*. **22**(4): 758-762.

Ghosh, S.K., Laskar, N. and senapati, S.K. (2005) Seasonal fluctuation of thrips on brinjal and field evaluation of pesticides. *Pest Management and Economic Zoology*. **13**(2): 185-190.

Ghosh, S.K., Laskar, N. and Senapati, S.K,(2004 a) Seasonal fluctuation of *Aphis gossypii* Glov. on brinjal,md field evaluation of pesticides from different origin against *A. gossypii* under tcrai region. *Indian J Agric. Res*. **38**(3): 171-177.

Ghosh, S.K. and Chakraborty, K. (2014) Bio-Efficacy of plant extracts against red spider mite (*Tetranychus* spp.) infesting brinjal (*Solanum melongena* L.). *Research journal of Agricultural and Environmental Sciences*. **1** (1): 26-31.

Ghosh, S.K., Laskar, N., Basak,S.N. and senapati, S.K. (2006 a) Seasonal fluctuation of spider on brinjal and efficacy of pesticides under terai region of West Bengal. *Orissa Iournal of Horticulture*.**34**(1): 86-91.

Ghosh, S.K., Laskar, N. and senapati, S.K. (2007) Seasonal incidence of predator *Menochilus sexmaculatus* Berliner on brinjal and harmful effect of insecticides on the predator. *Indian J. Agric. Res*.**41**(2): 102-106.

Ghosh, S.K. (2016) Harmful effect of insecticides against predator, Coccinella sp. (Lady bird beetle) on eggplant (*Solanum melongena* L.). *Uttar Pradesh J. Zool*. 36(1): 17-23.

Ghosh, S.K. and Senapati, S.K(2001a) Seasonal incidence and biology of brinjal shoot and fruit borer (*Leucinodes orbonalis* Guen.) under terai region of West Bengal, India. *Annals of Ent*. **19**(1): 13-18.

Ghosh, S.K. and Senapati, S.K (2009) Seasonal fluctuation in the population of *Leucinodes orbonalis* Guen. Under the sub-himalayan resion of West Bengal, India and its control on eggplant (*Solanum melongena* L.) *Precision Agriculture*, **10**: 443-449. Springer

Ghosh, S.K. (2019 a) Sustainable Management of aphid (*Aphis gossypii* Glover) infesting brinjal (eggplant), *Solanum melongena* L.) Book Edited by Sachin Tyagi, Book title "Hi-Tech Horticulture, Volume 4: Plant Protection and Stress

Management" *pp. 77-99, 2018.* (New India Publishing Agency, New Delhi ISBN: 978-93-87973-40-4).

Gupta, S.K. (1985). Hand Book of plant mites in India. *Zoological Survey of India. Calcatta,* pp 520.

Kumar V., Chandraskekar, K. and Sidhu. O. P. (2007) Synergistic action of neem and karanj to aphid and mite. *J. Entomol. Res.* **31** (2): 86 (Users online).

Srivastava, R.P. (2003) Neem tree (*Azadiracta indica* A Juss) and insect pest management. Biopesticides and Bioagents in Integrated Pest Management of Agricultural Crops. *International book publishing Co., Lucknow,* pp. 31-220.

Chapter 21

Application of Nanoherbicides in Weed Management: A Recent Approach

N. Prathap Reddy*, R. Karthik and Rajap Shiva Kumar

Department of Agronomy, College of Agriculture, Professor Jayashankar Telangana State Agricultural University, Rajendranagar, Hyderabad – 500030, Telangana

The demand for food production is increasing to meet the needs of increasing global population. Weeds are the plants which interact with the crop and effect the productivity. Herbicides are used to control the weeds. The herbicides are to be used repeatedly to control the weeds and this repeated use of herbicides cause problem to the environment, non target organisms and herbicide drifts also harm the neighbouring crops. Some of the herbicides have long residual effect and limits the choice of next crop in the rotation. In this situation the nanotechnology has shown potential for efficient delivery of chemical and biological pesticides using nanosized preparations. Herbicides are loaded on nanomaterials to increase the bioavailability and ensure better eradication of weeds. Nanoherbicides can also helps to increase wettability and dispersion of formulations and restrict the movement of herbicides to other areas. The nanoherbicides also have higher affinity to target plants due to higher specific surface area. Different carrier systems like chitosan, alginate, poly epsilon caprolactone are used for formulating the nanoherbicides. The nano herbicides plays an important role in exhausting weed seed bank,degrading germination inhibitor, exhausting food reserve, slow release-season long weed control and rapid degradation of herbicide residue.

Keywords: *Weed management, Weed seed bank, Nanotechnology, Nanoherbicides.*

Introduction

The population of India and world is increasing every year which increases the demand for food production. This possess a great challenge to feed this growing population with the limited cultivable area. The major factors which reduce the yield are declining in the soil fertility, competition form weeds, incidence of pest and disease. Out of all these abiotic and biotic factors weeds cause greater damage to the yield of crops. Herbicides are used to control weeds and maximize crop productivity. The intensive use of herbicides has caused increase in the level of herbicide residues in natural water sources, soil, and foodstuff (Shezad *et al.*, 2011). Herbicide contamination is one of the significant environmental pollution in the recent years. The conventional herbicide application methods are the major cause for herbicide dissipation from the applied soil surface to the water bodies which eventually reaches the non-target organisms and the humans. The use of nanoparticles as carriers of active chemicals in agriculture results in increased biological activity of the active substances, reductions in the quantities of herbicide required and reduced contamination of water resources due to reduction in leaching (Cea *et al.*, 2010). Nanotechnology, the term used to describe the processes of generation, manipulation, and use of nanomaterials, has emerged as a promising field that could help in alleviating problems associated with agrochemicals.

Nanotechnology

Nanotechnology is science of manipulating materials at nanoscale. It has many applications in all stages of production, processing, storing, packaging and transport of agricultural products. Nanotechnology deals with the production and stabilization of various types of nanoparticle. It provides an excellent platform and as vehicle for delivery of active materials to the targeted site without effecting the activity of active material. Nano-formulations are generally expected to increase the apparent solubility of poorly soluble active ingredients, to release the active ingredient in a slow or targeted manner, and to protect against premature degradation.

Benefits of Utilizing Nanotechnology in Preparation of Conventional Herbicides

1. Nanoherbicide control the release speed of its active ingredient
2. The nanoformulation may increase the efficiency of herbicide and also reduce the dosage of the herbicide used.
3. Encapsulation of herbicide with semi-permeable membrane coating controls the release rate and release pattern of herbicide
4. Nanoherbicides can increase effective duration of the herbicide active ingredient in soil
5. Nanoherbicides reduce the runoff and leaching losses into the soil

Role of Nanoherbicides in Weed Control

A nanoherbicide consists of either minute particles of active ingredients of herbicide or other minutely engineered structures possessing herbicidal properties.

Nanoherbicides also offer large specific surface area and hence increased affinity to the target. Nanoemulsions, nanoencapsules, nanocontainers and nanocages are some of the nanoherbicide delivery techniques capable of controlling or delaying the delivery. Nanocapsules may reduce herbicide loss due to lixiviation, volatilisation, drift and degradation soils, reduce phytotoxicity, facilitate herbicide management and safer applications, reducing toxicity in humans, offer the possibility to act more selectively on the intended target and increase the duration of activity and the effectiveness of the active ingredient, in soil as well as at the foliar level (Sopena *et al.*, 2009). The release mechanisms of nano encapsulation of herbicides include diffusion, dissolution, biodegradation and osmotic pressure with specific pH. (Preet kaur *et al.*, 2011)

The physicochemical properties of nanomaterials are different from the same materials at macroscopic level. Carrier systems based on nanoparticles provide modified release of bioactive agents, extending their duration of action and directing them more efficiently to target organisms In the case of herbicides, the use of nanoparticles could be beneficial in many respects, such as decreased contamination of water resources by leaching and less risk of harming non target organisms, including humans

Nanoherbicide in 1–100 nm range will try to mingle with the soil particles and destroy weed seeds and weeds through their roots. Most of the herbicides available in the market are designed in such a way that either they kill or control the above-ground part of the weed plants, none of them inhibits the activity of viable underground plant parts like rhizomes or tubers, which act as a source for new weeds in the following season

Carrier Systems Employed in Formulating Nanoherbicides

Chitosan

Chitosan is a natural product derived from chitin. It is a cationic polysaccharide composed of β-(1-4)- linked D-glucosamine and *N*-acetyl-Dglucosamine units. Generally, the use of chitosan as stabilizer for the synthesis of Cu-NPs is gaining momentum because of their availability, biocompatibility, hydrophilicity, nontoxicity, biodegradability, adsorption properties bioactive polymer and drug absorption enhancement, environmental-friendly material with many superior properties with momentous amounts of amine and hydroxyl groups that can be easily engineered. Itodo *et al.* (2017) proved that pendimethalin encapsulated with copper chitosan nanoparticle exhibited controlled release of herbicide than the commercial formulation.

Alginate

Alginate gels are used as matrices in agricultural applications because of their biodegradability and the ease of incorporation of pesticides using an aqueous system at ambient temperatures. The leaching of pesticides during the preparation of alginate beads has been improved by synthesizing bipolymeric beads of alginate with other natural polysaccharides. Shu *et al.* (2000) proved that it is needed to modify chitosan with alginate due to the weak mechanical strength of chitosan. Alginate has open lattice structure which forms porous beads which help in stabilization of chitosan. Starch and chitosan were reinforced with alginate and used in preparation of slow release formulations of imazaquin herbicide. The addition of alginate improved matrix strength and prevented leakage of encapsulated herbicide. (Namonu *et al.*, 2012)

Poly Epsilon Caprolactone

Poly epsilon caprolactone is used as carrier due to its good physico–chemical properties, biodegradability, and biocompatibility (Sinha *et al.*, 2004). The poly epsilon carpolactone show good colloidal stability and a high capacity for the encapsulation of bioactive compounds (Grillo *et al.*, 2012), making them potentially useful in agricultural applications. Pereira *et al.* (2014) reported that the encapsulation of atrazine with poly epsilon carpolactone was efficient and the formulation was stable over time. The encapsulated herbicide did not cause any damage to a non-target organism and were more effective against target organisms compared to use of the free herbicide indicating that the nanoparticles enhanced the bioavailability of the active principle. The mobility of atrazine in the soil column was increased due to reduced soil sorption, which also increased the effectiveness of the herbicide against the target organism. The encapsulated atrazine was less genotoxic, compared to the free herbicide, which would improve the level of safety in agricultural applications.

Rice Husk

In India, rice husk which is agro-industrial waste generated in large quantities of about 18– 22 million tonnes as a rice-milling by-product causing environmental problems (Chowdary *et al.*, 2011). It consists of cellulose (32 per cent), hemicellulose (21 per cent), lignin (21 per cent), silica (20 per cent) and crude protein (3 per cent). It is granular in structure, insoluble in water with high chemical stability and mechanical strength as well possess functional groups like carboxyl, hydroxyl and amidogen which accounts for its excellent adsorption capacity (Bansal *et al.*, 2009) The enormously generated rick husk is usually disposed *via* landfill, leading to serious water and air pollution. Due to its excellent adsorption characteristics, mechanical strength and chemical stability.

Rice husk has been converted to nano rice husk for increasing the adsorption capacity of 2,4-D. The reduction in size results in increased adsorption capacity due to higher surface area. The Dn-RH formulation prepared using rice husk by adsorption phenomenon showed high and irreversible sorption of 2,4-D and this proves that rick husk can be used as a carrier for formulating nanoherbicides. Thus rice husk serves as a herbcidie carrier and also the soil amendments. The DnRH nanoformualtion was checked for phytotoxic effect against target plant (*Brassica* sp.) and non-target plant (*Zea mays*). In case of *Zea mays*, an germination index of 92 per cent was obtained in case of free 2,4-D and the DnRH nanoformulation. On the other hand, the germination index of *Brassica* sp. exposed to technical grade 2,4-D and DnRH formulation caused 100 per cent mortality but the activity was found enhanced, in case of the 2,4-D nanoformulation The results reveal the enhanced potential of 2,4-D nanoformulation in target species while the non-target plant was not affected at all. (Abigail *et al.*, 2016)

Advantanges of using Nanoherbicides

1. Nanoparticles helps in exhausting weed seed bank and food reserve
2. Nanoherbicides release the herbicide slowly and result in the season long weed control
3. Nanoherbicide degrade the germination inhibitors
4. Nanoparticles helps in rapid degradation of herbicide residue
5. Reduce the dose of herbicide required to control the weeds
6. Reduce the leaching losses, volatilization losses, cytotoxicity effects, genotoxicity effects

Exhausting Weed Seed Bank

The purple nutsedge (*Cyperus rotundus*) is one of the world worst weed. It propagates mainly by producing rhizomes, basal bulbs and tubers. The tubers remain viable for more than three years. The bud dormancy is due to the presence of inhibitors in the tuber. The presence of phenolic compounds and abscisic in the tubers reduce their sprouting. (Jangaard *et al.*, 1971)

Degrading the phenol breaks the dormancy and induce the germination of all buds present in the tubers at once. As the tubers are germinate, they can be controlled by using herbicides or cultural methods which results in reduced weed seed bank size. Advanced oxidation processes (AOPs) have been realized as efficient technologies for phenol degradation (Bach *et al.*, 2010).

Viji *et al.* (2015) reported that iron oxide nanoparticles at the concentration of 3.0 g kg tubers^{-1} recorded higher percentage of phenol degradation (89 per cent) over control, which was on par with 2.5 g kg tubers^{-1} (87.1 per cent) over control. Advanced oxidation processes (AOP) are widely used for the removal of recalcitrant organic constituents such as phenols. In the case of the AOPs, the generation of hydroxyl radicals takes place through a catalytic mechanism in which the iron oxide nanoparticles play a very important role in phenol degradation. As a result the dormancy of tubers is broken and all the buds germinate at once which provides an opportunity to eradicate all the germinated buds by using chemicals and reduce the weed seed bank.

Exhausting Food Reserve

Brindha *et al.* (2017) reported that the presence of higher amount of starch (142 mg g^{-1}) phenol (10.5 mg g^{-1}) in the control treatment inhibited germination of tubers and no germination (0 per cent) was recorded. Whereas, the ZnO treated tubers at the rate of 1500 mg kg^{-1} registered lower content of starch (102 mg g^{-1}) phenol (6.0 mg g^{-1}) and higher per cent of germination (80 per cent). ZnO nanoparticles act as a power house of electron donor possessing ability to degrade organic and inorganic compounds present in the tubers and thus starving the tubers which results in the germination of these tubers.

Rapid Degradation of Herbicide Residue

Excessive use of herbicides leave residue in soil and cause damage to the succeeding crops. Atrazine, an s-triazine-ring herbicide, is used globally for the control of pre and post-emergence broadleaf and grassy weeds, which has high persistence (half-life, 125 days) and mobility in soils. Residual problems due to the application of atrazine herbicide pose a threat towards widespread use of herbicide and limit the choice of crops in rotation.Application of silver modified with nanoparticles of magnetite stabilized with Carboxy Methyl Cellulose (CMC) nanoparticles recorded 88 per cent degradation of herbicide atrazine residue under controlled environment

Ali *et al.* (2015) have showed that iron nano particles have capacity to degrade atrazine from water up to 95 per cent. Iron nano particles have positive charges. Atrazine has lone pair of nitrogen which coordinated to the positive charges of the adsorbent. This result in the formation of coordinate bond with composite iron nanoparticle. Hence the removal of atrazine herbicide is fast. The regenerated composite INPs can be used for seven cycles for removal of atrazine, with 90–98 per cent removal capacities.

Satapanajaru *et al.* (2008) showed that nano zerovalent iron can be used to degrade the atrazine contamination in soil and water. The major process involved in the degradation of atrazine by nano zerovalent iron is reductive dechlorination. The dechlorination product was 2-ethyl-amino-4-isopropylamino- 1,3,5-triazine.

Preisler *et al.* (2019) reported that from 17 to 60 days, atrazine levels in the soil decreased regardless of the applied formulation. At 17 and 60 days the atrazine content was higher in the soil exposed to nanocapsule + atrazine relative to atrazine, whereas it did not differ among soil samples treated with nanocapsule + atrazine 1/10 and atrazine 1/10. The treatment with 10-fold diluted formulation led to lower atrazine levels compared with nanocapsule + atrazine and atrazine. Atrazine nano encapsulation could potentially reduce the effective residual effects in agricultural lands, as lower atrazine dosage would be applied for weed control resulting in reduced herbicide content in the soil.

Slow Release-Season Long Weed Control

Pradeesh kumar *et al.* (2017) proved that nano encapsulated pendimethalin using solvent evaporation method released the herbicide over a period of 40 days under controlled conditions.The outer covering and porous nature of polymer determine the release pattern of encapsulated herbicide. When irrigation is applied to the field the entrapped herbicide inside the carrier is slowly released and the left over herbicide inside the carrier will be released during the subsequent irrigations which helps in controlling the germination of weeds over the entire crop growing period and also the biosafety studies proved that nanoencapsulated pendimethalin by solvent evaporation was nontoxic to earthworms.

Maruyama *et al.* (2016) reported that the release percentages of 55 and 97 per cent were obtained for free Imazapic and Imazapyr, respectively, after 300 min, while in the presence of the nanoparticles, the values decreased to 30 per cent (Imazapic) and

20 per cent (Imazapyr) after 300 min. It showed that encapsulation of the combined herbicides resulted in slower release, compared to the free compounds.

Nanoherbiides Helps to Decrease the Dose of the Herbicide Required to Control the Weeds

Preisler *et al.* (2019) reported that all treatments with atrazine enhanced the mortality percentage compared with the control, whereas nanocapsule did not affect the mortality. In the nanocapsule + atrazine and atrazine treatments, the mortality percentage was above 80 per cent, which did not differ from that of nanocapsule + atrazine 1/10. However, seedlings grown in the soil treated with Atrazine 1/10 had lower mortality (44.8 per cent) than those exposed to other atrazine-containing formulations. The improved pre emergence herbicidal activity can be related to the decrease in atrazine mobility in the soil caused by nanoencapsulation. It would thereby result in greater atrazine levels in the upper soil layers that are mainly explored by roots of germinating seeds, thus increasing the availability of the herbicide to the seedlings.

Jhones *et al.* (2014) proved that solid lipid nanoparticles (SLN) containing herbicides caused greater effects on both the aerial parts and roots of plants, compared to the commercial formulation. This provided evidence of the effectiveness and agricultural potential of SLN containing herbicides, since when applications were performed at the recommended concentrations, all the formulations caused inhibition of growth, while at 10 times lower concentrations, the effect of the commercial formulation decreased, while SLN containing herbicides remained effective which results in the lower doses of herbicide application.

Kumar *et al.* (2016) proved that at 30 DAT weed control efficiency and weed control index was highest with pretilachlor microemulsion @ 600 g ha^{-1} compared to pretilachlor monolithic dispersion and pretilachlor @ 600 g ha^{-1}. WCI and WCE with pretilachlor microemulsion @ 450 g ha^{-1} was highest when compared to pretilachlor @ 600 g ha^{-1} which proves that pretilachlor microemulsion is superior to pretilachlor. Highest grain yield was achieved with weed free conditions (4.8 t ha^{-1}) followed by pretilachlor miroemulsion (4.5 t ha^{-1}), pretilachlor monolithic dispersion (4.3 t ha^{-1}) and pretilachlor (4.1 t ha^{-1}) used @ 600 g ha^{-1}. However the grain yield with pretilachlor microemulsion @ 450 g ha^{-1} is similar to pretilachlor @ 600 g ha^{-1} which proves that nanoformulations at lower doses can results in better weed management than commercial formulations. In general microemulsion gave better results due to monolithic dispersion due to the smaller average particle size of miroemulsion

Vimalraj *et al.* (2018) reported that application of nanoparticle herbicide combination significantly influenced the plant height of blackgram. Weed free plot produced taller plants at all the three stages. Application of H_2O_2 fb pendimethalin + ZnO Nps equally produced higher plant height and comparable with weed free plot. Increased plant height is due to the weed free environment accelerated the enzyme activity and auxin metabolism in plants, which promote the growth of the crop. It can also be attributed due to improved absorption of ZnO nanoparticles influenced the plant growth.

Nanoherbicides Reduce the Leaching Losses, Volatilization Losses, Cytotoxicity Effects, Genotoxicity Effects

Oliver *et al.* (2004) conducted the experiment for 32 days in which all formulations were incorporated in the soil, evolved alachlor from the alachlor and Lasso 4EC formulations substantially exceeded that from the polymeric formulations only through the first 4 days could indicate increased degradation of the microcapsules with passage of time. This suggest that microencapsulation of alachlor with selected polymers can reduce volatilization, particularly when the formulation is incorporated in the soil.

The leaching of 2,4-D was evaluated at different depths of soil column 0–4,4–8, 8–12 and 12–16 cm for DnRH formulation and the control (2,4-D technical). Highest concentration of 2,4-D was found with the top segment which was found reduced in the following bottom segments. The greater retention of 2,4-D on the top most segment could be due to the presence of colloids in the soil. The results indicate the bioavailability of the DnRH formulation in field conditions where increased persistence of the herbicide could have enhanced the bioavailability of the active herbicide to the target plants for prolonged duration. But in the control column with technical grade 2,4-D, the residual herbicide concentration was detected in almost all the four segments, which depicts the reason for its ground water contamination potential. The lowest residual 2,4-D concentration was noticed in the lowest 12–16 cm column segment. (Abigail *et al.*, 2016)

Lidong *et al.* (2017) reported that the total amount of 2,4-D sodium salt leached was clearly lower in the CRFs (48.4 per cent) than in the free system (97.3 per cent). The soil column leaching test confirmed controlled release of 2,4-D sodium salt from formulations based on MSN-TA nanoparticles. This retarded the vertical movement of the herbicide through soil and reduced the leaching potential.

Grillo *et al.* (2014) proved that IC_{50}value for free paraquat obtained from the cytotoxity assays was approximately 0.12 mg/mL, while the nanoparticles(with or without paraquat) showed virtually no toxicity at the concentrations tested for 24 h. Comparison of NP with NP:PQ revealed small decrease in cellular viability when paraquat was associated with the nanoparticles, although the viability values remained close to 100 per cent, indicating a good ability of the nanoparticles to reduce the toxicity of herbicide in this CHO cell. The genotoxicity data showed that there was increased DNA damage for all treatments. Greater damage was caused when free paraquat or paraquat associate with the particles was used. The association of the herbicide with the nanoparticles reduced the degree of damage, compared to the use of free paraquat, suggesting that the particles provided a degree of protection against toxic effects

Tong *et al.* (2017) reported that in the control group, mPEG–PLGA nanoparticles without metolachlor exhibited negligible inhibition to plants. The seedling height and root length in the MNPs treatment groups exhibited higher inhibition on *Digitaria sanguinalis* seedlings than L-MNPs and MMPs under relatively low concentrations. However, it showed no significant inhibition or even less inhibition comparing to L-MNPs and MMPs treatment groups under relatively high concentrations. The controlled release maintained the persistence effectiveness of MNPs during the experiment time. MNPs exhibited high-activity owing to its absorption into plants at low concentrations. This process along with water absorption guaranteed the utilization of metolachlor

Conclusion

Weeds are the pioneer plants which compete with crops for nutrients, light and water. Herbicides are used intensively to control the weeds. The half life of herbicides is very short ranging from few hours to couple of weeks. The intensive use of herbicides cause several problems like soil, ground water and food contamination. Nano-herbicides helps to reduce this problems. They are required in low amount with high bioavailability to the problematic weeds and also reduce the phytotoxcitiy to the non target organisms. Nano herbicides persist in soil where they are applied through out the entire growing season and take of the germinating weeds and leaves no reside at the end of the growing season allowing for taking up the next crop. Even the one tenth of the recommended does of nanoherbicide can control the weeds effectively than the complete recommended dose of commercial formulation.

REFERENCES

Abigail, E.Y., Samuel, M.S., Chidambaram.R. Application of rice husk nanosorbents containing 2,4-dichlorophenoxyacetic acid herbicide to control weeds and reduce leaching from soil. *Journal of the Taiwan Institute of Chemical Engineers*. 2016, 63, 318-326.

Ali, I. ALOthman, Z. A., Al-Warthan. A. Sorption, kinetics and thermodynamics studies of atrazine herbicide removal from water using iron nano-composite material. *International Journal of Environmental Science and Technology*. 2015, 13, 733-742.

Bach, A., Shemer,H., Semiat. R. Kinetics of Phenol Mineralization by Fenton-Like Oxidation. *Desalination*. 2010, 264 (3), 188–192.

Bansal, M., Garg, U., Singh, D., Garg, V.K. Removal of Cr (VI) from aqueous solutions using pre-consumer processing agricultural waste: A case study of rice husk. *Journal of Hazardous Mater*. 2009,162,312–20.

Brindha, K., Chinnamuthu, C. R. Zinc oxide nanorods to degrade phenolics and stored starch of *Cyperus rotundus* tubers management. *Journal of Crop and Weed. 2017*, 13(3), 184-188.

Cea, M., Cartes, P., Palma, G., Mora, M.L. Atrazine efficiency in an andisol as affected by clays and nanoclays in ethylcellulose controlled release formulations. Revista de la Ciencia del Suelo Nutricion Vegetal. 2010, 10(1), 62–77.

Chowdhury, S., Mishra, R., Saha, P., Kushwala, P. Adsorption thermodynamics, kinetics and isosteric heat of adsorption of malachite green onot chemically modified rice husk. *Desalination* 2011,265,159–68.

Grillo, R., dos Santos, N.Z.P., Maruyama, C.R., Rosa, A.H., de Lima, R., Fraceto L.F. Poly (caprolactone) nanocapsules as carrier systems for herbicides: physico–

chemical characterization and genotoxicity evaluation. *Journal of Hazardous Mater*. 2012, 231–232, 1–9.

Grillo, R., Rosa, A. H., Fraceto. L.F. Poly(e-caprolactone) nanocapsules carrying the herbicide atrazine: effect of chitosan-coating agent on physico-chemical stability and herbicide release profile. *International Journal of Environmental Science Technology*. 2014, 11, 1691–1700.

Itodo, H.U., Nnamonu, L.A., Wuana, R. A. Green Synthesis of Copper Chitosan Nanoparticles for Controlled Release of Pendimethalin. *Asian Journal of Chemical Sciences*. 2017, 2(3), 1-10.

Jangaard, N.O., Sckerl, M. M., Schieferstein. R. H. The role of phenolics and abscisic acid in nutsedge tuber dormancy. *Weed Science*. 1971,19 (1), 17-20.

Jhones,L.D.O., Estefania,V.R.C., Camila, M.G.D.S., Tatiane, P., P Renata, L., Leonardo, F.F. Solid Lipid Nanoparticles Co-loaded with Simazine and Atrazine: Preparation, Characterization, and Evaluation of Herbicidal Activity. *Journal of Agricultural and Food Chemistry*. 2014, 63 (2), 422–432.

Kumar, N., Kuamr, R., Shakil, N,A., Das, T.K. Nanoformulation of pertilachlor herbicide : Preparation, characterization and activity. *Journal of Scientific and Industrial Research*. 2016, 75, 676-680.

Lidong Cao, Zhaolu Zhou, Shujun Niu, Chong Cao, Xiuhuan Li, Yongpan Shan, and Qiliang Huang. Positive-Charge Functionalized Mesoporous Silica Nanoparticles as Nanocarriers for Controlled 2,4-Dichlorophenoxy Acetic Acid Sodium Salt Release. *Journal of Agricultural and Food Chemistry*. 2018, 66, 6594–6603.

Maruyama, C R., Guilger, M., Pascoli, M., Bileshy, J.N., Abhilash, P.C., Fernandes, L. F., Lima, R. D. Nanoparticles Based on Chitosan as Carriers for the Combined Herbicides Imazapic and Imazapyr. *Scientific Reports. 2016*, 6(1), 1-2.

Muhammad, S.U, Nor, AI., Kamyar, S., Norhazlin, Z., Wan, M.D., Zin, W.Y. Copper nanoparticles mediated by chitosan: Synthesis and characterization via chemical methods. *Molecules*. 2012, 17,14928-14936.

Nnamonu,L.A., Sha, R. A., Onyido. I. Alginate Reinforced Chitosan and Starch Beads in Slow Release Formulation of Imazaquin Herbicide—Preparation and Characterization. *Materials Sciences and Applications*. 2012, 3, 566-574.

Oliver, D. D.J.R. Volatilization of Alachlor from Polymeric Formulations. *Journal of Agricultural and Food Chemistry*. 2004, 52 (22), 6742-6746.

Pereira, A. E. S., Renato Grillo, Nathalie, F.S., Melloa, Andre Rosab, H., Leonardo Fraceto, F. Application of poly(epsilon-caprolactone) nanoparticles containing atrazine herbicide as an alternative technique to control weeds and reduce damage to the environment. *Journal of Hazardous Materials* 2014, 268, 207–215.

Pradeesh Kumar., Chinnamuthu. Assembly of Nanoencapsulated Pendimethalin Herbicide using Solvent Evaporation Method for Season Long Weed Control under Irrigated Ecosystem. *International. Journal of Pure and Applied. Bioscience*. 2017, 5 (1), 349-357.

Preisler, A. C., Anderson, E, S, P., Campos E,V.R., Dalazen, G., Leonardo F, F.,Oliveira, H, C. Atrazine nanoencapsulation improves pre-emergence herbicidal activity against *Bidens pilosa* without enhancing long-term residual effect on *Glycine max*. 2019. *Pest Management Science*. wileyonlinelibrary.com/journal/ps. (published online May 12, 2019).

Preet Kaur, S., Rao, R., Hussain, A., Khatkar, S. Preparation and characterization of rivastigmine loaded chitosan nanoparticles. *Journal of Pharmaceutical sciences and Research*. 2011, 3(5), 1227-1232.

Satapanajaru, P., Anurakpongsatorn, P., Pengthamkeerati, P., Boparai, H. Remediation of Atrazine contaminated Soil and Water by Nano Zerovalent Iron.*Water Air Soil Pollution*. 2008, 192, 349-359.

Shehzad, F., Shah, J., Jan, M.R. Determination of pendimethalin herbicide in water and vegetable samples by microwave-assisted solvent extraction and HPLC method. *Pakistan Journal of Weed Science and Resources*. 2011, 17 (2), 175-185.

Shu, X. Z., Zhu, K. J. A Novel Approach to Prepare Tripolyphosphate/Chitosan Complex Beads for Controlled Release Drug Delivery. *International Journal of Pharmaceutics*. 2000, 201 (1), 51-58.

Sinha, V.R., Bansal, K., Kaushik, R., Kumria, R., Trehan, A. Poly-caprolactone microspheres and nanospheres: an overview. *International Journal of Pharmaceutics*.2004, 278 (1), 1–23.

Sopena, F., Maqueda, C., Morillo, E. Controlled release formulations of herbicides based on microencapsulation. *Ciencia e investigación agrarian*. 2009, 36(1), 27-42.

Tong, Y., Yan, Wu., Zhao, C., Yong, Xu, Jianqing, Lu, Xiang, S., Zong, F and Xuemin, Wu. Polymeric Nanoparticles as a Metolachlor Carrier: Water-Based Formulation for Hydrophobic Pesticides and Absorption by Plants. *Journal of Agricultural and Food Chem*istry. 2017, 65 (34), 7371–7378.

Vijji, N., Chinnamuthu, C.R. Iron Oxide Nanoparticles to break the tubers dormancy of the world's worst weed the *cyperus rotundus*. *International Journal of Agricultural Science and Research*. 2015, 5 (3), 259-266.

Vimalrajiv, B., Chinnamuthu,C.R., Subramanian, E., Senthil, K. Effect of nanoparticles in combination with pendimethalin and hydrogen peroxide on growth parameters and nodulation of blackgram (*Vigna mungo L*). *International Journal of Chemical Studies*. 2018, 6(3), 2816-2819.

Chapter 22

Role of Secondary Plant Metabolites against Insects

S. Lekha Priyanka[1], R. Nagasri Navya[1], M. Chandrasekaran[2] and Wajid Hasan[3]*

[1]Department of Agricultural Entomology, Tamil Nadu Agricultural University, Coimbatore – 641003, Tamil Nadu
[2]Department of Plant Protection, Horticultural College and Research Institute for Women, TNAU, Tiruchirappalli – 620027, Tamil Nadu
[3]Krishi Vigyan Kendra, Jehanabad, BAU, Bihar

Plants are having several morphological and biochemical phenomenon to withstand insect pest damages; however, insects have developed several adaptive mechanisms to withstand these defense mechanisms. Plants use a number of morphological, chemical, and biochemical defenses against insect herbivores. Among plant defensive traits, the chemical barriers, both inducible and constitutive, and the nutritional content of a plant are considered to be more essential in terms of increasing defense against insects. While plants withstand the mechanical barrier by means of unique behavioural and life cycle adaptations, adaptation to chemical defense is not easy as it is highly dynamic. Plants synthesize variety of compounds distinct from the intermediates called secondary metabolites.

Introduction

Secondary plant metabolites occur in plants or produced in response to herbivory. Naturally occurring secondary plant alkaloid based metabolites are generally known as phyto - anticipins, while the induced ones are known as phytoalexins (Howe and Jander, 2008). Secondary metabolites are not stringently need for the plant growth and reproduction (Howe and Jander, 2008), but they play a vital role in conferring the mechanism of defense against insects (Bennett and Walls Grove, 1994), microbial infections (Wink, 2003) and other roles such as protection from UV radiation– have been identified (Schafer *et al.*, 2009).

Apart from line of defense, some of these compounds are utilized by plants to attract pollinators and seed dispersal (Wink, 1988). For centuries, Secondary metabolites secretions are used by the mankind to improve their health, nutrition and enhancing agricultural productivity in a positive way (Wink, 2003).

Classification of Secondary Plant Metabolites

Secondary plant metabolites are classified based on their structure, diversity in functional biosynthesis.

The main categories of secondary metabolites synthesized in plants are,

A. Terpenes
B. Phenolics
C. Nitrogen and
D. Sulphur containing compounds (Grayson, 1998).

Terpenes

These are the primary group of metabolites derived from acetyl Co-A or glycolytic intermediates (Gershenzon and Croteau, 1991), which involved in defence as toxins, feeding deterrents to insects and mammals. These are further divided into five subclasses.

Source: These are synthesized from shikimic acid pathway.

Monoterpenes

The monoterpene esters which are present in different parts of chrysanthemum plant exhibit insecticidal response to various group of insects belongs to coleoptera, hymenoptera, and lepidoptera *etc.* In gymnosperms *viz.*, fir, pine *etc.*, these monoterpenes compounds are occur in resin ducts, twigs and trunks as α - pinene, β - pinene, myrecene and limonene that are toxic to serious beetle pest of barks in conifers (Turlings *et al.*, 1995).

Sesquiterpenes

Sesquiterpene compounds such as costunolides are important anti - herbivore agents characterized by five membered lactone rings shown higher feeding repellency against mammals and numerous herbivorous insects (Wink, 2007).

Diterpenes

Abietic acid, a diterpene found in the trees comes under leguminous group and pine are secreted as resin compounds through pore canals of the trunk region when injured by insects. The flow of resin substances blocks insect feeding and act as a deterrent (Bradley *et al.*, 1992).

Triterpenes

In plant system, sterols are the major components confined with cell plasma membranes, which regulates and maintain the permeability by altering the fatty acid chains. The milkweeds, *Asclepias* spp. produces the sterols like bitter tasting glucosides against cattle and insects (Cianfrogna *et al.*, 2002). Azadirachtin, a complex limonoid compound posses various toxic effects derived from the neem plant, *Azadirachta indica* exerts deterrence to some insects (Mordue and Blackwell, 1993).

Polyterpenes

These are terpene compounds with high molecular weight occur in plants. Carotenoids are the primary pigmental tetraterpenes (Li *et al.*, 2002). Rubber is the another polyterpene found in long vessels with 1500 – 15000 isopentenyl units called laticifers, which provide protection mechanism to heal injured wounds of herbivores as reported by Eisner and Meinwald, 1995.

Phenolic Compounds

Plants produce multiple numbers of products containing phenols, which is an integral component of conferring the defence mechanism against insect pests, diseases and the parasitic nematodes which infesting the root regions(Wuyts *et al.*, 2006).

Source: These are synthesized from shikimic acid pathway.

Coumarin

Coumarins are the simple phenolic defense compounds that are wide spread in the vascular bundles and acts against herbivores (Brooker *et al.*, 2008). Coumarins are the main derivative compounds of shikimic acid pathway (Wittstock *et al.*, 2004) in plants, bacteria and fungi but not produced in mammals. It is suspected that coumarins serves as natural defence compounds for plants and constitute a starting point to focus on new derivatives with a range of improved antifungal activity.

Furano Coumarins

It is a type of coumarin, abundant in the members of umbelliferae responsible for phytotoxicity. In general, these compounds become toxic when they get activated by light (UV – A). Psoraline, basic furano-coumarin very rarely found within the SO_2 treated plants and expresses its activity against fungal compounds (Ali *et al.*, 2008).

Lignin

It is one of the polymer of phenyl - propanoid group, derived by oxidation of alcohols like coniferyl, sinapyl and coumaryl by a plant enzyme-peroxidase to form lignin. The proportions of the monomeric units in lignin vary among species, plant organs and even layers of a single cell wall (Stevens *et al.*, 2000). Lignin compounds with high chemical durability elicits physical toughness in plants. Hence, the herbivorous and animals finds difficulty in digestion (Mader and Fisher, 1982).

Flavonoids

In plants flavonoids play a vital role in many biological processes like seed development (Song *et al.*, 2016), pollen tube germination (Ylstra *et al.*, 1992), transport of hormones (Peer and Murphy, 2007) and fruit ripening (Andreotti *et al.*, 2008). Flavonoids prevent the damages induced by the viruses, bacteria, fungi and herbivores but acts as a attractant to pollinating agents (Wink, 2010). They are also responsible for colour differences in fruits, flowers and seeds (Falcone *et al.*, 2012). New pesticides are being synthesized by using the identified flavonoid compounds as an good alternative to the synthetic pesticides because of increasing demand to plant products. Flavonoids inhibit activity of enzymes to arrest the larval growth of insect species in early instars (Kim *et al.*, 2000). Flavonoids compounds *viz.*, naringin, rutin, and quercetin are used as an insecticidal elements in the management of immature and mature stages of the aphid, *Eriosoma lanigerum* Hausmann (Ateyyat, 2012).

Tannins

Most tannin has molecular masses between 600 and 3000. Tannins are toxins which retard the survivorship, growth and development of herbivore insects and animals. Tanins are the polymers of phenolic compounds with pesticidal properties. The properties of tannins are mainly attributed with its ability to bind on the insect proteins. These are proved feeding deterrents to phytophagous insects. The deterrent effect against feeding is undoubtedly a mechanism of plants to protect against insects. Effects of tannins on behaviour and physiology of herbivores are affected by the nutritional context of the tannins. Tanniferous plants with low protein levels are showing high deterrence against herbivores than the plant with high nutrition with same level of tannin compounds. These tannin compounds induce an astringent excitement in the mouths of insects due to the adherence of salivary proteins. Hence, the mammals deer, cattle, and apes generally avoid the plants with high tannin content (Oates *et al.*, 1980).

Sulphur Containing Compounds

Sulphur containing compounds include

- ☆ Glutathione Synthetase (GSH)
- ☆ Glucosinolates (GSL)
- ☆ Phytoalexins
- ☆ Thionins
- ☆ Defensins and
- ☆ Lectins

Source: These are synthesized from amino acids and they play key role in governing the defence of plants (Saito, 2004).

GSH (Glutathione Synthetase)

GSH is a sulphur containing glucosides which is produced by plants when the plant suffers sulphur deficiency. This mobile and easily soluble form of sulphur is readily assimilated by plant to combat sulphur shortage (Kang *et al.*, 2007). Thus it regulates the growth of the plant (Noctor *et al*., 1998). They also provoke resistance of plants during microbial attack and act as cellular antioxidants during stress (Kang and Kim, 2007).

GSL (Glucosinolates)

GSL is a nitrogen and sulphur containing plant glucosides and are produced by plants to induce resistance against insect pests (Siemans *et al*., 2009). When the plant is subjected to insect attack GSL breaks down to isothiocyanates which activates the antioxidant defence system in plants (Renwick *et al*., 1992). This protects the plant against insect damage (Lipka *et al*., 2010). The leaves of a resistant lines of *Brassica* which conferred the tolerance to the fungal pathogen, *Leptosphearia maculans* had higher levels of GSL than the susceptible line (Mithen and Magrath, 1992; Halkier and Gershenzon, 2006).

Phytoalexins

Response of plants to bacterial or fungal invasion is mainly due to the synthesis of phytoalexins (Van Etten *et al*., 1994). Phytoalexins are group of disparate secondary plant metabolites with high antimicrobial nature, which accumulate around the site of external infection (Grayer and Harborne, 1994). These phytoalexins are usually undetectable in the host plant before the external attack. But, very rapidly synthesized immediately after microbial/insect pest attack (Monde *et al*., 2000). The point of attack for the activation of these biosynthetic pathways is usually the initiation of gene transcription (Harborne, 1999). Thus the plants do not have storing mechanism of any enzymatic machinery required for phytoalexin synthesis (Mehta *et al*., 1995). But, soon after invasion of infective microbials, they begin to start the transcribing and translating process of the appropriate mRNAs and need based synthesize of the enzymes *de novo* (Grubb and Abel, 2006).

The production of phytoalexin compounds in plants might be a common mechanism of resistance to insects or pathogenic microbes (Siemans *et al*., 2009). The nature of production of secondary phytoalexin compounds may vary plant to plant depends upon its family. For example, isoflavonoids are the common in leguminous plants such as soybean and alfalfa. But, sesquiterpenes are main phytoalexins compounds in solanaceous plants *viz*., potato, tobacco and tomato.

Keen and Kennedy (1974) documented the common phytoalexin compounds synthesized in plants *viz*., Phaseolin in the pulse, *Phaseolus vulgaris*, glyceollins in soybean, *Glycine max*, Ipomeamarone in sweet potato, Pistin in the pods of *Pisum sativum*, Trifolirhizin in red clover and Orchinol in tubers of orchids. Some evidenced reports are mention that the production of phytoalexins leads to death of plant cells in few situations and it is known as the hypersensitive response of the host plants (HR) or Apoptosis (Hammond – Kosack and Jones, 1996).

Defensins

As the name indicates defensins are antifungal and anti-bacterial (Thomma *et al*., 2002). These are also pathogen-inducible and are expressed in higher amounts when subjected to pathogen attack (Gu *et al*., 1992; Terras *et al*., 1995).

Thionins

Thionins are also involving in the strengthening and conferring the natural defense mechanism in plants against insects, micro-organisms and mammals. Infected wheat spikes with higher amount of thionins were resistant to *Fusarium culmorum* (Kang and Buchenauer, 2003).

Lectins

Lectins are kind of defensive proteins which bind with carbohydrate or carbohydrate containing proteins. Lectin binds in to the epithelial cell lining present in the digestive system and disrupt the nutrient absorption, after being ingested by many herbivores (Peumans and Van Damme, 1995).

Nitrogen Containing Compounds

Nitrogen containing compounds include:

- Alkaloids
- Cyanogenic glucosides and
- Non protein amino acids

Alkaloids

These are nitrogen containing secondary metabolites present abundantly in vascular tissues of plants (Hegnauer, 1988). Alkaloids are abundantly found in dicots compared to monocots and gymnosperms (Harborne, 1989). These include Pyrolizidine alkaloids (PAs) which confers defence against microbial infection and insect herbivory (Robinson, 1980). They are synthesized from the common amino acids namely aspartic acid, lysine, tyrosine and tryptophan (Pearce *et al.*, 1991).

Cyanogenic Glucosides

These are a group of nitrogen containing compounds. They are commonly present in *Graminae, Rosaceae* and *leguminosae* (Seigler, 1991).When the plant is subjected to damage, they readily break down to emit volatile poisonous gas HCN and H_2S. When the plant is damaged due to herbivore feeding, the cell contents of different tissues mix and form HCN. The emission of this gas deters the insects and acts as an anti-feedant. Cyanogenic glucosides also repels snails and slugs (Taize and Zeiger, 1995). Examples are Amygdalin in seeds of almonds, apricot, cherries and peaches, Dhurin in *Sorghum bicolar* (Poulton, 1990).

Non-protein Amino Acids

Non-protein amino acids found in free forms and are mistakenly incorporated for proteins (Johnson *et al.*, 1989). They serve as barriers against insect pests, for example azetidine and canavanine are analogues of proline and arginine, respectively. These block the intake of amino acids by insects. Canavanine binds to the enzyme to which arginine normally bind, thus arginine transferred RNA molecule is incorporated with canavanine in place of arginine. This process leads the formation of a non-functional protein in place of arginine. Hence, the metabolism of arginine in insects is disrupted (Funck *et al.*, 2009). Plants which synthesize the non-protein amino acid might provide resistance against herbivorous animals, insects and pathogenic microbes (Rosenthal, 1991).

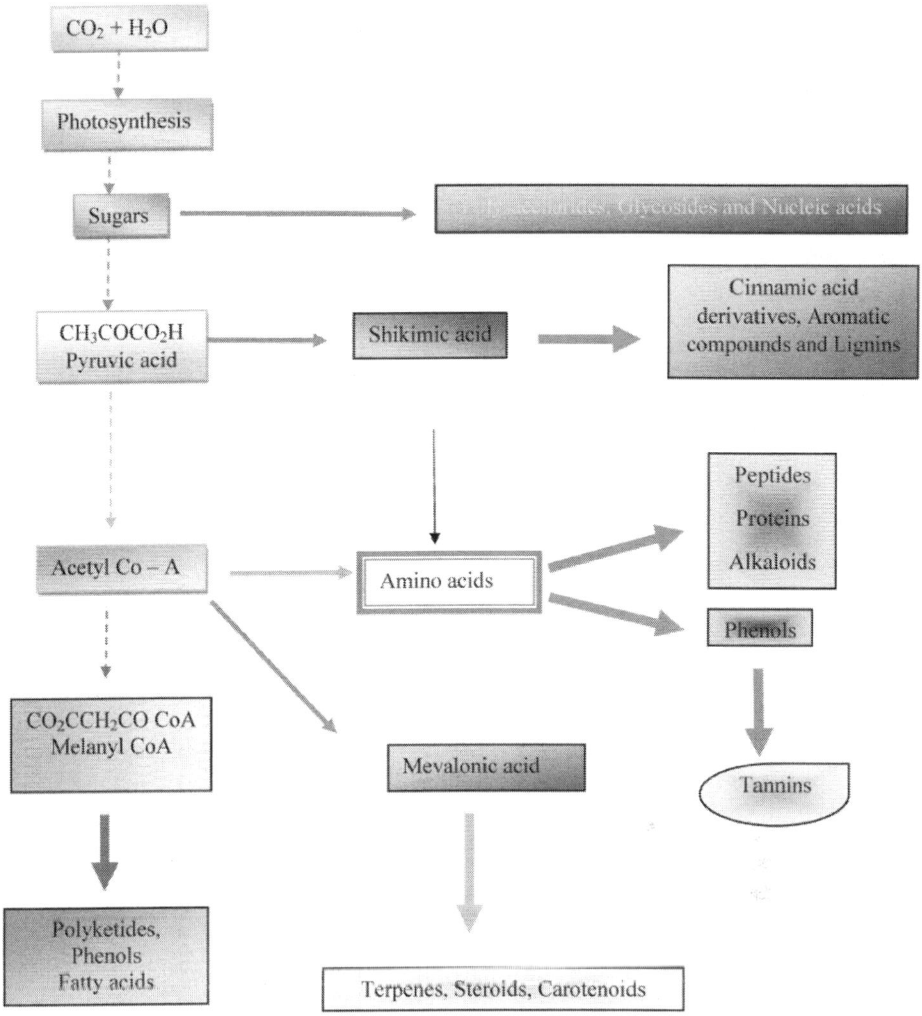

Figure 22.1: Relationship between the Primary and Secondary Metabolites.

Secondary Plant Metabolites - Mode of Action

The secondary plant metabolites *viz.*, alkaloids, nicotine, tubocourarine, ergot alkaloids, agroclavine, muscarine, caffeine, theobromine, theophylline (Wink, 2012) modulates neuronal signal transduction.

Terpenes

Secondary plants metabolites of are usually lipophilic in nature, like monoterpenes, diterpenes, triterpenes, sesquiterpenes, phenyl propanoids, steroids and mustards oils. Further, lipophilic metabolites attack the bio membranes surrounding the living cells and intracellular compartments. Apart from changing the structure of proteins, these compounds change the permeability of bio membranes by being trapped inside them.

Phenolics

Secondary metabolites interact with cytoskeleton of the cells, thus interfering with cell division. However, most of secondary plant metabolites involves and modulates the 3D structure and activity of the protein by forming complex ionic bonds (Wink, 2008).

Alkaloids

Alkaloids are highly toxic to the insect pests. These compounds affect the ionic pathways, neuro-transmitter receptors, transporters and enzymes. Modulation of neuronal signal transduction components, concentration of neurotransmitters, function of neurotransmitter receptors or their expressions are altered may lead to significant changes in the physiology and behavior of the insect.

Saponins

Saponins contain a lipophilic steroid or triterpene moiety with hydrophobic nature, form complexes with membrane cholesterol. It involves in modulating the neuronal signal transduction, inhibiting protein synthesis, altering the protein structures, and interacting with bio membranes, some of the plant secondary metabolites interfere with metabolizing nucleic acids and enzymes (Wink, 2010), while some are involved in intercalating the DNA. The interspersing of DNA by saponin compounds stabilizes the DNA sequences during its replication processes and it prevents the activities of helicases and RNA, by the way inhibiting the intermediately steps in DNA replication (Wink, 2018).

Secondary metabolites serve as protein inhibitors while others alter protein structure and function. Specific protein inhibitors like podophyllotoxin, colchicine, vinblastine, sanguine, maytansine and rotenone inhibits the microtubule assembly necessary for mitotic spindle assembly during cell division.

Secondary plant metabolites contains active functional groups that interact with amino, sulfhydryl, or hydroxyl groups of residues including protein amino acids, thereby altering their structure and functional properties (Wink, 2015).

The frameshift mutations and deletions by plant secondary metabolites lead to cell death. The alkaloid plant secondary metabolites with intercalating properties have been detected in alkaloids like quinoline, emetine, furano-quinoline, anthra-quinones, furano-coumarins, and beta-carboline (Wink, 2010).

Insect Adaptation to Plant Secondary Metabolites

Insects have developed adaptations to overcome the toxic secondary plant metabolites through alteration in its behavioural and biochemical adaptations or mechanisms like detoxification, degradation, excretion and sequestration (War and Sharma, 2014).

Detoxification

A number of enzymes are involved in the detoxification process of plant toxins by insect pests. These toxic secondary metabolites are detoxified or less toxified by insect pests after ingestion with the help of detoxifying enzymes (Ferreres *et al.*, 2008). The function of these enzymes may vary based on diet composition of the host plant, insect species depends upon the glycosylation, glutathionation, sulfation and deacylation processes.

Detoxifying enzymes occur in the cytoplasmic cells and midgut lumen. Metabolism of induced phyto-toxins occurs in cytoplasm and midgut lumen before entering into the cells (Halkier and Gershenzon, 2006). Detoxification enzymes like cytochrome P450 monooxygenases, glutathione S-transferases (GSTs), and the carboxylesterases (COEs) are present in insects in low concentrations. Their level increases when insect feeds on the toxic metabolites. This increased concentration of detoxifying enzymes converts the toxic metabolites into least toxic form.

Detoxification by P450

P450s are the primarily used by insects against plant allelochemicals (Feyereisen, 1999). Examples are In *S. frugiperda,* the toxic metabolites like 2 – phenylethyl isothiocyanate, indole – 3 – carbinol, and indole – 3 – acetonitrileare detoxified in insect midgutby P450 (Yu, 2000). In cotton, one of the important secondary metabolite gossypol is detoxified by P450 monooxygenase. (Mao *et al.*, 2007). In the bark beetles, *Ips paraconfusus* Lanier and *Ips pini* Say the secondary plant terpene metabolites are detoxified by P450s (Seybold *et al.*, 2006).

Glutathionation

Another important detoxifying enzyme is GST (Glutathione S-transferases). Detoxification by GSTs occurs in insect haemolymph, fat body and midgut. Detoxification using Glutothion S-transferases reported by Enayati *et al.* (2005) in many lepidopteran insects. Insect GS-transferases catalyze the conjugation of glutathione to electrophilic toxic molecules, which leads to the formation of water-soluble glutathione S-conjugates that are easily degraded and eliminated by the insect (Schramm *et al.*, 2011). In *Myzus persicae,* high level of GST was documented. Hence they were able to feed on brassicaceous plants with metabolites, glucosinolates and isothiocyanates which can be detoxified by *Myzus persicae* (Francis *et al.*, 2005).

Excretion

Esterases are another group of metabolic enzymes involved in the metabolization of toxic compounds. Esterases detoxify toxic compounds by creating cleavage enzymatic pathway or sequestration of the plant toxic metabolites. The compounds are hydrolyzed into non-toxic polar compounds that are easily excreted from the insect body.

Degradation

UDP Glycosyl-Transferases (UGTs) is an important enzyme involved in the degradation of secondary metabolites (Mackenzie *et al.*, 2005). These enzymes catalyze the conversion of a glycosyl group from UGTs acceptor molecules. In *Manduca sexta*, degradation of plant compounds occurs by UGTs. Gene coding of BmUGT1 in the silkworm, *Bombyx mori* (L.) has been reported for the degradation of coumarins and flavonoids (Luque *et al.*, 2002).

Sequestration

Sequestration is defined as the uptake and accumulation of selective and specific toxins of insect pests, which determines their growth and development (Vanveen, 2015). In milkweed bugs, cardenolides are synthesized by sequestration (Bramer *et al.*, 2015). In monarch butterfly, *Danaus plexippus* (L.) the sequestration of cardenolides taken place in target sites and substitute the histidine and valine in place of asparagine and leucine, respectively.

Conclusion

Plants produce a multifariousness of secondary plant metabolites which confers protection against its predators pathogens. They also enhance the protection against biotic and abiotic stresses (Rosenthal *et al.*, 1991). The unique groups of secondary metabolites *viz.*, phenolic compounds, terpenes, nitrogen and sulphur containing compounds exhibits feeding deterrence to many herbivores. They act as attractants (colour, taste and odour) to group of pollinators and seed dispersing organisms. They serve as the agents of plant-plant interactive competition and plant-microbe symbiosis. The capability of plants to survive and compete is greatly affected by the functions of secondary metabolites. Recent biotechnological approaches are also intricate in the production of secondary plant metabolites; *in vitro* experiments also confirm the defensive roles of secondary plant metabolites (Wink, 1988).

Compounds with ecological, chemical and defensive roles, pave way for a new area of science development called it as ecological biochemistry (Harborne, 1989). Therefore, additional research in area of natural pesticides development is needed in the current scenario. In the longer run, it will be possible to generate geneswhich could be used for production of valuable defensive secondary metabolites in bioreactors or for metabolic engineering of crop plants. This will improve their resistance against herbivores, pathogens and various environmental stresses.

This chapter is incorporated with the major aspects of plant secondary-metabolites and their importance in conferring defence mechanisms in plants and its ecological adaptations. Wide range of SPS compounds present in plants, with a vast distribution. Hence research works to be focussed to on secondary metabolites and to use them in crop protection measures.

REFERENCES

Ali, S. T., Mahmooduzzafar – Abdin, M. Z. and Iqbal, M. (2008). Ontogenetic changes in foliar features and psoralen content of *Psoraleacorylifolia* Linn. exposed to SO_2 stress. *Journal of Environmental Biology*. 29(5): 661 – 668.

Andreotti, C., Ravaglia, D., Ragaini, A. and Costa, G. (2008). Phenolic compounds in peach (*Prunuspersica*) cultivars at harvest and during fruit maturation. *Annals of Applied Biology*. 153: 11 – 23. doi: http: //doi.org/10.1111/j.1744-7348.2008.00234.x.

Ateyyat, M. (2012). Impact of flavonoids against woolly apple aphid, *Eriosoma lanigerum* (Hausmann) and its sole parasitoid *Aphelinusmali* (Hald.). *Journal of Agricultural Science*. 24: 227 – 236.

Bennett, R. N. and Wallsgrove, R. M. (1994). Secondary metabolites in plant defence mechanisms. *New Phytologist*. 127(4): 617 – 633.

Bradley, D. J., Kjellborn, P. and Lamb, C. J. (1992). Elicitor and wound induced oxidative cross linking of a proline rich plant cell protein: A novel rapid defence response. *Cell*. 70(1): 21 – 30.

Bramer, C., Dobbler, S., Deckert, J., Stemmer, M. and Petschenka, G. (2015). Na/K ATPase resistance and cardenolide sequestration: basal adaptations to host plant toxins in the milkweed bugs (Hemiptera: Lygaeidae: Lygaeinae). *Proceedings of Biological Science*.282: 1805.

Brooker, N., Windorski, J. and Blumi, E. (2008). Halogenated coumarins derivatives as novel seed protectants. *Communication in Agriculture and Applied Biological Sciences*. 73(2): 81 – 89.

Cianfrogna, J. A., Zangeri, A. R. and Berenbaum, M. R. (2002).Dietary and developmental influences on induced detoxification in an oligophage.*Journal of Chemical Ecology*. 28: 1349 – 1364.

Eisner, T. and Meinwald, J. (1995). In: Chemical ecology: The chemistry of biotic interaction.

Enayati, A. A., Ranson, H. and Hemingway, J. (2005).Insect glutathione transferases and insecticide resistance.*Insect Molecular Biology*. 14: 3 – 8.

Falcone, F. M., Rius, S. P. and Casati, P. (2012). Flavonoids: Biosynthesis, biological functions, and biotechnological applications. *Frontiers in Plant Science*. 3: 1 – 16. doi: http: //doi.org/10.3389/fpls.2012.00222.

Ferreres, F., Valentao, P., Pereira, J. A., Bento, A., Noites, A. and Seabra, R. M. (2008). HPLC – DAD MS/MS – ESI screening of phenolic compounds in *Pierisbrassicae* L. reared on *Brassica rapa* var. *rapa* L. *Journal of Agriculture and Food Chemistry*. 56: 844 – 853.

Feyereisen, R. (1999). Insect P450 enzymes.*Annual Review of Entomology*.44: 507 - 533.

Francis, F., Vanhaelen, N. and Haubruge, E. (2005).Glutathion S – transferases in the adaptation to plant secondary metabolites in the *Myzus persicae* aphid. *Archives of Insect Biochemistry and Physiology*. 58: 166 – 174.

Funck, D., Stadelhofer, B. and Koch, W. (2009). Ornitine – delta – aminotransferase is essential for arginine catabolism but not for proline biosynthesis. *BMC Plant Biology*. 8: 40 – 45.

Gershenzon, J. and Croteau, R. (1991).Terpenoids. In Herbivores their interaction with secondary plant metabolites, Volume I: The chemical participants, II edition, G.A. Rosenthal and M. R. Berenbaum, editions, Academic press, San Diego: 165 – 219.

Grayer, R. J. and Harborne, J. B. (1994).A survey of antifungal compounds from higher plants 1982 – 1993.*Phytochemistry*. 37: 19 – 42.

Grayson, D. H. (1998).Monoterpenoids.*Natural Product Reports*. 5: 497 – 521.

Grubb, C. and Abel, S. (2006). Glucosinolate metabolism and its control.*Trends in Plant Science*. 11: 89 – 100.

Gu, Q., Kawata, E. E., Morse, M. J., Wu, H. M. and Cheung, A. Y. (1992). A flower specific cDNA encoding a novel thionin in tobacco.*Molecular and General Genetics*. 234: 89 – 96.

Halkier, B. A. and Gershenzon, J. (2006).Biology and biochemistry of glucosinolates. *Annual Review of Plant Biology*. 57: 303 – 333.

Hammond – Kosack, K. E. and Jones, J. D. G. (1997).Plant disease resistance genes. *Annual Review of Plant Physiology and Plant Molecular Biology*. 48: 575 – 607.

Harborne, J. B. (1989). Recent advances in chemical ecology, *Natural Products Reports*, 6: 85 – 109.

Harborne, J. B. (1999). The comparative biochemistry of phytoalexin induction in plants.*Biochemical Systematics and Ecology*. 27: 335 – 367.

Hegnauer, R. (1988). Biochemistry, distribution and taxonomic relevance of higher plant alkaloids.*Phytochemistry*.27: 2423 – 2427.

Howe, G. A. and Jander, G. (2008) Plant immunity to insect herbivores.*Annual Review of Plant Biology*. 59: 41 – 66.

Johnson, R., Narvaez, J., An, G. and Ryan, C. (1989). Expression of proteinase inhibitors I and II in transgenic tobacco plants: Effects on natural defence against *Manduca Sexta* larvae. *Proceedings of the National Academy of Sciences of the USA*.86: 9871 – 9875.

Kang, S. Y. and Kim, Y. C. (2007).Decursinol and decursin protect primary cultured rat cortical cells from glutamate – induced neurotoxicity. *Journal of Pharmacy and Pharmacology*. 59(6): 863 – 870.

Kang, Z. and Buchenauer, H. (2003).Immonocytochemical localization of cell wall – bound thionins and hydroxyproline – rich glycoproteins in *Fusarium culmorum* – infected wheat spikes. *Journal of Phytopathology*. 151(3): 120 – 129.

Keen, N. T. and Kennedy, B. W. (1974).Hydroxyphaseollin and related isoflavonoids in the hypersensitive resistance reaction of soybeans to *Pseudomonas glycinea*. *Physiological Plant Pathology*. 4: 173 – 185.

Kim, J. S., Kwon, C. S. and Son, K. H. (2000). Inhibition of Ü – glucosidase and Ü – amylase by luteolin, a flavonoid. *Bioscience, Biotechnology, and Biochemistry*.64: 2458 – 2461.

Li, X., Berenbaum, M. R. and Schular, M. A. (2002). Plant allelochemicals differentially regulate *Helicoverpa zea* cytochrome P450 genes. *Insect Molecular Biology*. 11: 343 – 351.

Lipka, U., Fuchs, R., Kuhns, C., Petutschnig, E. and Lipka, V. (2010). Live and let die –Arabidopsis non – host resistance to powdery mildews. *European Journal of Cell Biology*. 89(2): 194 – 199.

Luque, T., Okano, K. and Reilly, O. D. R. (2002).Characterization of a novel silkworm (*Bombyx mori*) phenol UDP – glucosyltransferase.*European Journal of Biochemistry*. 269: 819 – 825.

Mackenzie, P. I., Bock, K. W., Burchell, B., Guillemette, C., Ikushiro S, Iyanagi T, Miners, J. O., Owens, I. S. and Nebert, D. W. (2005). Nomenclature update for the mammalian UDP Glycosyl Transferase (UGT) gene superfamily. *Pharmacogenetics and Genomics*. 15: 677 – 685.

Mader, M. and Fisher, V.A (1982). Role of peroxidase in lignifications of tobacco cells, Oxidation of nicotinamide adenine dinucleotide and formation of hydrogen peroxide by cell wall peroxidises. *Plant physiology*. 70: 1128 – 1131.

Mao, Y. B., Cai, W. J., Wang, J. W., Hong, G. J., Tao, X. Y., Wang, L. J., Huang, Y. P. and Chen, X. Y. (2007). Silencing a cotton bollworm P450 monooxygenase gene by plant – mediated RNAi impairs larval tolerance of gossypol. *Nature Biotechnology*. 25: 1307 – 1313.

Mehta, R. G., Liu, J., Constantinou, A., Thomas, C. F., Hawthorne, M., You, M., Gerhauser, C., Pezzuto, J. M., Moon, R. C. and Moriarty, R. M. (1995). Cancer chemopreventive activity of brassinin, a phytoalexin from cabbage. *Carcinogenesis*. 16(2): 399 – 404.

Mithen, R. and Magrath, R. (1992).Glucosinolates and resistance to *Leptosphaeria maculans* in wild and cultivated *Brassica* species.*Plant Breeding*. 108: 60 – 68.

Monde, K., Osawa, S. M., Harada, N., Takasugi, M., Suchy, M., Kutschy, P., Dzurilla, M. and Balentova, E. (2000).Synthesis and absolute stereochemistry of a cruciferous phytoalexin, (–) – Spirobrassinin.*Chemistry Letters*. 598: 886 – 887.

Mordue, A. J. and Blackwell, A. (1993). Azadirachtin: an update. *Journal of Insect Physiology*. 39: 903 – 924.

Noctor, G. and Foyer, C. H. (1998).Ascorbate and glutathione: Keeping active oxygen under control. *Annual Review of Plant Physiology and Plant Molecular Biology*. 49: 249 – 279.

Oates, J. F., Waterman, P. G. and Choo, G. M. (1980).Food selection by the south Indian leaf – monkey, *Presbytisjohnii*, in relation to leaf chemistry.*Oecologia*. 45: 45 – 56.

Pearce, G., Strydom, D., Johnson, S. and Ryan, C. A. (1991). A polypeptide from tomato leaves induces wound inducible proteinase inhibitor proteins. *Science*. 253: 895 – 898.

Peer, W. A. and Murphy, A. S. (2007). Flavonoids and auxin transport: Modulators or regulators? *Trends in Plant Science*. 12: 556 – 563. doi: http: //doi.org/10.1016/j.tplants.2007.10.003.

Peumans, W. J. and Van Damme, E. J. M. (1995). Lectins as plant defence proteins. *Plant Physiology*.109: 347 – 342.

Poulton, J. E. (1990). Cyanogenesis in plants.*Plant Physiology*. 94: 401 – 405.

Renwick, J. A. A., Radke, C. D., Gupta, S. K. and Staedler, E. (1992). Leaf surface chemicals stimulating oviposition by *Pierisrapae* (Lepidoptera: Pieridae) on cabbage. *Chemoecology*. 3: 33 – 38.

Robinson, T. (1980). In: The organic constituents of Higher Plants.

Rosenthal, G. A. (1991). The biochemical basis for the deleterious effects of L – canavanine.*Phytochemistry*. 30: 1055 – 1058.

Saito, K. (2004). Sulphur assimilatory metabolism. The long and smelling road.*Plant Physiology*.136: 2443 – 2450.

Schafer, H. and Wink, M. (2009). Medicinally important secondary metabolites in recombinant microorganisms or plants: progress in alkaloid biosynthesis. *Biotechnology Journal*. 4(12): 1684 – 1703.

Schmelz, E. A., Carroll, M. J., LeClere, S., Phipps, S. M., Meredith, J., Chourey, P. S., Alborn, H. T. and Teal, P. E. A. (2006). Fragments of ATP synthase mediate plant perception of insect attack. *Proceedings of the National Academy of Sciences*.103: 8894 – 8899.

Schramm, K., Vassao, D. G., Reichelt, M., Gershenzon, J. and Wittstock, U. (2011). Metabolism of glucosinolates – derived isothiocyanates to glutathione conjugates in generalist lepidopteran herbivores. *Insect Biochemistry and Molecular Biology*. 42: 174 – 182.

Seigler, D. S. (1981).Secondary metabolites and plant systematic. In: *The biochemistry of plants*. 7: 139-176.

Seybold, S. J., Huber, D. P. W., Lee, J. C., Graves, A. D. and Bohlmann, J. (2006). Pine monoterpenes and pine bark beetles: a marriage of convenience for defense and chemical communication. *Phytochemistry Review*. 5: 143 – 178.

Siemans, D. H., Haugen, R., Matzner, S. and Vanasma, N. (2009). Plant chemical defense allocation constrains evolution of local range. *Molecular Ecology*. 18(23): 4974 – 4983.

Song, C., Xiang, D. B., Yan, L., Song, Y., Zhao, G. and Wang, Y. H. (2016). Changes in seed growth, levels and distribution of flavonoids during tartary buck wheat seed development. *Plant Production Science*. 1008: 1 – 10. doi: http: //doi.org/10.1080/1343943X.2016.1207485.

Stevens, J. L., Snyder, M. J., Koener, J. F. and Feyereisen, R. (2000).Inducible P450s of the CYP9 family from larval *Manduca sexta* midgut.*Insect Biochemistry and Molecular Biology*. 30: 559 – 568.

Taize, L. and Zeiger, E. (1995). In: Plant Physiology Edition.

Terras, F. R., Eggermont, K., Kovaleva, V., Raikhel, N. V., Osborn, R. W., Kester, A., Rees, S. B., Torrekens, S., Van – Leuven, F. and Vanderleyden, J. (1995). Small cysteine – rich antifungal proteins from radish: their role in host defense. *Plant Cell*. 7: 573 – 588.

Thomma, B. P. H. J., Cammue, B. P. A. and Thevissen, K. (2002).Plant defenses. *Planta*. 216(2): 193 – 202.

Turlings, T. C. J., Loughrin, J. H., Mccall, P. J., Roese, U. S. R., Lewis, W. J. and Tumlinson, J. H. (1995). How caterpillar damaged plants protect themselves by attracting parasitic wasps. *Proceedings of the National Academy of Sciences*. 92 (10): 4169 – 4174.

Van Etten, H. D., Mansfield, J. W., Bailey, J. A., Farmer, E. E. (1994). Two classes of plant antibiotics: Phytoalexins versus "phyto anticipins". *Plant Cell*. 6: 1191 – 1192.

Vanveen, F. J. F. (2015). Plant – modified trophic interactions. *Current Opinion in Insect Science*. 8: 29 – 33.

War, A. R. and Sharma, H. C. (2014). Induced resistance in plants and counter – adaptation by insect pests.*Short views on Insect Biochemistry and Molecular Biology*. 1 – 16.

Wink, M. (1988). Plant breeding: Importance of plant secondary metabolites for protection against pathogens and herbivores. *Theoretical and Applied Genetics*.75(2): 225 – 233. doi: http: //doi.org/10.1007/BF00303957.

Wink, M. (2003). Evolution of secondary metabolites from an ecological and molecular phylogenetic perspective.*Phytochemistry*. 64: 3 – 19. doi: http: //doi.org/10.1016/S0031-9422(03)00300-5.

Wink, M. (2007). Molecular modes of action of cytotoxic alkaloids from DNA intercalation, spindle poisoning, topoisomerase inhibition to apoptosis and multiple drug resistance.*Alkaloids*. 64: 1 – 48.

Wink, M. (2008). Evolutionary advantage and molecular modes of action of multicomponent mixtures used in phytomedicine.*Current Drug Metabolism*. 9: 996 – 1009. doi: https: //doi.org/10.2174/138920008786927794.

Wink, M. (2010). Introduction: Biochemistry, physiology and ecological functions of secondary metabolites. *Annual Plant Reviews*. 40: 1 – 19.

Wink, M. (2012). Medicinal plants: a source of anti – parasitic secondary metabolites. *Molecules*. 17: 12771 – 12791.

Wink, M. (2015). Modes of action of herbal medicines and plant secondary metabolites. *Medicines*. 2: 251 – 286. doi: https: //doi.org/10.3390/medicines2030251.

Wink, M. (2018). Plant secondary metabolites modulate insect behaviour steps towards addiction? *Frontier Physiology*.9: 364.doi: https: //doi.org/10.3389/fphys.2018.00364

Wittstock, U., Agerbirk, N., Stauber, E. J., Olsen, C. E., Hippler, M., Mitchell – Olds, T., Gershenzon, J. and Vogel, H. (2004).Successful herbivore attack due to metabolic diversion of a plant chemical defense.*Proceedings of National Academic Sciences*.101: 4859 – 4864.

Wu, J. Q., Hettenhausen, C., Meldau, S. and Baldwin, I. T. (2007). Herbivory rapidly activates MAPK signaling in attacked and unattacked leaf regions but not between leaves of *Nicotianaattenuata*. *Plant Cell*. 19: 1096 – 1122.

Wuyts, N., Waele, D. D. and Swennen, R. (2006).Extraction and partial characterization of polyphenol oxidase from banana (*Musa acuminate* - granaine) roots.*Plant Physiology and Biochemistry*. 44: 308 – 314.

Ylstra, B., Touraev, A., Moreno, R. M., Stoger, E., Van, T. and Vicente, O. (1992). Flavonols stimulate development, germination, and tube growth of tobacco pollen. *Plant Physiology*. 100: 902 – 907. doi: http: //doi.org/10.1104/pp.100.2.902.

Yu, S. J. (2000). Allelochemical induction of hormone – metabolizing microsomal monooxygenases in the fall armyworm.*Zoology Studies*. 39(3): 243 – 249.

Chapter 23

An Insight into Science of Insect Neuropeptide and its Role in Insect Pest Management: A Review

***Pankaj Sharma*[1] *and Pratibha Sharma*[2]**

[1]Department of Entomology, Dr. Yashwant Singh Parmar University of Horticulture and Forestry, Nauni, Solan, Himachal Pradesh
[2]Department of Plant Pathology, CSK Himachal Pradesh Agricultural University, Palampur, Himachal Pradesh

Neuropeptides are the neurohormones which are synthesised in the neurons or neuro endocrine cells and are released in the haemolymph. They coordinate physiological functions like mating, oviposition, moulting, water balance, fat mobilization and also act as neuromodulators in circuits of the central nervous system. Neurohormone and neuropeptide differ in the fact that the former is released in the neuro-muscular or neuro-neuro junction while the later in the haemolymph. Most of the neuropeptide receptors belong to the G-protein coupled receptors (GPCRs). However, there are some exceptions like the prothoracicotropic hormone (PTTH) which acts through the activation of receptor tyrosine kinase (RTK). The first neuropeptide isolated was proctolin from cockroach (Periplanata americana L.) in 1975. Currently, there are about 4782 insect neuropeptides grouped in 54 insect neuropeptide families, covering 23 insect orders. Among them greater emphasis is given to four neuropeptides (proctolin, kinin, pheromone biosynthesis activating neuropeptide-PBAN and allatostatin). Advanced mechanisms like synthetic ligands and peptide analog mimetics [BackBone Cyclic Neuropeptide-Based Antagonist (BBC-NBA)] or RNAi induced gene silencing can help in utilising neuropeptides for pest control. The target specific and eco-friendly nature of neuropeptides unfold their potential in the management of insect pests in an effective manner.

Keywords: *Neuropeptides, PBAN, Proctolin, Kinin, Allatostatin.*

Introduction

To meet the requirements of growing population in India green revolution started which resulted in the use of toxic chemicals and with time insects too gained resistance to overcome these toxic substances. Continuous use of the toxic substances in turn resulted in the degradation of environment. To overcome those difficulties, strategies based on integrated pest management were approached which include the characteristics like **insect specificity, non-toxicity, compatibility with other insect control agents,** *etc.* Also , tactics such as use of bio control agents and transgenic plants is gaining importance. All these methods are effective but either there application is limited or insect-pests can find some ways to escape the effects of these control measures and cause crop damage. Neuropeptide- based pest control is an innovative and novel approach for pest management which possess great potential to address several problems of the current pest management strategies.

Neuropeptides are the neurohormones which are synthesised in the neurons or neuro endocrine cells and are released in the haemolymph. Complex physiological functions like **moulting, mating, fat mobilization, oviposition and water balance** ***etc.*** are controlled directly or indirectly by neuropeptides (Yeoh *et al.*, 2017). The larger precursor proteins from which neuropeptides get synthesized are called **Prepropeptides**. These comprise of a **signal peptide** (which directs the protein to the secretary pathway), **spacer peptides** (peptide fragments with no known biological function and non-conserved sequences), **progenitors of mature peptides** (the biologically active peptides) and **cleavage sites** (monobasic and dibasic) (Yeoh *et al.*, 2017).

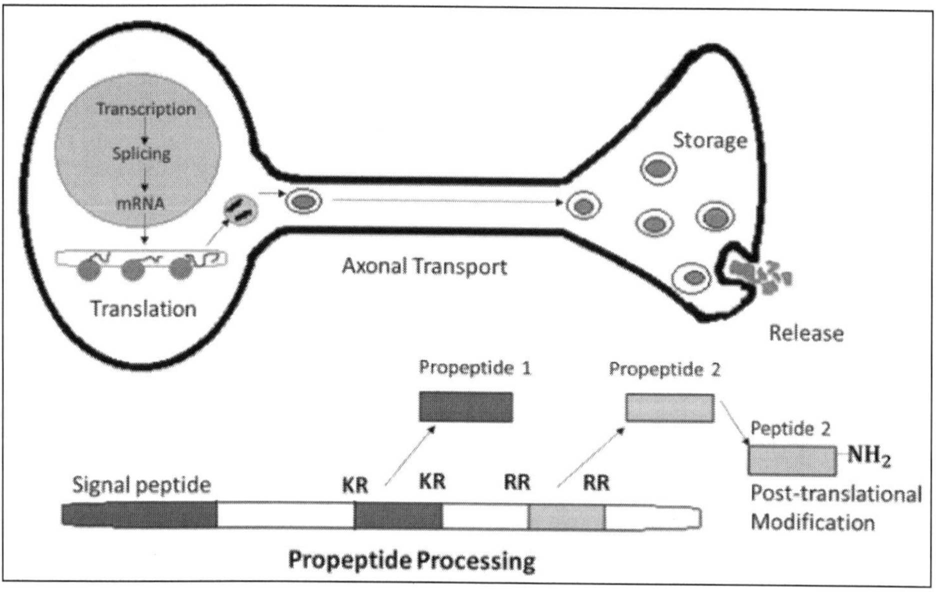

Figure 23.1

The prepropeptides (precursor) can produce more than one bioactive peptide. The genes for neuropeptide production are transcribed in the nucleus and after splicing, mRNA is translated on ribosome followed by the signal peptide aiding the incorporation of immature prepropeptide in the secretory pathway which finally ends up in the vesicle. The vesicles are then transported to the axon termination and simultaneously the precursor is also processed. Each bioactive peptide is surrounded by mono/dibasic **cleavage sites like KR or KK**, which administer the peptidases to release the peptides. The **white boxes depict the non-conserved sequences (spacing regions) between** peptide progenitors. In the final step post translational modifications like C-terminal amidation ($-NH_2$), as shown in the diagram, may take place. These fully-fledged neuropeptides are then deposited in dense core vesicles in the axon termination. The release of these peptides is triggered by the depolarization of axon termination accompanied by Ca^{2+} influx.

[This figure was redrawn and modified from Nassel and Larhammar (2013)]

Naming and Classification of Neuropeptides

The literature on naming of neuropeptides is disconcerting. Conventionally, the naming of neuropeptides in many cases was done according to their preliminarily reported functions. However, neuropeptides can have multiple functions and therefore, same family is likely to attract several names. For example, the first allatostatin identified from **cockroach** were named after their inhibitory action on juvenile hormone production (Woodhead *et al.*, 1989). But later similar neuropeptides with alike inhibitory properties and slightly different functional groups were discovered in cricket by Lorenz *et al.* (1995). This new group was named **cricket** type allatostatin or allatostatin B (AstB), and the allatostatin from **cockroaches** were designated allatostatin A (AstA) or FGLa type allatostatin (FGLa/AST). However, recently, myoinhibitory properties were discovered in AstB type orthologs of the migratory **locust**, *Locusta migratoria* and were therefore renamed as Myoinhibitory Peptide (or MIP) (Schoofs *et al.*, 1991).

Although there is a huge amount of 'roughly arranged data' on insect neuropeptides, their nomenclature and functions still there is a dearth of exclusive source for insect neuroendocrinology. The currently available databases either focus more on amino acid sequences from higher animals or lack the attempts to capture functional interpretation. Here are some of the available online data sources which throw some light on the science of neuropeptides:

- ☆ **'Neuropeptides'** (http://www.neuropeptides.nl) is a single table resource, which lists information on gene families, mainly for **vertebrate neuropeptides** (Burbach, 2010).
- ☆ **'NeuroPedia'** (http://proteomics.ucsd.edu/Software/NeuroPedia) is a downloadable database of neuropeptide sequence and spectra data libraries which aid in identifying neuropeptides from mass spectrometry, available at (Kim *et al.*, 2011). However, it lacks information about **arthropod neuropeptides.**
- ☆ **'NeuroPep'** (http://isyslab.info/NeuroPep/ home.jsp), contains 21 **arthropod neuropeptide** families, less than half those known (Wang *et al.*, 2015).

- **DINeR** (http://www.neurostresspep.eu/diner/) is a web-based database application dedicated to **insect neuropeptide** information of multiple insect species detailing their peptide sequences, physiological functionality and images of their receptor-binding sites.
- For standardization of neuropeptide classification, they were classified into different families on the basis of homology of amino acid sequence. This nomenclature proposed by Coast and Schooley (2011) is adopted by DINeR (Database for Insect Neuropeptide Research) for naming insect neuropeptide families.
- **Families of neuropeptide**

  Presently, about 54 insect neuropeptide families are identified covering 23 insect orders. Four of (**Proctolin, kinin, pheromone biosynthesis activating neuropeptide and allatostatin)** these 54 neuropeptide families are studied in more detail for their bioassay activity against various insects.

Table 23.1: Families of Neuropeptide

Families of Insect Neuropeptide	*Isolated from*	*Function*	*References*
Adipokinetic hormone	*Locusta migratoria*	May play a part in development and ecdysis	Siegert, 1999
Anti-diuretic Factor	*Tenebrio molitor*	Inhibit fluid secretion in Malpighian tubules	Eigenheer *et al.*, 2002.
Allatostatin A,B,C	Cockroach, Cricket, Moth	Inhibits JH synthesis	Woodhead *et al.*, 1989; Lorenz *et al.*, 1995.
Allatotropin	*Manduca sexta*	Stimulates JH biosynthesis	Kataoka *et al.*, 1989
Bursicon	*Drosophila melanogaster*	Cuticle tanning	Luo *et al.*, 2005; Mendive *et al.*, 2005
Capability	*Manduca sexta*	Impacts desiccation and cold stress tolerance	Huesmann *et al.*, 1995
Crustacean Cardio-Active Peptide	*Locusta migratoria*	Initiating the ecdysis	Stangier *et al.*, 1989
CCHamide	*Bombyx mori*	Increased the motivation to feed	Roller *et al.*, 2008
Pigment-dispersing factor	*Romalea microptera*	Pigment movements in response to light	Rao and Riehm., 1988
Corazonin	*Periplaneta americana*	Initiating ecdysis	Veenstra, 1994
Diuretic Hormone 31, 44	*Diploptera punctata, Manduca sexta*	Fluid secretion in MTs	Furuya *et al.*, 2000.
Eclosion hormone	*Manduca sexta and Bombyx mori*	Ecdysis behavior	Kataoka *et al.*, 1987.
Ecdysis-triggering hormone	*Manduca sexta*	Triggers ecdysis	Zitnan *et al.*, 1996
FMRFamide	*Drosophila melanogaster*	Ecdysis, myostimulatory in action	Nambu *et al.*, 1988.
GP2, 5	*Drosophila melanogaster*	Anti-diuresis	Hsu *et al.*, 2002.

Families of Insect Neuropeptide	Isolated from	Function	References
Insulin-like Peptide	*Bombyx mori*	Growth, metabolism and reproduction.	Nagasawa *et al.*, 1986
Kinin	*Leucophaea maderae*	Myotropic, diuretic activities.	Holman *et al.*, 1986
Limostatin	*Drosophila melanogaster*	Regulate production and release of DILPs in Drosophila	Alfa *et al.*, 2015
Myosuppressin	*Leucophaea maderae*	Inhibit heart and visceral muscle	Holman *et al.*, 1986
Neuropeptide F	*Drosophila*	Helps in foraging and feeding	Brown *et al.*, 1999
Neuropeptide-like precursor	*Drosophila*	Role in development	Baggerman *et al.*, 2002, 2005
Natalisin	*Drosophila*	Plays role in reproduction	Jiang *et al.*, 2013
Orcokinin	*Bombyx mori*	Play roles in gut function.	Yamanaka *et al.*, 2011
Pheromone Biosynthesis Activating Neuropeptide (PBAN)	*Leucophaea maderae*	Regulating pheromone biosynthesis	Holman *et al.*, 1986
Diapause hormone	*Leucophaea maderae*	Regulation of insect diapause	Holman *et al.*, 1986
Pyrokinin	*Leucophaea maderae*	Myostimulatory activity and regulates hormone biosynthesis	Holman *et al.*, 1986
Proctolin	*Periplaneta americana*	Stimulate muscle contractions	Starratt and Brown, 1975
Prothoracicotropic hormone	*Bombyx mori*	Regulates molts and metamorphosis	Kataoka *et al.*, 1991
Sulfakinin	*Leucophaea maderae*	Induces hyperactivity and aggression	Nachman *et al.*, 1986
Short neuropeptide F	*Drosophila*	Roles in feeding, growth.	Nässel and Wegener 2011.
Vasopressin	*Locusta migratoria*	Triggers diuresis	Proux *et al.*, 1987
Trissin	*Drosophila melanogaster*	Regulation of foregut-midgut contractions and food intake	Ida *et al.*, 2011
SIFamide	*Sarcophaga bullata*	Involved in control of sexual behaviour	Janssen *et al.*, 1996
Tachykinin-related peptide	*Locusta migratoria*	regulation of release of adipokinetic hormone from the corpora cardiac	Schoofs *et al.*, 1990
Calcitonin	*Locust and termite*	Not known	Veenstra, 2014
CNMamide	*Drosophila melanogaster*	Function not available	Jung *et al.*, 2014
RY amide	*Nasonia vitripennis*	Not known	Hauser *et al.*, 2010

Important Families of Neuropeptide

Proctolin

In locusts, proctolin is produced by motor neurons and in *Drosophila,* it influences corpora cardiaca centred crawling behaviour (Clark and Undem, 2006). The biologically active sequence of proctolin is arginine, tyrosine, threonine and leucine. Slight modification between Tyr2 and Leu3 (by replacing just an amino group with $O_{2)}$ was found to inactivate the muscle contraction in *Locusta migratoria* at a very low concentration (1 mmol/L). While modification between Arg1 and Tyr2 (replacing an amino group with $O_{2)}$ was helped to retain the activity of muscle contraction significantly. On the other hand, studies (Scherkenbeck and Zdobinsky, 2009) have shown that cyclization of proctolin (Cycloproctolin) help in formulating a potent antagonist of proctolin-induced production of the second messengers InsP3 (insulin P3) and InsP4 (insulin P4).

Kinins

The first member of the kinin family was isolated from *Leucophaea maderae* also found to be present in nematodes, annelids and molluscs (Radford *et al.,* 2018) where it triggers ecdysis behaviour. It has a common C- terminal pentapeptide sequence *i.e.,* Phe-Phe-Aib-Trp-Gly-NH_2.

Pyrokinins/Pheromone Biosynthesis Activating Neuropeptides (PBANs)

These are the peptides of pyrokinin type regulating pheromone biosynthesis and are named after their functions *i.e.,* Pheromone Biosynthesis Activating Neuropeptides (PBAN), Diapause Hormone (DH), Melanization and Reddish Coloration Hormone (MRCH) and so on. DH and PBAN were found primarily in drosophila and Lepidoptera. This group of neuropeptides have been extensively explored to develop biostable analogs to be used in insect pest control (Raina and Menn, 1993). Phe-Ser-Pro-Arg-Leu-NH_2 is the biologically active c-terminal pentapeptide sequence of PBAN and found homologous to *Mythimna sapareta* as pheromonotropin and to *Bombyx mori* as diapause hormone. H-Tyr-Phe-Ser-Pro-Arg-Leu-NH_2 is the c-terminal hexapeptide, active sequence of PBAN. In a study the peptide H-Arg-Tyr-Phe-D-Phe-Pro-Arg-Leu-NH_2 was observed to exhibit the highest antagonistic activity and it caused jitters action in insect. Administration of this peptide at 100 pmol, inhibited sex pheromone biosynthesis by 63 per cent after 2h (Gilon *et al.,* 1997). Further, substitution of l-amino acids with d-Phe followed by backbone cyclization led to the discovery of several highly potent linear and selective, conformationally constrained, metabolically stable backbone cyclic (BBC) pure antagonists for PBAN.

Allatostatin –A, B, C

The first member of the insect AST family was isolated in the year 1989 from brain extracts of the cockroach *Diploptera punctata* (Woodhead *et al.,* 1989). AST is found to be three types. **AST-A as cockroach type, AST- B as cricket type and AST C as moth type**. The name allatostatin is derived from their ability to inhibit juvenile hormone (JH) biosynthesis by the corpora allata. It also regulates various aspects of feeding and metabolism in several species (Yeoh *et al.*, 2017). Allatostatins contains 8-13 amino acids and are amidated. The biologically active peptide sequences of allatostatin are Ala-Pro-Ser-Gly-Ala-Gln-Arg-Leu-Tyr-Gly-Phe Gly-Leu- NH_2.

Different Neuroreceptors

The various physiological functions performed by neuropeptides are exerted by interaction with specific receptors, termed as **G-protein coupled receptors-GPCRs** (Most of the neuropeptides belong to this family *i.e.*, the largest family of cell surface proteins) and the receptors of respective neuropeptides like **AKH receptor, allatotropin receptor, allatostatin receptor** ***etc.*** Some common neuroreceptors are:

- ☆ G Protein-Coupled Receptors
- ☆ Adipokinetic Hormone Receptors
- ☆ Allatostatin Receptors
- ☆ Capa Receptors
- ☆ Kinin (Mykonin) Receptors
- ☆ Proctolin Receptors
- ☆ Pyrokinin Receptors

Utilising Insect Neuropeptides in Pest Control

Neuropeptides are important to carry out various physiological activities in insects, modification of these neuropeptides paves the way for the novel pest control strategy. The insect specific and environmentally safe nature of neuropeptides, enable their use as a component in integrated pest management and give an alternative way of pest management reducing their dependency on toxic pesticides (Karuppaiah and Sujithra, 2013). However, the thorough knowledge of specific protein sequence in relation to its function and isolation is essential to exploit this area. Therefore, by targeting the precursors, or by employing peptidase induced degradation highly insecticide agent can be produced which can fit well in pest management programs. According to the common functions performed by neuropeptides their role in pest management and possible future functionalities are summarised as-

Growth and Development

In developing new active compounds, in preventing the production or release of moulting hormone, *ecdysone* the *allatoregulatory* (*allatotropins*/AT/ and *allatostatins*/ AST/) have a great potential. The endocrine gland corpora allata (CA) does not store **juvenile hormone** in appreciable amounts and the titre of the hormone in the haemolymph is determined by the rate of biosynthesis or inhibition in corpora allata, along with the activity of JH-esterases (JHE). Since JHs are not only crucial during the fine tuning of development, but in adult females they are responsible for vitellogenesis for example, therefore allatoregulation is important throughout insect life. Till date only few ATs, compared with several ASTs have been isolated from nervous tissues of various species. Furthermore, CA of the beetle, *Tenebrio molitor*, the locust, *Schistocerca gregaria* and the cockroach *P. americana* were not activated by synthetic AT, whereas the CA of the moth, *H. virescens* were stimulated, suggesting order specificity. Studies on the amino-terminal truncated fragments of synthetic peptide indicate that the 6–13 amino acids constitute the biological active core. It was also found that isolated peptides structurally belonging to ATs showed other *pleiotropic* effects including *myotropic* activity.

Chemists have developed special pseudopeptides introducing sterically bulky components like indane ring or cyclopropyl, into the active core region to make them resistant to both haemolymph as well as tissue bound catabolic peptidases, while maintaining the biological activities (Nachman *et al.*, 1999).

Reproduction

The use of synthetic species blending several hundred known pheromones may be effective in agriculture (forecasting, reduction by trapping, *etc.*). Also, the employment of latest results regarding the delicate mechanisms controlling pheromone production (by PBAN, occasionally supplemented by JHs and *ecdysone*) has a great potential. The interest towards PBANs and other neuropeptides bearing the common c- terminal pentapeptide FXPRLamide (active sequence) like DH or *myotropic* PKs is well established. Synthetic PBAN (or other PKs) injected *in vivo* into females maintain their activity and induce intensive pheromone production, therefore designing and production of PBAN/PK analogs, antagonists/agonists, which may be practically very promising **(**Fonagy A., 2006**)**.

In some *Helicoverpa* species (Rafaeli, 2002) the involvement of cAMP in inducing the rate limiting enzyme prior to fatty acid synthesis is strongly supported. To date, however, no convincing data on cAMP dependent protein kinases have been presented, therefore the complete model has not yet been established.

Based on recent advances there are two major concepts in development of analogs or mimetics of PBANs for insect control.

- The first strategy is based on the analogy of Substance *P* where the backbone cyclisation to impose conformational constraint on peptide was implemented (Gilon *et al.*, 1991). The method involves the cyclisation of the active hexapeptide of Hez-PBAN28–33 (YFSPRLamide) called Backbone cyclic neuropeptide-based antagonist (BBC-NBA) approach (Altstein *et al.*, 2000).

Backbone cyclization of PBAN

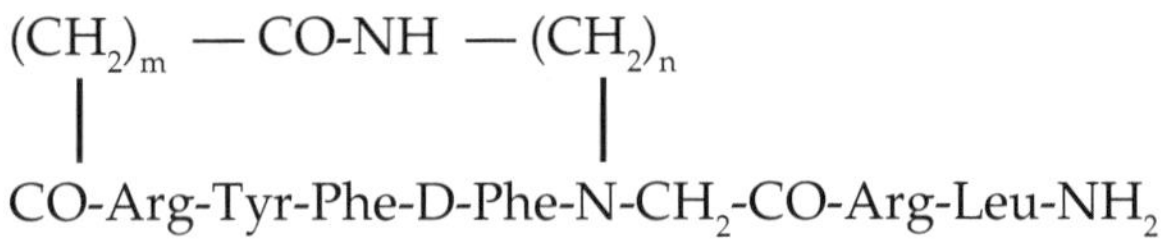

- According to the second strategy the development of PBAN/PK pseudopeptides is a good alternative (Nachman *et al.*, 1996a). The basic concept is to provide a protection for the synthetic active core peptide, which allows easy penetration of the polar molecule through the apolar matrix of the cuticle, while it resists attacks of endopeptidases in the gut and haemolymph.

Metabolism and Homeostasis

A number of different AKH mimetics have been designed and synthesized like the ones where the C (Lee *et al.*, 1996) or N termini were modified (Lee *et al.*, 1997), or it can be achieved by replacing just one amino acid (Ziegler *et al.*, 1998). The regulation of the water and ion balance of insect body for haemolymph homeostasis and interference with these processes is particularly an attractive as well as effective target. Agonists of *diuretic* peptides (DP) or antagonists of *antidiuretic* hormones could induce dehydration leading to death. Both act on the main excretory organ, the Malpighian-tubules producing the primary filtrate. The CRF-DP acts via the cAMP signal transduction cascade stimulating urine secretion.

Muscle Movement

Proctolin, the first identified neuropeptide (Starratt and Brown, 1975) and the *cardioactivating peptides* (CAP) stimulating heartbeat frequency compose a separate family and both are involved in muscle movement regulation. Studies related to the pentapeptide *proctolin* and possible analogues are still in the focus of interest. Konopiñska (1997) summarized the results derived from analysis (*in vitro* insect and *in vivo* rat toxicology) of about 80 different molecules in which amino acids were replaced in 1–5 position and cycloanalogues and analogues with a truncated or elongated structure were synthesized. The *proctolin* receptor has also been characterized pharmacologically, but its isolation and cloning are still lacking. All the above results however, have theoretical significance, are so far suitable, stable and easily penetrating analogues have yet not been found.

Besides the *myoinhibitory* peptides (MIP) from *Locusta*, the majority of inhibitors belong to large family of *FMRFamide related peptides* (FaRPs) which is characterized by its C-terminus (Orchard *et al.*, 2001). A non-peptide *myoinhibitor* antagonist, the benzethonium chloride (Bztc) is capable to mimic physiological activity of *myosuppressins* (MS) belonging to FaRPs (Nachman *et al.*, 1996b).

Isolation and modification of neuropeptides to obtain desired antagonistic activity:

- First **Sequencing of protein** is done by isolating the specific gene of interest using PROCHECK-NMR and AQUA (Laskowski *et al.*, 1996).
- Then **artificial synthesis of peptide** is done by the condensation of the carboxyl group of one amino acid to another amino acid group using Fmoc (fluorenylmethoxycarbonyl) resin (Shin *et al.*, 1999).
- After isolation, the ***in vitro* and *in vivo* protein docking** is done using various softwares like DIVALI, ADAM, AutoDock, DOCK, DockVision, EUDOC, FlexX, FLOG, FTDOCK, SANDOCK, GOLD, Hammerhead, ICM, LUDI, MCDOCK, Prodock, Proleads, QXP *etc.* (Sousa *et al.*, 2006).
- Structural modification of synthesised protein is done in order to obtain antagonist activity either by **Linear replacement/side chain modification (Figure 23.2) or by backbone cyclization (Figure 22.3)** (Gilon *et al.*, 1997).

☆ The **final product thus formed is in the form of dry powder** (Alstein, 2003).

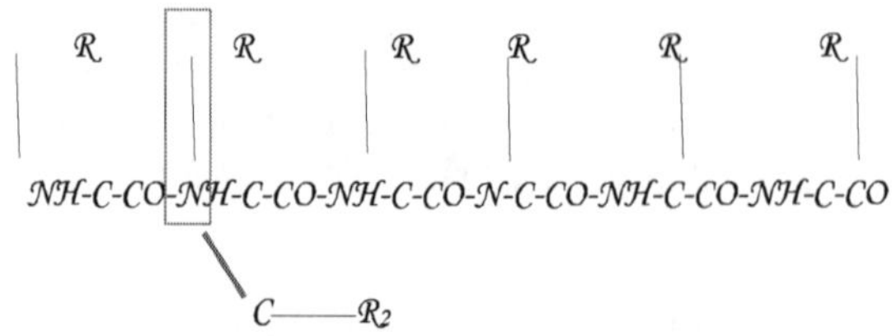

Figure 22.2: Linear Replacement of a Peptide Chain.

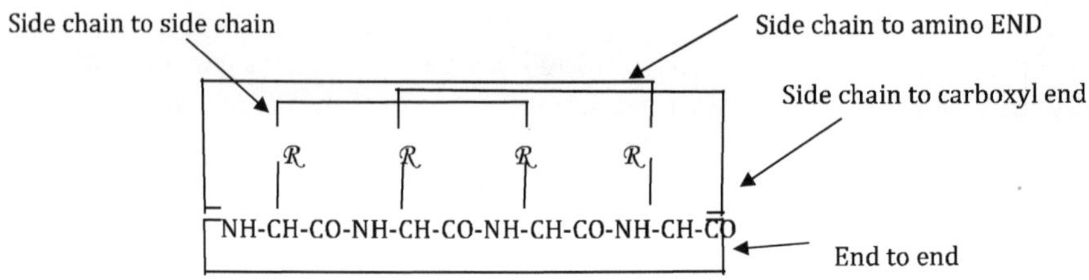

Figure 23.3: Backbone Cyclization of a Peptide Chain

Backbone cyclic neuropeptide-based antagonist (BBC-NBA) approach (Altstein *et al.*, 2000).

1. Identification of the neuropeptide that controls the required function
2. Elucidation of the shortest sequence which constitutes the active site
3. Discovery of peptidic linear lead agonist on the basis of the sequence found in (2)
4. Discovery of a potent BBC antagonist devoid of agonistic activity based on (3)
5. Determination of the structural requirements for antagonistic activities as determined in (4);
6. Designing and synthesizing an insecticide prototype (*i.e.*, small, metabolically and structurally stable, with selective bioavailability, cost-effective peptidomimetic compound).

BBC-NBA approach for generation of antagonists and insecticide prototypes.

HTSA: High throughput screening assay; **Np**: neuropeptide; **SAR**: structure-activity relationship; **SML**: small molecules

RNAi induced gene silencing (using modified dsRNA)

It is a biological process involving introduction of small RNA molecules to silence gene expression at the transcriptional or post-transcriptional level. This can occur by directing chromatin modifications, affecting the stability or translation

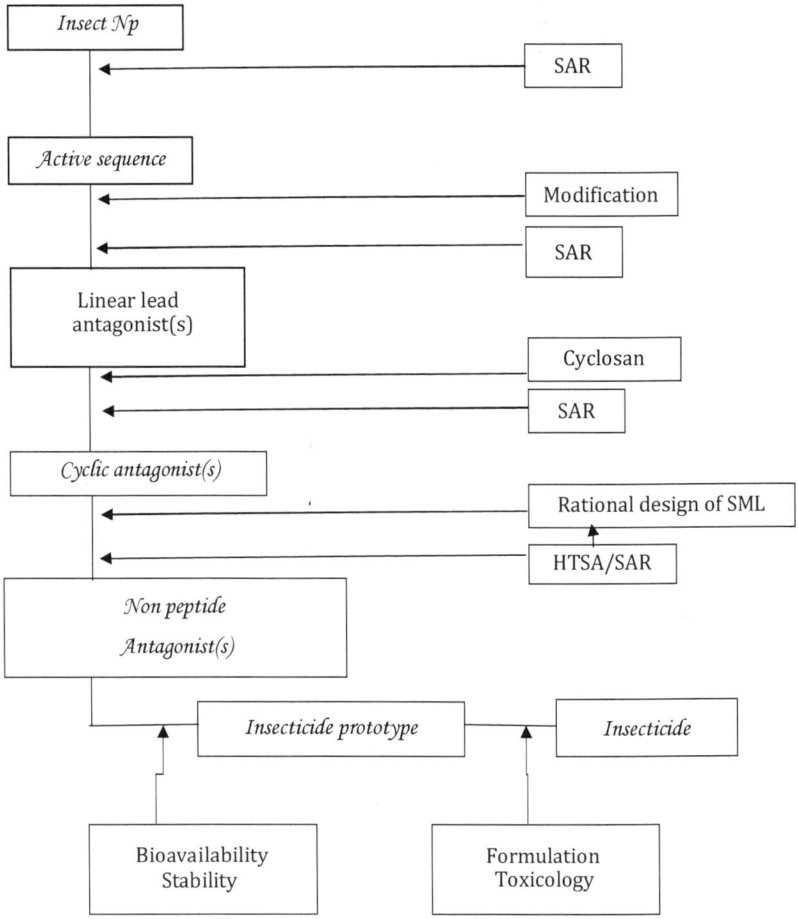

potential of the targeted mRNA. A double -stranded RNA (dsRNA) is used as a trigger for gene silencing (Liu *et al.*, 2020). In 2006, studies suggested that insect pests are sensitive to oral administration of dsRNA (Turner *et al.*, 2006; Araujo *et al.*, 2006). Since then, much attention has been focused on developing dsRNA application strategies for insect control in field test conditions such as genetically modified organism (GMO) based strategies, in which a dsRNA that silences a critical insect gene is produced as a result of ingestion and direct application of dsRNA onto plants (Rosa *et al.*, 2018; Gordon and Waterhouse, 2007).

RNAi induced gene silencing is a novel approach and holds great potential for use in pest control, as by gene silencing the neuropeptide production or function can be altered. Time-course analysis indicates that the RNAi effect is highly persistent in some insects, lasting for weeks in *Tribolium* and for multiple generations in grain aphids (Bucher *et al.*, 2002; Abdellatef *et al.*, 2015). However, in these insects, as well as in green peach aphid and milkweed bug, the strength of pRNAi (parental RNAi) decreases over time and with successive generations (Liu and Kaufman,

2004; Coleman *et al.,* 2015). Currently, the mechanism of action of pRNAi is not well understood more research is needed in this field to utilize this novel technique efficiently in pest control.

Conclusion

The usefulness of research on insect neuropeptide signalling systems goes beyond the world of insects as several mammalian neuropeptides and/or their receptors have orthologs in insects. It is clear that the neuropeptides and their receptors in insect physiology is immensely important. The variety among the neuropeptides and their receptors between species makes them potential targets for the development of a new generation of insecticides with high species specificity. Insecticides developed in this way would ideally only be harmful for target pest while beneficial insects would remain unharmed. The development of such novel controlling agents is need of the hour because of the detrimental effects of the currently used insecticides on the environment and their toxicity to non-target species. The increasing resistance of pest insects against the used insecticides is also an expanding problem. As such, more and more research work is performed on neuropeptides in order to develop synthetic ligands that can disturb the proper functioning of neuropeptide signalling systems, provoking detrimental effects on the insect's fitness.

Physiological functions of peptides can be manipulated with use of antagonist or inhibitors of biosynthetic enzymes. These could be achieved at any level of synthesis, release and biological activity. Further, diverse nature of neuropeptides and its receptors among the insect species provides enormous scope to use them as potential targets for development of new insecticides with specific selectivity. These selective insecticides are environmentally safer and thus, play an alternative to the broad-spectrum chemicals, where insecticide resistance is a challenge. Neuropeptides signalling system can be disturbed by developing synthetic ligands and peptide analog mimetics and pesticides, which interfere with their regular functions. Employing peptidase induced degradation and down regulation of synthesis and receptors using RNAi induced gene silencing of specific neuropeptides would give some new area in this field. However, it needs thorough knowledge of synthesis pathways which regulates the physiological functions, characterization of receptor sites and peptide signalling systems. Presently no neuropeptide-based insecticides are in commercial use, but a lot of progress is made on a number of interesting neuropeptides. But insect neuropeptides and its receptors are for sure the candidates to develop the selective and effective insecticide, which would fit well in future pest management- programmes.

REFERENCE

Abdellatef E, Will T, Koch A, Imani J, Vilcinskas A and Kogel KH. 2015. Silencing the expression of the salivary sheath protein causes transgenerational feeding suppression in the aphid *Sitobion avenae. Plant Biotechnology Journal* 13: 849 -57.

Alfa RW, Park S, Skelly KR, Poffenberger G, Jain N, Gu X, Kockel L, Wang J, Liu Y, Powers AC and Kim SK. 2015. Suppression of insulin production and secretion by a decretin hormone. *Cell metabolism* 21: 323-333.

Alstein M. 2003. Novel Insect Control Agents Based on Neuropeptide Antagonists. *Journal of Molecular Neuroscience* 147-157pp.

Altstein M, Ben-Aziz O, Schefler I, Zeltser I and Gilon C. 2000. Advances in the application of neuropeptides in insect control. *Crop Protection* 19: 547–555.

Araujo RN, Santos A, Pinto FS, Gontijo NF, Lehane MJ and Pereira MH. 2006. RNA interference of the salivary gland nitrophorin 2 in the triatomine bug *Rhodnius prolixus* (Hemiptera: Reduviidae) by dsRNA ingestion or injection. *Insect Biochemistry and Molecular Biology* 36: 683 -93.

Baggerman G, Boonen K, Verleyen P, De Loof A and Schoofs L. 2005. Peptidomic analysis of the larval *Drosophila melanogaster* central nervous system by two-dimensional capillary liquid chromatography quadrupole time-of-flight mass spectrometry. *Journal of mass spectrometry* 40: 250-260.

Baggerman G, Cerstiaens A, De Loof A and Schoofs L. 2002. Peptidomics of the larval *Drosophila melanogaster* central nervous system. *The Journal of biological chemistry.* 277: 40368-40374.

Brown MR, Crim JW, Arata RC, Cai HN, Chun and Shen P. 1999. Identification of a *Drosophila* brain-gut peptide related to the neuropeptide Y family. *Peptides* 20: 1035-1042.

Bucher G, Scholten J and Klingler M. 2002. Parental RNAi in *Tribolium* (Coleoptera). *Current Biology* 12: R85 -6.

Burbach JPH. 2010. Neuropeptides from concept to online database. *European Journal of Pharmacology* 626: 27-48.

Clark TT and Undem BJ. 2006. Transduction mechanisms in airway sensory nerves. *Applied Physiology* 101: 950–959.

Coast GM and Schooley DA. 2011. Toward a consensus nomenclature for insect neuropeptides and peptide hormones. *Peptides* 32: 620-31

Coleman AD, Wouters RHM Mugford ST and Hogenhout SA. 2015. Persistence and transgenerational effect of plant -mediated RNAi in aphids. *Journal of Experimental Botany* 66: 541 -48.

Eigenheer RA, Nicolson SW, Schegg KM, Hull JJ and Schooley DA. 2002. Identification of a potent antidiuretic factor acting on beetle Malpighian tubules. *Proceedings of the National Academy of Sciences* 99: 84-89.

Fonagy A. 2006. Insect Neuropeptides and their Potential Application for Pest Control. *Acta Phytopathologica et Entomologica Hungarica* 41, 137–152pp. DOI: 10.1556/APhyt.41.2006.1-2.14

Furuya K, Milchak RJ, Schegg KM, Zhang J, Tobe SS, Coast GM and Schooley DA. 2000. Cockroach diuretic hormones: Characterization of a calcitonin-like peptide in insects. *Proceedings of the National Academy of Sciences* 97: 6469–6474.

Gilon C, Halle D, Chorev M, Selinger Z and Byk G. 1991. Backbone cyclization: a new method for conferring conformational constraint on peptides. *Biopolymers* 31, 745–750pp.

Gilon C, Zeltser I, Daniel S, Aziz O, Schefler I and Alstein M. 1997. Rationally designes neuropeptide antagonists: A Novel Approach for Generation of Environmentally Friendly Insecticides. *Invertebrate Neuroscience*, 245-250pp.

Gordon KHJ and Waterhouse PM. 2007. RNAi for insect -proof plants. *Nature Biotechnology* 25: 1231 - 1232.

Hauser F, Neupert S, Williamson M, Predel R, Tanaka Y and Grimmelikhuijzen CJ. 2010. Genomics and peptidomics of neuropeptides and protein hormones present in the parasitic wasp *Nasonia vitripennis*. *Journal of proteome research* 9: 5296-5310.

Holman GM, Cook BJ and Nachman RJ. 1986. Isolation, primary structure and synthesis of leuco myosupressin, an insect neuropeptide that inhibits spontaneous contractions of the cockroach hindgut. *Comparative Biochemistry and Physiology* 85: 329-333.

Holman GM, Cook BJ and Nachman RJ. 1986. Isolation, primary structure and synthesis of two neuropeptides from *Leucophaea maderae*: members of a new family of cephalotropins. *Comparative Biochemistry and Physiology* 84: 205-211.

Holman GM, Cook BJ and Nachman RJ. 1986. Primary structure and synthesis of a blocked myotropic neuropeptide isolated from the cockroach, *Leucophaea maderae*. *Comparative Biochemistry and Physiology* C 85: 219-224.

Hsu SY, Nakabayashi K and Bhalla A. 2002. Evolution of glycoprotein hormone subunit genes in bilateral metazoa: identification of two novel human glycoprotein hormone subunit family genes, GPA2 and GPB5. *Molecular Endocrinology* 16: 1538–51.

Huesmann GR, Cheung CC, Loi PK, Lee TD, Swiderek KM and Tublitz NJ. 1995. Amino acid sequence of CAP2b, an insect cardio acceleratory peptide from the tobacco hawkmoth *Manduca sexta*. *FEBS Letters* 371: 311–4.

Ida T, Takahashi T, Tominaga H, Sato T, Kume K and Ozaki M. 2011. Identification of the novel bioactive peptides dRYamide- 1 anddRYamide-2, ligandsfor a neuropeptide Y- like receptor in *Drosophila*. *Biochem. Biophys.*

Janssen I, Schoofs L, Spittaels K, Neven H, Broeck J and Devreese B. 1996. Isolation of NEB-LF amide, a novel myotropic neuropeptide from the grey flesh fly. *Molecular and cellular endocrinology* 117: 157-165.

Jiang H, Lkhagva A, Daubnerova I, Chae HS, Simo L, Jung SH, Yoon YK, Lee NR, Seong JY, Zitnan D, Park Y and Kim YJ. 2013. Natalisin, a tachykinin-like signaling system, regulates sexual activity and fecundity in insects. *Proceedings of the National Academy of Sciences* 110: 3526-3534.

Jung SH, Lee JH, Chae HS, Seong JY, Park Y, Park ZY and Kim YJ. 2014. Identification of a novel insect neuropeptide, CNMa and its receptor. *FEBS Letters* 588: 2037–2041.

Karuppaiah V and Sujithra M. 2013. Insect Neuropeptides: As Potential Targets for Pest Control. *Double Helix Research* 10: 2250 –3668.

Kataoka H, Nagasawa H, Isogai A, Ishizaki H and Suzuki A. 1991. Prothoracicotropic hormone of the silkworm, *Bombyx mori*: amino acid sequence and dimeric structure. *Agricultural* and Biological *Chemistry* 55: 73-86.

Kataoka H, Toschi A, Li JP, Carney RL, Schooley DA and Kramer SJ. 1989. Identification of an allatotropin from adult *Manduca sexta*. *Science*. 243: 1481–3.

Kataoka H, Troetschler RG, Kramer SJ, Cesarin BJ and Schooley DA. 1987. Isolation and primary structure of the eclosion hormone of the tobacco hornworm, *Manduca sexta*. *Biochemical* and *Biophysical Research Communications* 146: 746–750.

Kim Y, Bark S, Hook V, Bandeira N. 2011. NeuroPedia: neuropeptide database and spectral library. *Bioinformatics* 27: 2772-2773.

Konopiñska D. 1997. Insect neuropeptide proctolin and its analogues. *The International Journal of Peptide Research and Therapeutics* 49: 457–466.

Laskowski RA, Rullmann JAC, Arthur MWM, Kaptein R and Thornton JM. 1996. AQUA and PROCHECK-NMR: Programs for checking the quality of protein structures solved by NMR. *Journal of Biomolecular NMR*. 8: 477-486.

Lee ML, Cusinato O, Luswata R, Wheeler CH and Goldsworthy GJ. 1997. N-terminal modifications to AKH-I from *Locusta migratoria*: assessment of biological potencies *in vivo* and *in vitro*. *Regulatory Peptides* 69: 69–76.

Lee ML, Goldsworthy GJ, Poulos CP and Velentza A. 1996. Synthesis and biological activity of adipokinetic hormone analogues modified at the C-terminus. *Peptides* 17: 1285–1290.

Liu PZ and Kaufman TC. 2004. Hunchback is required for suppression of abdominal identity, and for proper germband growth and segmentation in the intermediate germband insect *Oncopeltus fasciatus*. *Development* 131: 1515 -27.

Liu S, Jaouannet M, Dempsey DA, Imani J, Coustau C and Kogel KH. 2020. RNA-based technologies for insect control in plant production. *Biotechnology Advances* 39: 107463 DOI: 10.1016/j.biotechadv.2019.107463

Lorenz MW, Kellner R and Hoffmann KH. 1995. Identification of two allatostatins from the cricket, *Gryllus bimaculatus* de Geer (Ensifera, Gryllidae): additional members of a family of neuropeptides inhibiting juvenile hormone biosynthesis. *Regulatory Peptides* 57: 227–236.

Luo CW, Dewey EM, Sudo S, Ewer J, Hsu SY, Honegger HW and Hsueh AJW. 2005. Bursicon, the insect cuticle-hardening hormone, is a heterodimeric cystine knot protein that activates G protein-coupled receptor LGR2. *Proceedings of the National Academy of Sciences* 102: 2820–5. DOI: 10.1073/pnas.0409916102.

Mendive FM, Van Loy T, Claeysen S, Poels J, Williamson M, Hauser F, Grimmelikhuijzen CJP, Vassart G. and Broeck J. 2005. *Drosophila* molting neurohormone bursicon is a heterodimer and the natural agonist of the orphan receptor DLGR2. *FEBS Letters* 579: 2171–6. DOI: 10.1016/j.febslet.2005.03.006.

Nachman RJ, Garside CS and Tobe SS. 1999. Hemolymph and tissue bound peptidase-resistant analogs of the insect allatostatins. *Peptides* 20: 23–29.

Nachman RJ, Holman GM, Haddon WF and Ling N. 1986. Leucosulfakinin, a sulfated insect neuropeptide with homology to gastrin and cholecystokinin. *Science*. 234: 71-73.

Nachman RJ, Olender EH, Roberts VA, Holman GM and Yamamoto D. 1996b. A nonpeptidal peptidomimetic agonist of the insect FLRFamide myosuppressin family. *Peptides* 17: 313–320.

Nachman RJ, Teal PEA, Radel P, Holman GM and Abernathy RL. 1996a. Potent pheromonotropic/myotropic activity of a carboranyl pseudotetrapeptide analogue of the insect pyrokinin/PBAN neuropeptide family administered via injection or topical application. *Peptides* 17: 747–752.

Nagasawa H, Kataoka H, Isogai A, Tamura S, Suzuki A, Mizoguchi A, Fujiwara Y, Takahashi SY and Ishizaki H. 1986. Amino acid sequence of a prothoracicotropic hormone of the silkworm *Bombyx mori*. *Proceedings of the National Academy of Sciences* 83: 5840-5843.

Nambu JR, Murphy-Erdosh C, Andrews PC, Feistner GJ and Scheller RH. 1988. Isolation and characterization of a *Drosophila* neuropeptide gene. *Neuron*. 1: 55–61.

Nassel DR and Larhammar D. 2013. Neuropeptides and peptide hormones. In: Galizia, C.G., Lledo, P.M. (Eds.), Neurosciences - from Molecule to Behavior: A University Text Book. Springer, Berlin, pp. 213-237

Nässel DR and Wegener C. 2011. A comparative review of short and long neuropeptide F signalling in invertebrates: Any similarities to vertebrate neuropeptide Y signalling *Peptides*. 32: 1335-1355.

Orchard I, Lange AB and Bendena WG. 2001. FMRFamide-related peptides: a multifunctional family of structurally related neuropeptides in insects. *Advances in Insect Physiology* 28: 267–329.

Proux JP, Millert CA, Lit JP, Girardie RLA, Delaages M and Schooley DA. 1987. Identification of an arginine vasopressin-like diuretic hormone from *Locusta migratoria*. *Biochemical and biophysical*

Radford JC, Davies SA and Dow AT. 2018. Systematic GPCR Analysis in *Drosophila melanogaster*. *The American Society for Biochemistry and Molecular Biology* 8: 1-22.

Rafaeli A. 2002. Neuroendocrine control of pheromone biosynthesis in moths. *International Review of Cytology* 213: 49–92.

Raina AK and Menn JJ. 1993. Pheromone Biosynthesis Activating Neuropeptide: From Discovery to Current Status. *Archives of Insect Biochemistry and Physiology* 22: 141-151.

Rao KR and Riehm JP. 1988. Pigment-dispersing hormones: A novel family of neuropeptides from arthropods. *Peptides* 1: 153-159.

Roller L, Yamanaka N, Watanabe K, Daubnerova I, Zitnan D, Kataoka H and Tanaka Y. 2008. The unique evolution of neuropeptide genes in the silkworm Bombyx mori. Insect Biochem. *Molecular Biology*. 38: 1147–1157.

Rosa C, Kuo YW, Wuriyanghan H and Falk BW. 2018. RNA Interference mechanisms and applications in plant pathology. *Annual Review of Phytopathology* 56: 581-610.

Scherkenbeck J and Zdobinsky T. 2009. Insect Neuropeptides: Structures, Chemical Modification and Potential for Insect Control. *Bioorganic and Medical Chemistry* 17: 4071–4084.

Schoofs L, Holman GM, Hayes TK, Nachman RJ and De Loof A. 1990. Locust a tachykinin I and II, two novel insect neuropeptides with homology to peptides of the vertebrate tachykinin family. *FEBS letters* 261: 397-401.

Schoofs L, Holman GM, HayesTK, Nachman RJ and DeLoof A. 1991. Isolation, primary structure, and synthesis of locusta pyrokinin – amyotropic peptide of *Locusta migratoria. General and Comparative Endocrinology* 81: 97–104.

Shin Y, Winans KA, Backes BJ, Kent SBH, Ellman JA and Bertozzi CR. 1999. Fmoc-based synthesis of peptide-thioesters: Applicaion to the total chemical synthesis of a glycoprotein by native chemical ligation. *Journal of the American Chemical Society* 121: 11684-11689.

Siegert KJ. 1999. Locust corpora cardiaca contain an inactive adipokinetic hormone. *FEBS Letters* 447: 237–240. DOI: 10.1016/S0014-5793(99)00299-9.

Sousa SF, Fernandes PA and Ramos MJ. 2006. Protein-ligand docking: Current status and future challenges. *Proteins: Structure, function and bioinformatics*. 65: 15-26.

Stangier J, Hilbich C and Keller R. 1989. Occurrence of crustacean cardioactive peptide (CCAP) in the nervous system of an insect, *Locusta migratoria. Journal of Comparative Physiology B* 159: 5–11. DOI: 10.1007/BF00692677.

Starratt AN and Brown BE. 1975. Structure of the pentapeptide proctolin, a proposed neurotransmitter in insects. *Life Sciences* 17: 1253-56.

Turner CT, Davy MW, MacDiarmid RM, Plummer KM, Birch NP and Newcomb RD. 2006. RNA interference in the light brown apple moth, *Epiphyas postvittana* (Walker) induced by double -stranded RNA feeding. *Insect Molecular Biology* 15: 383 -91.

Veenstra JA. 1994. Isolation and structure of the *Drosophila* corazonin gene. *Biochemical and Biophysical Research Communications* 204: 292-296.

Veenstra JA. 2014. The contribution of the genomes of a termite and a locust to our understanding of insect neuropeptides and neurohormones. *Frontiers in Physiology* 5: 1–22. DOI: 10.3389/fphys.2014.00454

Wang Y, Wang M, Yin S, Jang R, Wang J, Xue Z, Xu T. 2015. NeuroPep; a comprehensive resource of neuropeptides. *Database: The Journal of Biological Databases and Curation* (article ID bav038).

Woodhead AP, Stay B, Seidel SL, Khan MA and Tobe SS. 1989. Primary structure off our allatostatins: neuropeptide inhibitors of juvenile hormone synthesis. *Proceedings of the National Academy of Sciences* 86: 5997–6001.

Yamanaka N, Roller L, Zitnan D, Satake H, Mizoguchi A, Kataoka H and Tanaka Y. 2011. *Bombyx* orcokinins are brain-gut peptides involved in the neuronal regulation of ecdysteroidogenesis. *The Journal of Comparative Neurology* 519: 238-246.

Yeoh JGC, Pandit AA, Zandawala M, Nassel DR, Davies SA and Dow AT. 2017. DINeR: Database for insect neuropepdide research. *Insect Biochemistry and Molecular Biology* 86: 9-19.

Ziegler R, Cushing AS, Walpole P, Jasensky RD and Morimoto H. 1998: Analogs of *Manduca* adipokinetic hormone tested in a bioassay and in a receptor-binding assay. *Peptides* 19: 481–486.

Zitnan DTG, Kingan JL, Hermesman and Adams ME. 1996. Identification of ecdysis-triggering hormone from an epitracheal endocrine system. *Science*. 271: 88–91. doi: 10.1126/science.271.5245.88.

Chapter 24

Trends in Biopesticides Formulations for Crop Protection

Vijay Laxmi Rai[1], *Sameer Kumar Singh*[1] *and Rudra Pratap Singh*[2]

[1]*College of Agriculture, Campus Azamgarh, Uttar Pradesh*
[3]*Krishi Vigyan Kendra, Kotwa, Azamgarh – 276001, ANDUA&T, Ayodhya, Uttar Pradesh*

Biopesticides are typically inherently less harmful than traditional pesticides. They often have an effect on solely the target pest and closely connected organisms, in distinction to broad spectrum, conventional pesticides that will affect organisms as completely different as birds, insects, and mammals. They often are effective in tiny quantities and infrequently decompose quickly. Once used as a part of Integrated Pest Management (IPM) programs, biopesticides will greatly decrease the use of conventional pesticides, whereas crop yields stay high. To use biopesticides effectively, however, users have to be compelled to grasp a good deal regarding managing pests. Choice of the suitable formulations which will improve product stability and viability may solve the issue of inconsistency of field performance of the many potential biological management agents. The fascinating feature of biopesticides is environment friendly and simple biodegradability. In terms of production and development also, biopesticides have a position over chemical pesticides as a result of they involve low analysis expenditure and have quicker rates of product development. Though there are about one hundred forty biopesticide production units existing within the country as on today, but they are not able to meet the demand of more than 1 per cent of cropped area. There exists a large gap, which might only be bridged by putting in place of additional and more units for production of biopesticides.

Introduction

Biopesticides have been gaining attention and interest among those concerned with developing environment friendly and safe integrated crop management practices for pest management. As biopesticides are generally less toxic than traditional synthetic chemical pesticides which generally effective against target pest and closely related organisms while broad spectrum, conventional pesticides that may affect organisms of a wide range that includes birds, beneficial insects, spiders as well as mammals. They often are effective in small quantities and often decompose quickly, thereby resulting in less exposure to non-target organisms and avoid the environmental pollution problems caused by conventional pesticides to a great extent. When biopesticides are taken as a curative or precautionary measures in any Integrated Pest Management (IPM) programs, it can greatly decrease the use of synthetic chemical pesticides, while ultimate crop yields remain high. Biopesticides can be used effectively, only if the farmers or users know a great deal about managing pests.

In most cases, the active ingredients of biopesticides are formulated in the same way as the synthetic pesticides and most convenient for farmers to use the same equipment for application. The basis for most of the biopesticides is living organisms and their viability has to be maintained during the formulation process and stored at acceptable levels. The organisms must revive from their dormant state in order to be active at the application time. Final product is maintained by mixing the microbial component with different carriers and adjuvants during formulation process for better protection from environmental factors, controlled rates, improved bioactivity and storage stability. To achieve the most important function of the developed biopesticides formulation such as easy handling and application of the product, high shelf life of the microbial strains during handling and storage, safeguarding of the bio-agent from adverse weather conditions, bio-agents activity can be improved by increasing contact and interaction with the target pest is necessary to ensure these.

Biopesticides are generally the formulated form of active ingredients based on microorganisms such as fungi, viruses, bacteria, nematodes or various botanicals and semio-chemicals. Biopesticides are highly target species-specific and their use is highly preferable but required formulations are very challenging to develop. Because there is need to maintain good physical properties and convenience in use, the formulated product must also keep its biological agent functional throughout storage and during application. Commercial biological products manufacturing should be economical, have high persistent during storage for certain period of time, should not be easily degradable, be easy to handle, mix and apply, and provide invariably effective control of target pests. Different formulations of biopesticides should be manufactured to neglect problems regarding their efficacy and their degradation and that should be easy during handling and application. Selection of the appropriate formulations that can improve product stability and viability may reduce inconsistency of field performance of many potential biological control agents (Boyetchko, 1996; Glass, 1993). A suitable formulation should provide a protective habitat for the introduced bacteria, thereby increasing then potential for survival and successful colonization (Elsas and Heijnen, 1990)

Scope and Potential of Biopesticide

The interesting feature of biopesticides is environment friendly and not persist in the environment for creating any kind of negative impact. For more production and commercialization biopesticides must involve high research expenditure and have more number of effective formulations.

Though there are about 140 biopesticide production units existing in the country as on today, they are able to meet the demand of only less than 1 per cent of cropped area. There exists a wide gap, which can only be bridged by setting up of more and more units for production of biopesticides. This requires large scale investment and private participation.

Few local small scale industries have started production and distribution of *Trichoderma viride* (against few fungal diseases) and *Trichogramma* (against sugarcane early shoot borer). There is a large scope to enhance production and use of biopesticides in near future as the demand is increasing every year.

Biopesticides cover only 2.89 per cent of the overall pesticides market in India and are expected to exhibit an annual growth rate of about 2.3 per cent in the forthcoming years. In India, till now 12 types of biopesticides have been registered under the Insecticide Act, 1968 (Gupta, 2006). Neem based pesticides, *Bt*, NPV and *Trichoderma* are the major biopesticides manufactured and efficiently used in our country.

Increasing demand for residue-free crop produce is one of the key reasons to flourish the biopesticide market (Pratap and Bhowmick, 2008). However, growing organic food market and easier registration than chemical pesticides are other important driving factors for the growing biopesticide market. Due to high cost of developing new synthetic molecules and most Indian companies have impotency to invest such huge amounts, there is chances of greater growth in the biopesticide sector. India's rich biodiversity provides plenty of scope for many micro- and macro-organisms as well as natural botanicals. Rich diversity of knowledge base of the diverse indigenous communities would also help in providing valuable clues for developing effective biopesticides.

Factors affecting Growth of Biopesticides

Some factors which have restricted the growth of biopesticides are:

1. Low reliability on biopesticides because of low stability or highly degradable in field condition, Thus, more number of biopesticides application is required
2. Target specificity means more work for the farmers as every major pest species require different management strategy

3. Slow action compared to synthetic pesticides which is unsuitable for the immediate treatment of crops
4. Irregular and low availability of biopesticide in the market

Biopesticides are formulated in various ways. Depending on the physical states of the biopesticide formulation as dry or liquid forms, the active ingredients are formulated/modified by addition of stabilizers, synergist, spreaders, stickers, surfactants, colouring agents, anti-freezing compounds, additional nutrients, and dispersants and melting agents. In general, biopesticides are usually formulated as dry formulation (for direct applications) and liquid formulations.

Dry Formulation for Direct Applications

1. **Dust Powders (DP):** Active ingredient concentration for dust formulations is usually 10 per cent and is generally formulated by adsorption of active ingredient on finely ground, solid mineral powder (talc, clay *etc.*) with particle size ranging from 50-100 mm. The inert ingredients for dust formulations are UV protectants, adhesive materials (*i.e.* stickers) to enhance adsorption and anticaking agent.
2. **Granules (GR):** Active ingredient concentration for granules ranges from 2-20 per cent and the active ingredients either cover the outside of the granule or are absorbed within the granules. To control the rate of effectiveness of active ingredients after application, granules can be coated with resins or polymers. Granules are mostly applied to control insects living in soils, weeds and nematodes for uptake by roots. Granules with coarse size particles range from 100-600 microns made from such materials such as kaoline, silica, starch, polymers, groundnut plant residue *etc.* Some granules release their active ingredients after exposure to soil moisture.
3. **Seed Dressing (SD):** In this type of biopesticide formulation active ingredient is generally mixed to the seed material so that the biopesticides can adhere to the seed coats. These seed dressing materials also contain colouring agents inform of red pigment as a safety maker for treated seed.
4. **Wettable Powders (WP):** These are also dry formulation ground finely and applied after suspension in water. Wettable Powders are formed by mixing active ingredients with synergist, melting and dispersing agents, surfactants, and inert fillers. Strict safety measures are usually taken because of their dustiness that can cause serious health problems to manufacturers and during application. Besides, WPs have long stability during storage, good miscibility with water and can be applied with conventional spraying equipment.
5. **Water Dispersible Granules (WDG):** It is designed to be suspended in water and to overcome problems associated with WPs, dust free and with good storage stability.

Liquid Formulations

1. **Emulsion:** Emulsion formulations are designed to be mixed with water and it could be normal emulsion which is oil in water (O/W) or an inert emulsion which is water in oil (W/O).
2. **Suspension Concentrate (SC):** These SC formed by mixing finely ground, solid active ingredient dispersed in water. Agitation is always a requisite before application to keep particles evenly distributed because the solid particles are not dissolved in liquid phase. The particle size distribution is 1-10 µm and these small particles size offers easier access of the active ingredients to plant tissue and improved bio-efficiency. It is a popular type of formulation because of safety to operator and environment.
3. **Suspo-Emulsion (SE):** It is a mixture of emulsion and suspension concentrate and highly demanding formulation, because it is necessary to develop a homogenous emulsion component with a particle suspension component so that the final product will remain stable. In addition, it is necessary to be carried out using storage stability resting.
4. **Oil Dispersion (OD):** The product of the formulation is produced in the same ways as suspension concentrate. Instability problems could be avoided by proper selection of inert ingredients.
5. **Capsule Suspension (CS):** In this, active ingredients are formulated in microencapsulated stable suspension which require dilution with water before use. Capsules made from gelation, starch, cellulose and other polymers are used to encapsulate the bio-agents and in this way the bioagents are protected from the harsh environmental conditions. Interfacial polymerization principle is the most frequent applied method of encapsulation which is used to give smaller size and highly efficient formulations typically fungal biopesticides.
6. **Ultra Low Volume Liquids (ULV):** Formulations not intended to be diluted in water before use and have concentration of active ingredients. It is easy to transport and can be formulated using a suspended bio-control agent as an active ingredient.

Methods of Biopesticides Application

Efficient pest management can be obtained by proper selection of application techniques/methods and an appropriate time and/or frequency of biopesticides application.

Seed Treatment is one way to apply biopesticides is by seed treatment and is the most effective method or technique. Powder formulations are applied on seeds by tumbling seed with the product that is designed to adhere to the seed. Another method is foliar application in which biopesticides application is done on leaf surface as sprays. For example application of *Bacillus thuringiensis* to Gram leaves reduced the incidence of Gram pod borer, *Helicoverpa armigera*. Seedling Dipping is another method which involves dipping roots of the seedlings in biopesticides suspension for some minutes or hours prior to transplanting. For example *Trichoderma* spp. is applied in this way.

Formulation of Natural Insecticides

The earliest natural formulation is by using water as it serves as a universal solvent. The crude extracts in fact are better home-made formulations. Although it does not require any scientific skills but even today aqueous extract of plant products are being used by the conventional farmers, or those who believe in their own make. But the introduction of commercial synthetic pesticides made the farmers to use them as it is readily available from the market in plenty. Formulation of any botanical insecticide helps in providing adherence to the target sites and controlled release of the bioactive compounds, depending on the type of formulation (Borges *et al.*, 2018) Neem seed extract is one of the commercially exploited botanical based formulations. They are not only effective against crop pests but also relatively safer to natural enemies and non-target organisms (Raguraman, 2006). Here, few common formulations are given which can be easily get prepared by farmers (Ignacimuthu and Jayaraj, 2006; Raguraman, 2007).

Neem leaf extract: About 5 kg of fresh neem leaves were chopped and soaked in 6 lit of water and made into paste. The suspension was boiled for 30 min and cooled overnight. Soap powder (150 g) was added to the mixture and the final solution was sprayed.

Calotropis leaf extract: Calotropis leaves (5 kg) were collected with latex and cleaned well from any impurities, chopped and ground into fine paste. The paste was mixed with 50 lit of water and the suspension was filtered and sprayed.

Tobacco waste extract: After the tobacco waste was collected from the field, about 5 kg of the material was cut into pieces and soaked in cow urine (5 lit) for 7 days. After 7 days, the soaked tobacco suspension was filtered and 150 g of soap powder was mixed in 250 ml of water. This was also added at the rate of 50 ml/lit of water and the mixture was sprayed.

Neem products: Freshly collected neem seed kernels were shade dried and ground into paste. The paste (1 kg) was dissolved in 4 lit of water, and heated at 60-700 for 4-5 h. The hot extract was kept overnight and filtered in the morning. About 100 ml of the extract mixed with 1 ml of teepol or 10 ml of soap solution (NirmaR) was sprayed. Neem oil (3 lit) mixed with 200 ml of soap solution was sprayed in an acre of rice.

Leaf mixture extract: Fresh leaves of neem, Adathoda and Datura were soaked in water and ground into a paste, mixed in 4 lit of water. The suspension was boiled for about half an hour and kept undisturbed overnight. In the morning the suspension was filtered and sprayed. Asafetida (200 g) was dissolved in mild hot water (1 lit) and the two mixtures were mixed together with 150 g of soap powder and sprayed in a ratio of 1:20 (solution: water).

Jatropha plant extract: Stems and leaves of the Jatropha plant were washed with distilled water and ground well as paste. About 5 kg of the paste was soaked in 5 lit cow urine overnight and boiled for 3-4 h by adding 2 lit of water. The material was concentrated to 5 lit and filtered through coarse cloth and the filtrate was used.

Ipomoea leaf extract: About 5 kg of fresh leaves of Ipomoea were chopped. Twigs, barks and woody and dried stems were eliminated. The chopped leaves were made into a fine paste, mixed with equal volume of water and boiled. Cow urine was added to the boiling mixture. When the mixture was one fourth of its original volume, it was removed from the flame and allowed to cool overnight. The solution was filtered and washed repeatedly with cow urine and the volume was made to 4.5 lit. As emulsifier, 250 g of soap powder was added to 500 ml of water and the plant extract was mixed with the soap solution. Finally, the martial was used @ 500 ml per 10 lit of water and sprayed.

Common ash: The fire wood ash was collected and sieved (0.5 mm) to remove the solid pieces and debris and directly dusted in the field. About 15 kg of the ash would be required for an acre of crop.

Commercial Natural Formulations

Of the several options of plant materials possessing insecticidal properties neem products were at least successful in the commercial formulations. To date there are more than 80 formulations available in the market. Of course when it comes to the chemical formulation it does not much differ from its synthetic counterparts. Many products of neem seed extracts or other neem products are now available in dust, emulsifiable concentrate, or wettable powders which are either oil based or active ingredient based.

Future Prospects

Biopesticides have long been attracting global attention as a safer strategy than chemical methods of pest management, with having comparatively low risk to non-target organisms and the environment. However it is only possible, when cooperation between the public and private sectors takes place to facilitate the development, manufacturing, and distribution of this eco-friendly option that is use of biopesticides. In this regard, finding of new biological substances and research on formulation and delivery would enhance commercialization and use of biopesticides. Moreover, there is an urgent need of regulations that promote registration of these low-risk compounds with provision of incentives in the form of subsidies could also facilitate commercialization and supply of biopesticides in the market. Newly discovered formulations could serve as an attractive option for use in pest management. Apart from that more field research is required in this aspect to determine the efficacy on specific pest problems in various cropping systems. Microencapsulation based on nanotechnology could enhance the residual action of biopesticides, and this could boost their field use.

REFERENCES

Borges D, Lopes E and Moraes A. (2018). Formulation of botanicals for the control of plant-pathogens: a review. *Crop Protect.* (110) pp 135-140.

Boyetchko SM. (1996). Impact of soil microorganisms on weed biology and ecology. *Phytoprotectzon* 77,41-56.

Elsas V and Heijnen CE. 1990. Methods for the introduction of bacteria into soil: A review. *Biol. Fertil. Soils* 10, pp 127-133.

Glass DJ. (1993). Commercialization of soil microbial technologies, *In Soil Microbial Ecology Applications in Agricultural and Environmental Management* Marcel Dekker, New York, pp 595-618.

Greaves MP. (1993). Formulation of microbial herbicides to Improve performance in the field, *In Proceedings of 8th EWRS Symposium "Quantitative Approaches in Weed and Herbicide Research and Their Practical Application*, Germany, pp. 219-225.

Gupta PK. (2006). Status of biopesticides-Indian scene. *Toxicol. Int.* 13: 65-73.

Ignacimuthu S and Jayaraj S. (2005). Green Pesticides for Insect Pest Management. Narosa Publishing House, New Delhi pp 324.

Pratap A and Bhowmick D. (2008). Pesticides as microemulsion formulations. *J. Dispersion Sci. Technol.* 29 pp 1325-1330.

Raguraman S. (2006). Neem and Entomophages: Impact and Strategies in Pest Management. *Proceedings of the 2006 International Neem Conference,* November 11-15, 2006, Kunming, China, pp 149-181.

Raguraman S. (2007). Diversity and Role of Neem in Insect Pest Management, Green Farming 1 (1): 45-49.

Index

S

T

U

V

W

X

Y

Z